T0181009

Sustainable Textiles: Production, Processing, Manufacturing & Chemistry

Series Editor

Subramanian Senthilkannan Muthu, Head of Sustainability, SgT and API, Kowloon, Hong Kong

This series aims to address all issues related to sustainability through the lifecycles of textiles from manufacturing to consumer behavior through sustainable disposal. Potential topics include but are not limited to: Environmental Footprints of Textile manufacturing; Environmental Life Cycle Assessment of Textile production; Environmental impact models of Textiles and Clothing Supply Chain; Clothing Supply Chain Sustainability; Carbon, energy and water footprints of textile products and in the clothing manufacturing chain; Functional life and reusability of textile products; Biodegradable textile products and the assessment of biodegradability; Waste management in textile industry; Pollution abatement in textile sector; Recycled textile materials and the evaluation of recycling; Consumer behavior in Sustainable Textiles; Eco-design in Clothing & Apparels; Sustainable polymers & fibers in Textiles; Sustainable waste water treatments in Textile manufacturing; Sustainable Textile Chemicals in Textile manufacturing. Innovative fibres, processes, methods and technologies for Sustainable textiles; Development of sustainable, eco-friendly textile products and processes; Environmental standards for textile industry; Modelling of environmental impacts of textile products; Green Chemistry, clean technology and their applications to textiles and clothing sector; Eco-production of Apparels, Energy and Water Efficient textiles. Sustainable Smart textiles & polymers, Sustainable Nano fibers and Textiles; Sustainable Innovations in Textile Chemistry & Manufacturing; Circular Economy, Advances in Sustainable Textiles Manufacturing; Sustainable Luxury & Craftsmanship; Zero Waste Textiles.

More information about this series at https://link.springer.com/bookseries/16490

Ali Khadir · Subramanian Senthilkannan Muthu

Editors

Polymer Technology in Dye-containing Wastewater

Volume 1

 Springer

Editors
Ali Khadir
Western University
London Ontario, ON, Canada

Subramanian Senthilkannan Muthu
SgT Group and API
Hong Kong, Kowloon, Hong Kong

ISSN 2662-7108   ISSN 2662-7116   (electronic)
Sustainable Textiles: Production, Processing, Manufacturing & Chemistry
ISBN 978-981-19-1518-5   ISBN 978-981-19-1516-1   (eBook)
https://doi.org/10.1007/978-981-19-1516-1

This Springer imprint is published by the registered company Springer Nature Singapore Pte Ltd.
The registered company address is: 152 Beach Road, #21-01/04 Gateway East, Singapore 189721,
Singapore

Contents

Polymer Technology Coupling with Physical, Chemical, and Biological Methods in Textile Wastewater 1
Muhammad Hamad Zeeshan, Umm E. Ruman, Gaohong He,
Aneela Sabir, Muhammad Shafiq, and Muhammad Zubair

Evaluation of Fe–Mn–Zr Trimetal Oxide/Polyaniline Nanocomposite as Potential Adsorbent for Abatement of Toxic Dye from Aqueous Solution 15
Bibek Saha, Animesh Debnath, and Biswajit Saha

Silica-Polymer Composite for Dyes Removal 39
Bouhadjar Boukoussa and Adel Mokhtar

Polymer-Based Photocatalysis for Remediation of Wastewater Contaminated with Organic Dyes 57
Doaa M. EL-Mekkawi

Application of Hybrid Polymeric Materials as Photocatalyst in Textile Wastewater 101
Hartini Ahmad Rafaie, Norshahidatul Akmar Mohd Shohaimi,
Nurul Infaza Talalah Ramli, Zati Ismah Ishak, Mohamad Saufi Rosmi,
Mohamad Azuwa Mohamed, and Zul Adlan Mohd Hir

Synthesis of Pillared Clay Adsorbents and Their Applications in Treatment of Dye Containing Wastewater 145
Desai Hari and A. Kannan

Versatile Fabrication and Use of Polyurethane in Textile Wastewater Dye Removal via Adsorption and Degradation 179
Muhammad Iqhrammullah, Rahmi, Hery Suyanto, Kana Puspita,
Haya Fathana, and Syahrun Nur Abdulmadjid

Application of Polymer/Carbon Nanocomposite for Organic Wastewater Treatment 199
Adane Adugna Ayalew

"Environmental Issues Concerned with Poly (Vinyl Alcohol) (PVA) in Textile Wastewater" 225
Muhammad Hamad Zeeshan, Umm E. Ruman, Gaohong He, Aneela Sabir, Muhammad Shafiq, and Muhammad Zubair

Nanoparticles Functionalized Electrospun Polymer Nanofibers: Synthesis and Adsorptive Removal of Textile Dyes 237
Shabna Patel, Sandip Padhiari, and G. Hota

About the Editors

Ali Khadir is an environmental engineer and a member of the Young Researcher and Elite Club, Islamic Azad University of Shahre Rey Branch, Tehran, Iran. He has published several articles and book chapters in reputed international publishers, including Elsevier, Springer, Taylor & Francis, and Wiley. His articles have been published in journals with IF of greater than 4, including the Journal of Environmental Chemical Engineering and the International Journal of Biological Macromolecules. He also has been the reviewer of journals and international conferences. His research interests center on emerging pollutants, dyes, and pharmaceuticals in aquatic media, advanced water and wastewater remediation techniques, and technology.

Dr. Subramanian Senthilkannan Muthu currently works for SgT Group as Head of Sustainability and is based out of Hong Kong. He earned his Ph.D. from The Hong Kong Polytechnic University and is a renowned expert in the areas of environmental sustainability in textiles and clothing supply chain, product life cycle assessment (LCA), and product carbon footprint assessment (PCF) in various industrial sectors. He has five years of industrial experience in textile manufacturing, research and development, and textile testing and over a decade of experience in life cycle assessment (LCA), carbon and ecological footprints assessment of various consumer products. He has published more than 100 research publications, written numerous book chapters, and authored/edited over 100 books in the areas of carbon footprint, recycling, environmental assessment, and environmental sustainability.

Polymer Technology Coupling with Physical, Chemical, and Biological Methods in Textile Wastewater

Muhammad Hamad Zeeshan, Umm E. Ruman, Gaohong He, Aneela Sabir, Muhammad Shafiq, and Muhammad Zubair

Abstract Wet treatment and finishing processes in the textile industry is the major consumer of freshwater in large amounts. Considerable amounts of pollutants in textile wastewater such as chemical oxygen demand (COD), biological oxygen demand (BOD), dyes, xenobiotic compounds, organic compounds, etc. are released in portable water without treatments, which creates serious environmental and health problems. Treatment of such polluted water has become complicated and a challenging task over the last few decades. Due to the strict regulations by the Environmental Protection Agency (EPA) and World Health Organization (WHO), reclamations of wastewater become more and more attractive. However, there are no single and economical treatment methods that can effectively remove these pollutants and make water reusable. In this chapter, different hybrid methods are presented which are very efficient to remove textile effluents from wastewater. These hybrid methods include membrane bioreactor technology (MBR), photocatalyst membrane reactor technology (PMR), mixed adsorbent fixed bed reactor, and other polymer technologies coupled with different convention methods such as biological, chemical, and physical methods as pretreatment or posttreatment are detailed. The advantages and limitations of each method are also discussed. In the end, this chapter also discusses some future perspectives of these methods to make them more efficient, less costly, lower space, low maintenance cost, and easy handling.

Keywords Azo dyes removal · Textile effluents · Hybrid treatment · Membrane bioreactor treatment · Photocatalyst bioreactor · Fixed bed reactor · Coagulation · Flocculation

M. H. Zeeshan (✉) · G. He
State Key Laboratory of Fine Chemicals, R&D Center of Membrane Science and Technology, School of Chemical Engineering, Dalian University of Technology, Dalian 116024, China
e-mail: hamad.xeeshan@mail.dlut.edu.cn

U. E. Ruman · M. Zubair
Department of Chemistry, University of Gujrat, Gujrat 57200, Pakistan

A. Sabir · M. Shafiq
Department of Polymer and Textile Engineering, University of the Punjab, Lahore 54590, Pakistan

1   Introduction

The textile industry is the most complicated and quickly emerging industrial chain in the manufacturing industry. This chain begins with the harvesting of natural resources. There are several stages of mechanical processing to transform raw material into a useful product via spinning, knitting, weaving, and garments production [17]. Ready thread and garments are subjected to the finishing processes which consume enormous water. The wet processes such as washing, bleaching, desizing, mercerizing, dyeing, coating, printing, and finishing operations transformed water into highly loaded wastewater [15].

The release of textile effluents has increased proportionality into the environment with the increased demand for textile products over the last few decades. The textile wastewater composition depends on the nature of dyes, chemicals, organic-based compounds, and wet processing steps [48]. The extreme fluctuating parameters such as pH, BOD, color, COD, and salinity characterize textile wastewater. Table 1 depicts the characteristics of textile wet process wastewater [40].

Desizing, washing, bleaching, mercerizing, and dyeing processes are the commonly established processing methods in the textile industry [14]. Polyvinyl alcohol (PVA), starch, and carboxymethyl cellulose are added during the sizing preparation step as sizing agents which enhanced the fiber strength and curtail breakage. During desizing, sizing materials are removed prior to weaving. The alkaline solution, commonly sodium hydroxide (NaOH), is used to remove impurities during the washing step to break down fats, surfactants, waxes, and natural oils. In a washing bath, impurities are emulsified and suspended through the scouring method [7].

In the bleaching step, unwanted color from the fiber is removed using chemicals such as hydrogen peroxide (H_2O_2) and sodium hypochlorite. Mercerizing is a continuous process step where a concentrated alkaline solution is applied to enhance dye functionality, luster, and fiber finishing [10]. Before the dyeing process, the fibers are washed with an acid solution. Then color is applied to the fiber in the dyeing process.

Table 1   Characterization of textile wet process wastewater [9, 12, 35]

Process	Biological oxygen demand (g/l)	Chemical oxygen demand (g/l)	pH	Color (ADMI)	Total solids (g/l)	Total dissolved solids (g/l)
Desizing	1.8–5.2	4.5–5.9	–	–	16.1–32.0	–
Washing	0.1–2.9	7.9	10–13	697	7.6–17.4	–
Bleaching	0.1–1.8	6.7–13.5	8.5–9.7	155	2.3–14.4	4.8–19.5
Mercerizing	0.05–0.1	1.6	5.5–9.5	–	0.6–1.9	4.3–4.6
Dyeing	0.01–1.9	1.2–4.7	5–10	1450–4750	0.5–14.1	0.05[a]

ADMI = American Dye Manufacturer Institute
[a]The salt concentration in the dye bath can reach up to 60–100 g/l due to some reactive dyes. Therefore, the value can vary depending on the type of fiber and the dye

In this step large volume of water is required in the dye bath and during the rinsing process. Dye adsorption onto the fiber is improved by adding many chemicals such as salts, metals, surfactants, formaldehyde, and sulfide [41].

The wet processes are responsible for the high demand for water and lead to wastewater production in the textile industry. This wastewater contains a lot of pollutants which are the main source of severe pollution problems [38]. The release of colored effluents is undesirable not for their color because of many dyes. The degraded products of these effluents are mutagenic and toxic to the environment. Most of the dyes are stable and persistent for a longer time [23].

Textile wastewater has high discharge materials that contain toxic chemicals such as starch, acids, enzymes, alkalis, bleaching chemicals, resins, waxes, solvent, high in pH, chemical oxygen demand (COD), turbidity, temperature, high in color, and biological oxygen demand (BOD), [16, 20]. Wastewater from the textile industry if discharged directly into water bodies such as rivers, lakes, etc. pollutes the water and is very harmful to the environment. Dyes present in the discharged water show low biodegradability. Color present in the water cannot be tolerated and hinders sunlight penetration, thus disturbs the ecosystem [32, 44].

Nowadays, due to water scarcity, the textile industry often faces a shortage of available water sources. Thus, in the future to meet the requirement of reusing water will significantly increase. For this purpose, many conventional and non-conventional treatment methods are widely being applied in the textile industry to treat wastewater. These methods are costly and the quality of treated water does not meet the requirement of reuse [43]. In this chapter, we will explore the possible polymer technologies coupled with other process intensification to improve the treated water quality and possibilities of recycling in order to reduce the cost and consumption of available resources.

2   Treatment Processes for Textile Wastewater

Direct release of untreated textile wastewater into major water sources leads to an imbalance in the environment. So, before the discharge of polluted water into the lakes or rivers, many physical, chemical, and biological treatment processes have been developed [51]. Physical methods include coagulation/flocculation, filtration, adsorption, etc. [19]. Enzymes and microorganisms are considered efficient biological methods but the investment cost is high depending on the equipment used [8]. Chemical oxidation and advanced oxidation methods are also being used to treat textile effluents but are expensive as they demand a significant amount of electrical energy or chemical reagents [30]. Figure 1 shows the most used convention technologies in textile industries. However, to treat all kinds of textile effluents, there are no suitable treatment methodologies or universally adopted. Thus, due to certain limitations the quality of treated water does not meet the WHO standard [13].

To improve the quality of treated water, more innovative approaches are required to further optimize the different subprocesses to improve the wastewater quality

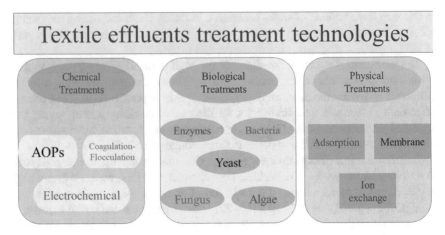

Fig. 1 Convention treatment technologies for textile effluents

and to reduce the consumption of freshwater. Therefore, a combination of several methods is being implemented for the treatment of textile wastewater [26]. Among all hybrid methods, polymer technologies coupled with other convention methods (physical, chemical, and biological) are more attractive, cost-effective, and efficient for the removal of textile effluents [3]. The selection of appropriate technologies depends on the type and quantum of pollution load.

3   Combined Polymer Technologies with Other Methods

From both economic and environmental points of view, polymer technologies coupled with other methods as pretreatment or posttreatment for the efficient removal of textile effluents are in critical need in the textile manufacturing processes as shown in Fig. 2. Such coupled technologies have major advantages including higher efficacy and lower cost. However, the selection of the process depends on the nature of effluents being treated during the wet process. Polymeric membrane technologies, polymer adsorbents, polymer flocculants–coagulants along with other physical, chemical, and biological methods are now being used in textile industries.

3.1   Polymer Membrane Technology Coupled with Biological, Chemical, and Physical Methods

Dyes are the major textile effluents that are discharged into major water sources. Most of the dyes remained unbound with the fiber. It is estimated that half of the unbound dyes are released directly and indirectly into the water bodies. To minimize

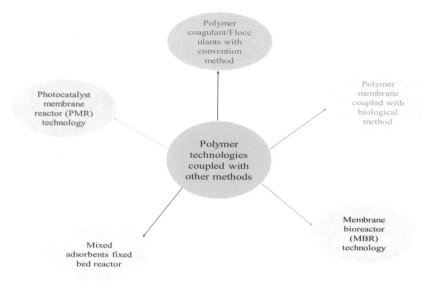

Fig. 2 Polymer technologies coupled with other methods

the adverse effect on the environment, Environmental Protection Agency (EPA) and other regulatory bodies have restricted the discharge of dye-containing wastewater in textile industries [37].

Biological treatment is one of the ecofriendly and low-cost processes to ensure dye control in the textile industry. This treatment process produces less sludge, energy savings, economical, and needs less amount of reagents. There are two biological treatment processes aerobic and anaerobic processes. In this method, dye solution is treated in an ecofriendly environment [2]. Bacteria, algae, fungi, yeast, and enzymes are used to biodegrade the dyes in wastewater. This method is used as the primary water treatment method. However, the use of microorganisms to degrade dyes is not an efficient process, thus limited the dye removal [21]. The most suitable method for dye removal is a chemical treatment regarding the chemistry of dyes. Chemical technologies are more expensive and unattractive in textile industries due to higher costs for equipments, a large amount of chemical reagents, and the high need for electrical energy for reactors [26]. However, advanced oxidation processes, coagulation–flocculation, and anodic oxidation are used in some industries as pretreatment and posttreatment methods to remove different dyes.

Physical methods such as adsorption are commonly used methods for dye removal and have 70–80% efficiency of dye removal. Simple design, cheapness, easy operation, and low cost have made physical techniques more productive with other polymer technologies [27]. In order to meet legislative requirements, this primarily treated water either with biological, chemical, or physical methods is further treated through polymeric membrane technologies as shown in Fig. 3.

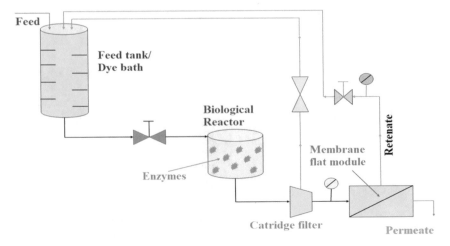

Fig. 3 Polymeric membrane technology coupled with the biological method

Effluents from textile dye baths are potentially removed through membrane processes and reusable water is produced. Microfiltration (MF) is a suitable membrane for eliminating colloidal dyes. This is a single-step treatment process for secondary textile wastewater. After MF, nanofiltration (NF) produces an appropriate softening effect by removing low molecular weight organic and divalent salts. Reverse osmosis (RO) is an appropriate membrane to remove ions and large species from dye bath [53]. Thus, polymeric membrane technology produces colorless and satisfactory water reuse in the textile industry.

3.2 Membrane Bioreactor (MBR) Treatment Technology

This hybrid technology is vastly being used for dye removal in the textile industry. This technology has the advantage of high dye removal efficiency, low maintenance, and less sludge formation [11]. This process has many stages. In the first stage, wastewater as feed is fed into the reactor to interact with biomass. In the next stage, wastewater and biomass mixture is transferred to the membrane unit as shown in Fig. 4. In the separation unit, the retentate is recycled. Membrane with 10^{-4}–10^{-3} μm pore size is used in MBR system and referred to as either microfiltration (MF) or ultrafiltration (UF) membrane [22]. In the MBR system, the membrane module is either placed in a separate stream system or in the submerged system. The shape of the membrane could be a hollow fiber or flat sheet.

Aerobic membrane bioreactor is the most used, attractive textile effluents treatment process. The removal efficiency of this process for the removal of (COD), color, and ammonium is reported more than 90%, 90%, and 80%, respectively [18].

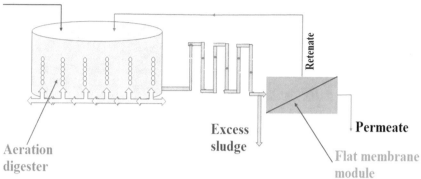

Fig. 4 Membrane bioreactor treatment technology

Anaerobic membrane bioreactor is also used for wastewater treatment. This method has many advantages such as low energy consumption, cost-effective, and low sludge generation. The significant advantage of an anaerobic membrane bioreactor is the production of biogas with a 100% dye removal efficiency [39].

3.3 Photocatalyst Membrane Reactor (PMR) Technology

In PMR technology, large dye molecules are broken down into smaller molecules by photocatalyst action. The products of this process are CO_2 and water [24]. In this process, by the action of high-energy photons, an electron is transferred to the conduction band, thus producing electron–hole pairs. Oxygen and hydroxyl groups present in water produces various reactive oxygen species such as OH, O_2^-, and H_2O_2 by the action of electron and holes [34]. The most important efficient factors of PMR technology are photocatalyst, membrane, light, temperature, pressure, and pH. These process parameters are crucial for the smooth operation and good performance of PMR technology. The most important factor of PMR technology is the selection of photocatalyst which is influenced by bandgap energy [52]. Figure 5 depicts the PMR technology dye removal process.

There are two PMR configurations, PMRs with suspended photocatalyst. This type of configuration has a strong active surface area, controls membrane damage by UV light, and generates hydroxyl in the suspended system. PMRs with immobilized photocatalyst membrane is the second type of configuration. In this system, the membrane and photocatalyst reactions are fused into one system. In the PMR process, crossflow and dead-end are the major operation ways with immobilized photocatalyst [36]. These processes have a high amount of color and chemical oxygen demand (COD) removal efficiency. Some typical polymeric membranes such as polyamide

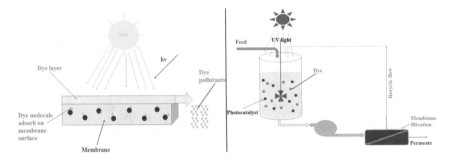

Fig. 5 Photocatalyst membrane reactor technology

(PA), polyvinylidene (PVDF), polyethersulfone (PES), polyurethane (PU), polyethylene terephthalate (PET), polyacrylonitrile (PAN), and cellulose acetate are widely used [6].

3.4 Mixed Adsorbents Fixed Bed Reactor Technology

To remove different types of dye pollutants from textile wastewater, physical adsorption is considered an efficient approach. This is a surface method in which molecules are attracted to the surface of adsorbents. The structure of adsorbents is porous that increases the total surface area. This method is considered cost-effective for the removal of dyes. Common adsorbents such as activated carbon, zeolite, alumina, and silica gel are widely used to remove dye from wastewater [1]. However, this technique has limited application and is unable to remove starch, PVA, and azo dyes from wastewater. Higher cost and less availability also hinder its vast practical application due to the high amount of adsorbents requirements to increase the rate of adsorption [50].

Thus to enhance the efficiency and to remove the carcinogenic dyes effectively, polymer adsorbents along with physical adsorbents or a mixture of both adsorbents are being widely used for the last two decades [29]. Figure 6 demonstrates this process. Azo dyes are carcinogenic stable organic substances and are difficult to effectively remove with one method. Polyaniline (PAni) is a polyaromatic amine and most potentially useful polymer for the adsorption of azo dyes. It is widely used because of its environmental stability, low cost, and strong adsorption application. More than 90% of azo dyes removal efficiency is obtained through PAni adsorbents and its removal efficiency is higher than single-activated carbon adsorbent [4, 5, 28].

To remove starch, PVA, and dyes, the crosslinked cationic starch/PVA composite is used in fixed bed adsorption along with the mixture of activated carbon for 100% pollutant removal. Fixed bed adsorption has a low cost of equipment, lower space, and lower operation cost. Desorption of polymer adsorbents is cost-effective and performed with a chemical with higher performance. Thus, high starch, PVA, and

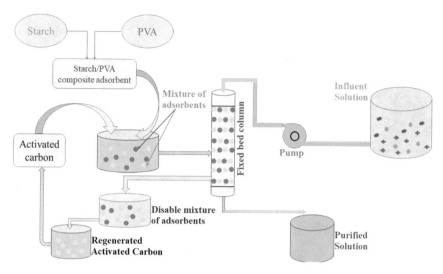

Fig. 6 Mixed adsorbents fixed bed reactor column

dyes removal from textile wastewater is obtained simultaneously with the mixture adsorbents composed of polymeric materials and activated carbon [49].

3.5 Blended Coagulation/flocculation Treatment Technology

Convention methods such as chemical coagulation/flocculation in textile wastewater treatment are regarded as the most successful pretreatment method. This method is used for the removal of suspended solids or colloidal which do not settle down easily [45]. This process encompasses the addition of chemicals that change the physical state of colloidal or suspended particles, thus assisting their confiscation by sedimentation. Due to low capital cost, coagulation is being used as a pretreatment method for many years in the textile industry. However, the application of this sole method has many limitations. Due to high solubility of coagulants, it is challenging to remove dyes. The selection of appropriate coagulants is difficult due to a large number of innovative dyes with complex structure. Sludge generation is also one of the major challenges in this treatment method. Thus, the efficiency of this convention method for the treatment of colored wastewater is very low due to the above-mentioned limitations [42].

In the last decade, the development of cationic polymer composite is one of the preferred technologies and is widely being used blended with convention methods for the advanced handling of textile effluents. In recent years, significant efforts have been exercised to synthesize less costly and high competent polymeric composite for the decolorization of textile wastes. The development and use of low-cost polymeric flocculants avoid the production of the extreme volume of sludge and is effective for

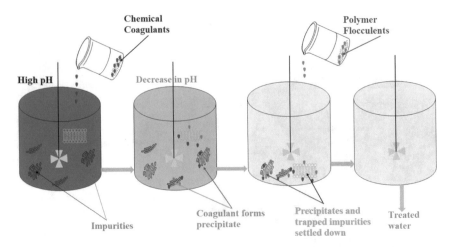

Fig. 7 Blended chemical coagulant and polymer flocculent treatment method

the remediation of reactive or azo dyes from wastewater [33]. The process is shown in Fig. 7. Formation of particle–polymer–particle bridge and adsorption phenomenon, cationic polymer flocculants destabilize the dye molecules and other suspended particles. Polyamine-based flocculants are found very operative over a wide range of pH. These polymeric coagulants/flocculants are excellent color removal with a lower dose and produce fewer volumes of sludge, thus making it an effective option as posttreatment technologies along with convention methods [46].

4 Conclusion

In each of the above-mentioned polymer hybrid technologies, several limitations and challenges still exist for their wide applications in the textile industry for effluents treatment and need to be addressed. The various membrane treatments such as NF, MF, UF, and RO are suitable for the specific cases as the final purification steps in the textile industry depend on the wastewater characteristics such as pH, types of dyes, BOD/COD ratio, alkalinity, salinity turbidity, TDS, organic contents, heavy metals, and toxic compounds. Membrane fouling is one of the major drawbacks of MBR and PMR technologies which reduce the permeation flux; therefore membrane washing is needed for long-term consistency and smooth performance. The recovery of redox mediators to accelerate decolorization rates in bioreactors still represents a challenge [47].

Desorption of mixed (polymer) adsorbents in fixed bed column is costly and is a major challenge nowadays. The performance of adsorbents is highly influenced by bed height and flow rate. The adsorption efficiency is also influenced by the pH, salinity, shape of adsorbent, swelling, and other toxic effluents. Therefore, there is still a major need for the introduction of suitable polymer adsorbents for the treatment of textile effluents [25].

Blended (polymer) coagulation–flocculation technology is found very efficient for dye removal but the efficiency of this process is controlled by its cationic potential, solution pH, and chemical characteristics of colloid which should be addressed further. Therefore, imminent exploration of these polymer-coupled technologies may overcome these challenges and may assist to get better performance and wider applications in the textile industry [31].

References

1. Afshari M, Dinari M (2020) Synthesis of new imine-linked covalent organic framework as high efficient absorbent and monitoring the removal of direct fast scarlet 4BS textile dye based on mobile phone colorimetric platform. J Hazard Mater 385:121514
2. Ahmad A, Mohd-Setapar SH, Chuong CS, Khatoon A, Wani WA, Kumar R, Rafatullah M (2015) Recent advances in new generation dye removal technologies: novel search for approaches to reprocess wastewater. RSC Adv 5(39):30801–30818
3. Al Prol AE (2019) Study of environmental concerns of dyes and recent textile effluents treatment technology: a review. Asian J Fisheries Aquatic Res 1–18
4. Ansari R (2006) Application of polyaniline and its composites for adsorption/recovery of chromium (VI) from aqueous solutions. Acta Chim Slov 53(1):88
5. Ansari R, Keivani M (2006) Polyaniline conducting electroactive polymers thermal and environmental stability studies. E J Chem 3(4):202–217
6. Argurio P, Fontananova E, Molinari R, Drioli E (2018) Photocatalytic membranes in photocatalytic membrane reactors. Processes 6(9):162
7. Arslan S, Eyvaz M, Gürbulak E, Yuksel E (2016) A review of state-of-the-art technologies in dye-containing wastewater treatment. Textile Ind Case
8. Bhatia D, Sharma NR, Singh J, Kanwar RS (2017) Biological methods for textile dye removal from wastewater: a review. Crit Rev Environ Sci Technol 47(19):1836–1876
9. Bhatia S, Devraj S (2017) Pollution control in textile industry. WPI Publishing
10. Bisschops I, Spanjers H (2003) Literature review on textile wastewater characterisation. Environ Technol 24:1399–1411. https://doi.org/10.1080/09593330309385684
11. Chang I-S, Le Clech P, Jefferson B, Judd S (2002) Membrane fouling in membrane bioreactors for wastewater treatment. J Environ Eng 128(11):1018–1029
12. Correia VM, Stephenson T, Judd SJ (1994) Characterisation of textile wastewaters-a review. Environ Technol 15(10):917–929
13. Crini G, Lichtfouse E (2019) Advantages and disadvantages of techniques used for wastewater treatment. Environ Chem Lett 17(1):145–155
14. Dos Santos AB, de Madrid MP, Stams AJ, van Lier JB, Cervantes FJ (2005) Azo dye reduction by mesophilic and thermophilic anaerobic consortia. Biotechnol Prog 21(4):1140–1145. https://doi.org/10.1021/bp050037t
15. Freeman N, Ikhu-Omoregbe D, Kuipa P, Muzenda E, Mohamed B (2009) Characterization of Effluent from textile wet finishing operations. Lecture Notes Eng Comp Sci 2178
16. Gao B-Y, Yue Q-Y, Wang Y, Zhou W-Z (2007) Color removal from dye-containing wastewater by magnesium chloride. J Environ Manage 82(2):167–172
17. Ghaly A, Ananthashankar R, Alhattab M, Ramakrishnan V (2014) Production, characterization and treatment of textile effluents: a critical review. J Chem Eng Process Technol 5(1):1–18
18. Hamedi H, Ehteshami M, Mirbagheri SA, Rasouli SA, Zendehboudi S (2019) Current status and future prospects of membrane bioreactors (MBRs) and fouling phenomena: a systematic review. Canad J Chem Eng 97(1):32–58

19. Holkar CR, Jadhav AJ, Pinjari DV, Mahamuni NM, Pandit AB (2016) A critical review on textile wastewater treatments: possible approaches. J Environ Manage 182:351–366
20. Hsu TC, Chiang CS (1997) Activated sludge treatment of dispersed dye factory wastewater. J Env Sci Health Part A Env Sci Eng Toxicol 32(7):1921–1932. https://doi.org/10.1080/109345 29709376655
21. Jafari N, Soudi MR, Kasra-Kermanshahi R (2014) Biodegradation perspectives of azo dyes by yeasts. Microbiology 83(5):484–497
22. Judd S (2010) The MBR book: principles and applications of membrane bioreactors for water and wastewater treatment. Elsevier
23. Kant R (2011) Textile dyeing industry an environmental hazard
24. Karabelas AJ, Plakas KV, Sarasidis VC (2018) How far are we from large-scale PMR applications? In: Current trends and future developments on (bio-) membranes. Elsevier, pp 233–295
25. Karcher S, Kornmüller A, Jekel M (2001) Screening of commercial sorbents for the removal of reactive dyes. Dyes Pigm 51(2–3):111–125
26. Katheresan V, Kansedo J, Lau SY (2018) Efficiency of various recent wastewater dye removal methods: a review. J Environ Chem Eng 6(4):4676–4697
27. Khan NA, Bhadra BN, Jhung SH (2018) Heteropoly acid-loaded ionic liquid metal-organic frameworks: effective and reusable adsorbents for the desulfurization of a liquid model fuel. Chem Eng J 334:2215–2221
28. Kitani A, Satoguchi K, Iwai K, Ito S (1999) Electrochemical behaviors of polyaniline/polyaniline-sulfonic acid composites. Synth Met 102(1–3):1171–1172
29. Klimaviciute R, Riauka A, Zemaitaitis A (2007) The binding of anionic dyes by cross-linked cationic starches. J Polym Res 14(1):67–73
30. Krull R, Döpkens E (2004) Recycling of dyehouse effluents by biological and chemical treatment. Water Sci Technol 49(4):311–317
31. Lee KE, Morad N, Teng TT, Poh BT (2012) Development, characterization and the application of hybrid materials in coagulation/flocculation of wastewater: A review. Chem Eng J 203:370–386
32. Maljaei A, Arami M, Mahmoodi NM (2009) Decolorization and aromatic ring degradation of colored textile wastewater using indirect electrochemical oxidation method. Desalination 249(3):1074–1078
33. Meimoun J, Wiatz V, Saint-Loup R, Parcq J, Favrelle A, Bonnet F, Zinck P (2018) Modification of starch by graft copolymerization. Starch Stärke 70(1–2):1600351
34. Mozia S (2010) Photocatalytic membrane reactors (PMRs) in water and wastewater treatment a review. Sep Purif Technol 73(2):71–91
35. Namasivayam C, Sumithra S (2005) Removal of direct red 12B and methylene blue from water by adsorption onto Fe (III)/Cr (III) hydroxide, an industrial solid waste. J Environ Manage 74(3):207–215
36. Ong C, Lau W, Goh P, Ng B, Ismail A, Choo C (2015) The impacts of various operating conditions on submerged membrane photocatalytic reactors (SMPR) for organic pollutant separation and degradation: a review. RSC Adv 5(118):97335–97348
37. Robinson T, McMullan G, Marchant R, Nigam P (2001) Remediation of dyes in textile effluent: a critical review on current treatment technologies with a proposed alternative. Biores Technol 77(3):247–255
38. Schoêberl P, Brik M, Braun R, Fuchs W (2005) Treatment and recycling of textile wastewater case study and development of a recycling concept. Desalination 171:173–183
39. Siddiqui MF, Singh L, Ab Wahid Z (2017) Treatment of dye wastewater for water reuse using membrane bioreactor and biofouling control. In: waste biomass management–a holistic approach. Springer, pp 121–136
40. Talarposhti AM, Donnelly T, Anderson GK (2001) Colour removal from a simulated dye wastewater using a two-phase anaerobic packed bed reactor. Water Res 35(2):425–432. https://doi.org/10.1016/S0043-1354(00)00280-3

41. Tunç YL, Yetis U, Çulfaz E (2014) Purification and concentration of caustic mercerization wastewater by membrane processes and evaporation for reuse. Sep Sci Technol 49. https://doi.org/10.1080/01496395.2014.914039
42. Tzoupanos N, Zouboulis A (2011) Preparation, characterisation and application of novel composite coagulants for surface water treatment. Water Res 45(12):3614–3626
43. Van der Bruggen B, Boussu K, De Vreese I, Van Baelen G, Willemse F, Goedeme D, Colen W (2005) Industrial process water recycling: principles and examples. Environ Prog 24(4):417–425
44. Vandevivere PC, Bianchi R, Verstraete W (1998) Review: Treatment and reuse of wastewater from the textile wet-processing industry: Review of emerging technologies. J Chem Technol Biotechnol 72(4):289–302. https://doi.org/10.1002/(SICI)1097-4660(199808)72:4<289::AID-JCTB905>3.0.CO;2-%23
45. Wang C, Alpatova A, McPhedran KN, El-Din MG (2015) Coagulation/flocculation process with polyaluminum chloride for the remediation of oil sands process-affected water: performance and mechanism study. J Environ Manage 160:254–262
46. Wang X, Jiang S, Tan S, Wang X, Wang H (2018) Preparation and coagulation performance of hybrid coagulant polyacrylamide–polymeric aluminum ferric chloride. J Appl Polym Sci 135(23):46355
47. Wang Z, Ma J, Tang CY, Kimura K, Wang Q, Han X (2014) Membrane cleaning in membrane bioreactors: a review. J Membr Sci 468:276–307
48. Weisburger JH (2002) Comments on the history and importance of aromatic and heterocyclic amines in public health. Mutation Res Fund Mol Mech Mutagen 506–507:9–20. https://doi.org/10.1016/S0027-5107(02)00147-1
49. Xia K, Liu X, Wang W, Yang X, Zhang X (2020) Synthesis of modified starch/polyvinyl alcohol composite for treating textile wastewater. Polymers 12(2):289
50. Ye B, Li Y, Chen Z, Wu Q-Y, Wang W-L, Wang T, Hu H-Y (2017) Degradation of polyvinyl alcohol (PVA) by UV/chlorine oxidation: radical roles, influencing factors, and degradation pathway. Water Res 124:381–387
51. Yukseler H, Uzal N, Sahinkaya E, Kitis M, Dilek F, Yetis U (2017) Analysis of the best available techniques for wastewaters from a denim manufacturing textile mill. J Environ Manage 203:1118–1125
52. Zheng X, Shen Z-P, Shi L, Cheng R, Yuan D-H (2017) Photocatalytic membrane reactors (PMRs) in water treatment: configurations and influencing factors. Catalysts 7(8):224
53. Zheng Y, Yao G, Cheng Q, Yu S, Liu M, Gao C (2013) Positively charged thin-film composite hollow fiber nanofiltration membrane for the removal of cationic dyes through submerged filtration. Desalination 328:42–50

Evaluation of Fe–Mn–Zr Trimetal Oxide/Polyaniline Nanocomposite as Potential Adsorbent for Abatement of Toxic Dye from Aqueous Solution

Bibek Saha, Animesh Debnath, and Biswajit Saha

Abstract The discharge of industrial effluents containing harmful synthetic dyes has gone up significantly since the last few decades and poses a severe environmental threat. Among a wide variety of treatment techniques available for the remediation of dye-laden wastewater, the adsorption process has been reported to be an effective method. Nanoscale adsorbents derived from metal oxides and their composites have recently attracted attention due to their enhanced ability to adsorb contaminants. The removal of congo red (CR) dye from dye solution was studied using a new adsorbent Fe–Mn-Zr trimetal oxide nanocomposite synthesized with polyaniline (PANI). The structural characterization of the Fe–Mn-Zr/PANI was investigated by X-ray diffraction (XRD) analysis, scanning electron microscopy (SEM) analysis and energy dispersive X-ray (EDX) analysis. It was observed that the adsorption process was substantially dependent on sonication time, the dose of adsorbent and the initial concentration of CR dye. Optimum solution pH was obtained as 5.0 from the effect of the pH study that has been performed for different initial concentrations of CR dye. The maximum percentage of adsorption efficiency for CR dye obtained was 89.25% at the initial concentration of 20 mg/L, with 0.2 g/L adsorbent dose and pH 5.0 within 15 min. In this work, kinetic analysis was performed utilizing pseudo-first-order, pseudo-second-order and intra-particle diffusion kinetic models, with the second-order kinetics model being shown to be the best fit. In comparison to pseudo-first-order kinetics, the experimental and calculated values of Q_e for the pseudo-second-order kinetics were in good agreement. The isotherm study was performed by varying the initial CR concentration from 5 to 70 mg/L with 0.1–0.4 g/L adsorbent dose. The adsorption experiments are best fitted with the Langmuir isotherm, giving a CR dye adsorption capacity of 111.111 mg/g at pH 5.0. The correlation coefficient value (0.991) indicates an accurate fit for the Langmuir model. The primary adsorption mechanism is the electrostatic interaction between the sulfonated

B. Saha · A. Debnath (✉)
Department of Civil Engineering, National Institute of Technology Agartala, Agartala, Jirania, Tripura (W) 799046, India

B. Saha
Department of Physics, National Institute of Technology Agartala, Agartala, Jirania, Tripura (W) 799046, India

© The Author(s), under exclusive license to Springer Nature Singapore Pte Ltd. 2022
A. Khadir and S. S. Muthu (eds.), *Polymer Technology in Dye-containing Wastewater*, Sustainable Textiles: Production, Processing, Manufacturing & Chemistry, https://doi.org/10.1007/978-981-19-1516-1_2

group ($-SO_3Na$) of negatively charged CR dye molecules and the positively charged amine group ($-NH_2$) of PANI surface. The response surface methodology (RSM) was studied to optimize the removal of the CR dye used with different experimental parameters like initial CR concentrations (5, 10, 15, 20 and 25 mg/L), with adsorbent dose (0.1–0.5 g/L), at reaction time (2 min, 4 min, 6 min, 8 min and 10 min) and solution pH: 5.0. Maximum CR dye removal of 97.78% was obtained at an optimum contact time of 15 min, with an initial CR dye concentration of 20 mg/L and adsorbent dose at 0.2 g/L.

Keywords Metal oxide nanoparticles · Polyaniline · Nanocomposite · Congo red · Dye adsorption · Response surface methodology · Adsorption mechanism · Kinetic study · Isotherm modelling · Central composite design · Optimization

1 Introduction

Society has become increasingly concerned about environmental sustainability during the past several decades. As a result of the challenges, people are now concerned about industrial wastewater's potential adverse effects on the environment. The textile industry is the second-largest source of water pollution after the agriculture industry [31, 47]. Dyes are coloured compounds widely used in textile printing, rubber, cosmetics, plastic, pharmaceutical and leather industries to colour their product, thus resulting in generating a large amount of coloured wastewater [12, 26]. Around 7×10^5 tonnes of dyes are produced annually by these industries, with 10–15% of the dyes being disposed of as industrial effluents into rivers and lakes [4]. Many of these dyes are carcinogenic, mutagenic and teratogenic, as well as harmful to humans, aquatic life and other living things. Hence their removal from wastewater becomes environmentally important. Congo red (CR) (the sodium salt of 3,3′-([1,1′-biphenyl]-4,4′-diyl) bis (4-aminonaphthalene-1-sulfonic acid) dye is a hazardous anionic azo dye that has been attributed to carcinogenicity, liver and kidney disorders, skin and eye disorders, and genotoxicity in a variety of human organs [33, 36]. CR dyes are mostly derived from effluents produced by sectors such as textiles, printing, paper production, pharmaceuticals and food processing [16, 37]. Nowadays, among the different dye removal technologies available such as electrochemical techniques, catalysis, photodegradation, biological treatments, adsorption and chemical oxidation/precipitation techniques, considering parameters such as initial cost, flexibility, design simplicity, ease of operation and sensitivity to toxic chemicals, adsorption technique has been reported to be superior to other technologies. Adsorption and ion exchange are the two mechanisms that cause decolourization [1, 17], and is affected by various physiochemical variables, including the interaction between sorbent and dye, the surface area of sorbent, particle size, temperature, contact time and pH [39]. Nanomaterials as synthetic adsorbents have recently attracted a lot of interest among researchers for dye adsorption studies due to their inexpensive cost, considerably simpler synthesis process, large surface area and large pore volume when compared to some bulk materials. The use of conducting polymer along with metal oxides has gained massive interest in the field of environmental research considering their different enhanced physicochemical properties and various potential uses [51]. Polyaniline/ZnO nanocomposite as prepared by Deb et al. [21] giving MO dye removal of 98.13% and nano-sized magnetic Fe0/polyaniline (Fe0/PANI) nanofibers as used by Das et al. [18] giving CR dye removal of 98% shows the effectiveness of using conducting polymer along with metal oxides for water purification. Polyaniline (PANI) used in this study is one type of polyaromatic amine and its synthesis process is much easier [5, 38]. Polyaniline is a nitrogen-containing conductive polymer with desirable characteristics such as ambient stability, a simple synthesis procedure, cheap monomer cost and great flexibility [49]. The surface area of PANI may decrease in water because of its weak dispersion and susceptibility for agglomeration in irregular morphology [9]. Hence, it is now focused on the assembly of PANI-multimetal-based nanocomposite having a larger surface area for efficient removal of toxic chemicals from water.

In light of these considerations, we synthesized a mixed phase of Fe–Mn-Zr/PANI using the oxidative polymerization method in the current study. Detailed characterization of synthesized nanoparticles was done by a variety of characterization techniques, including X-ray diffraction (XRD) pattern, scanning electron microscopy (SEM) and energy dispersive X-ray (EDX) analysis to study their crystalline nature, surface characteristics and morphology. Kinetic, isotherm and thermodynamic studies were performed to investigate the underlying process of dye ion adsorption onto fabricated nanoparticles. To analyse the adsorption behaviour, mathematical models based on response surface methodology (RSM) were constructed utilizing adsorption experimental data for precise prediction of CR dye removal efficiency [13, 14].

2 Materials and Methods

2.1 Chemicals and Reagents

CR dye stock solution (50 mg/L) was prepared by dissolving analytical grade CR dye powder which was purchased from Merck (India) in de-ionized water. Hydrochloric acid (HCl) (98% purity) and sodium hydroxide (NaOH) pellets were purchased from Merck (India) to increase or decrease the pH of a solution. A sonicator (Revotek Ultrasonic cleaner) was used to mix the adsorbent with dye solution, and a research centrifuge (Tanko KT-182A) was used to separate the adsorbent from the dye solution. Aniline monomer which was double-distilled before polymerization was purchased from Merck (India). Ammonium peroxodisulphate (APS) used for the polymerization of aniline was purchased from Sigma-Aldrich. Analytical grade ferric chloride anhydrous ($FeCl_3$), manganese chloride tetrahydrate ($MnCl_2 \cdot 4H_2O$) and zirconium oxychloride octahydrate ($ZrOCl_2 \cdot 8H_2O$) used as Fe, Mn and Zr combined trimetal form were purchased from Sigma-Aldrich. Using a UV–Vis spectrophotometer (Hach, DR-5000), the availability of CR dye concentration in effluent solutions was assessed spectrophotometrically [35, 44] at $\lambda_{max} = 500$ nm.

2.2 Synthesis of Fe–Mn-Zr Trimetal Oxide Nanoparticles

Initially, 32 g of $FeCl_3$ with 20_g of $MnCl_2 \cdot 4H_2O$, and 32_g of $ZrOCl_2 \cdot 8H_2O$ were dissolved in 500 mL of de-ionized water and rapidly agitated until a homogeneous solution was obtained [13, 20, 27]. After that, NaOH solution was formed by mixing 40 g of NaOH pellets with 200 mL de-ionized water. Following that, the NaOH solution which was formed was gradually added to the Fe–Mn-Zr solution while stirring was done slowly, and the pH of the solution was kept around 10. A heavy dark-brownish precipitate was produced, which was carefully separated by filtering and was rinsed multiple times with de-ionized water in order to remove excess alkalinity and contaminants. After washing the remaining precipitate was dried at 110 °C before being calcined in the muffle furnace for 8 h at 300 °C temperature. The sample was crushed to a fine powder when cooled to room temperature.

2.3 Synthesis of Fe–Mn-Zr Trimetal Oxide/PANI Nanocomposite

In a typical experiment, 150 ml of distilled water was poured into a 500 ml beaker and 12.25 g of APS was dissolved in it. In another beaker 150 ml of 1.0 M HCl solution was taken where 5 ml of aniline was added. Then both the solutions were stirred properly to remove all the precipitation. As for oxidative polymerization of aniline to poly-aniline highly acidic condition is favourable [15], and some more amount of HCl was added. The beaker having a mixed solution of aniline and HCl is put into an ultrasonicator. Under ultrasonication, Fe–Mn-Zr powder is added to the beaker and proper dispersion was attained by ultrasonication for 5 min. Then APS solution is added slowly to the beaker under ultrasonication and was kept for 15 min. The formation of PANI can be seen by greenish colour in the beaker [25]. The beaker was kept for 2 days in undisturbed condition to complete the reaction. After 2 days the precipitate was collected, and to remove unreacted aniline it was first washed with 0.5 N HCl solution and then with distilled water until a colourless solution appeared. The washed precipitate was dried for 8 h in a muffle furnace at 600 °C. After cooling to room temperature, the material was ground to a fine powder and used as an adsorbent in this study.

2.4   Ultrasound-Assisted Adsorption of CR Dye

The adsorption experiments assisted by ultrasonication were conducted to investigate the effect of various parameters on trioxide nano-adsorbent (Fe–Mn-Zr) with PANI composite for the removal of CR dye. For each experiment, 100 ml of 20 mg/L of CR dye was used which was prepared from the stock solution and a certain amount of nano-adsorbent was added to the solution. The mixture was sonicated in an ultrasonic bath for different time intervals to find the equilibrium time. By using HCl and NaOH solution, the pH of the dye solution was adjusted in this procedure. After sonication was done for a designed time interval, it has been centrifuged @6000 rpm for 15 min and Fe–Mn-Zr/PANI composite was parted from the solution and the residuals were analysed to determine the maximum absorbance of CR dye using UV-spectrophotometer. The removal efficiency (R) expressed in terms of percentage and equilibrium adsorption capacity, Q_e (mg/g), were calculated by the following equations [8]:

$$R = \frac{C_0 - C_e}{C_0} \times 100\% \tag{1}$$

$$Q_e = \frac{(C_0 - C_e) \times V}{m} \tag{2}$$

where C_0 and C_e are the initial and equilibrium dye concentrations in solution (mg/L), m is the adsorbent dose (g) and V is the dye solution volume (L).

2.5   Statistical Modelling by Response Surface Methodology (RSM)

Response surface methodology (RSM) can be considered as one of the most important design experimental techniques useful for developing, improving and optimizing processes and de-colourization of the dye solution. RSM is also defined as a set of mathematical and statistical skills used to fit a polynomial equation in which any data can be entered and is influenced by a number of process variables [2]. The fundamental goal of RSM is to find a region that meets the operating specifications. The most popular design of experiment (DOE) under RSM is the central composite design (CCD) which was employed in this study to optimize dye removal efficiency [6, 7]. This tool was used to explore how different experimental parameters (adsorbent dose, time and CR dye initial concentration) affected responses (CR dye removal efficiency). CCD has been used to construct a three-factor design of experiment (DOE) with five coded levels ($-\alpha$, -1, 0, $+1$ and $+\alpha$) for contact duration (X_1), adsorbent dose (X_2) and initial dye concentration (X_3). There were 20 experimental runs in total, containing eight factorial points, six axial points and six central points.

The following equation was used to calculate the total number of experiments [11].

$$N = 2^n + 2n + n_c = 2^3 + 2 \times 3 + 6 = 8 + 6 + 6 = 20 \tag{3}$$

In Eq. (3) N is the total number of experiments, n is the number of experimental parameters and n_c is the number of central point replicates. Design Expert (8.0.6.1 version from Stat Ease Inc., USA) was used in this study to prepare the design of experiment, evaluate the model and optimize the experimental findings of CR dye removal values using the RSM model which is shown in Table 1. The mathematical relationship between process variables and CR dye removal (percentage) was determined using the results of 20 experimental runs, and is as shown below [11]:

$$Y = \beta_0 + \sum_{j=1}^{k} \beta_j X_i + \sum_{j=1}^{k} \beta_{jj} X_j^2 + \sum_{j=1}^{k} \sum_{i=1}^{k} \beta_{ji} X_j X_i \tag{4}$$

where Y is the predicted response (CR dye removal efficiency); $X_i \ldots X_k$ are the independent variables (in this case $k = 3$); β_0 is constant or zero-order coefficient; β_i, β_{ii} and β_{ij} are the coefficients of linear, quadratic and interaction terms, respectively.

3   Results and Discussions

3.1   Characterization of Fe–Mn-Zr Trimetal Oxide/PANI Nanocomposite

The XRD pattern of the synthesized Fe–Mn-Zr/PANI nanocomposite is shown in Fig. 1a. The XRD pattern was obtained using the source of Cu Kα radiant of wavelength 8.389 Å. In the XRD patterns the appearing peaks at 2θ value of $25.26°$, $30.31°$, $35.26°$, $50.77°$ and $60.21°$ were observed. The XRD pattern exhibits with its crystalline behaviour the diffraction peak at $25.26°$ corresponding to the (082) crystal plane of pure PANI [29] which is confirmed as it creates PANI molecular chain with each molecule having an amorphous structure. The peak at $30.31°$ indicates to (110) miller plane of ZrO_2 [24, 45]. Another peak appearing at $35.26°$ corresponds to (051) miller plane of $MnFe_2O_4$ nanoparticles [10]. The remaining two peaks occurring at $50.77°$ and $60.21°$ resemble reflections from (054) and (039) miller planes of α-Fe_2O_3 nanoparticles with rhombohedral structure [28]. The presence of Fe_2O_3, $MnFe_2O_4$ and ZrO_2 nanoparticles creates dissimilarity in their lattice structure which results in smaller size crystalline with an increase in effective surface area, consequently resulting in an increase in the number of effective sites nearer to the edge of the crystalline [45]. Hence, the XRD patterns ensure the effective synthesis of Fe–Mn-Zr/PANI nanoparticle composite. SEM is also commonly employed in morphological studies to determine phases based on quantitative chemical analysis

Table 1 Variables in the adsorption process and their levels in the CCD, as well as observed and projected CR dye removal (%) from the RSM model

Parameter	Levels				
	$-\alpha$	Low(-1)	Central(0)	High($+1$)	$+\alpha$
X_1: Sonication time (min)	2	4	6	8	10
X_2: Adsorbent dose (g/L)	0.1	0.2	0.3	0.4	0.5
X_3: Initial CR conc. (mg/L)	5	10	15	20	25

Run	Factors			CR Removal (%)	
	X_1	X_2	X_3	Experimental	Predicted
1	6	0.3	5	89.15	92.59
2	8	0.4	20	86.96	88.05
3	8	0.2	10	96.41	91.60
4	6	0.3	15	89.26	89.51
5	4	0.4	20	81.55	85.30
6	6	0.3	15	89.26	89.51
7	2	0.3	15	93.75	93.01
8	8	0.4	10	97.85	97.41
9	6	0.3	15	88.93	89.51
10	6	0.5	15	95.68	93.21
11	10	0.3	15	95.91	97.72
12	6	0.3	25	76.79	74.41
13	6	0.3	15	89.18	89.51
14	6	0.1	15	79.38	82.91
15	4	0.4	10	96.51	95.97
16	6	0.3	15	90.44	89.51
17	6	0.3	15	88.93	89.51
18	8	0.2	20	84.62	84.09
19	4	0.2	10	91.79	89.63
20	4	0.2	20	81.44	80.81

or crystalline structure. Figure 1b shows a FESEM picture of the synthesized Fe–Mn-Zr/PANI, which demonstrates that the sample is made up of grains with spherical, cylindrical and rod-like shapes having some porosity, as well as some grain moulds. The grains have an average diameter of 20–60 nm, which is appropriate for an adsorbent since its small size offers a large surface area. The elemental composition and chemical characterization of a sample can be determined using EDX. To investigate the chemical composition of the prepared nanocomposite, it was proposed to analyse the EDX spectrum as shown in Fig. 1c, which shows the EDX spectrum of

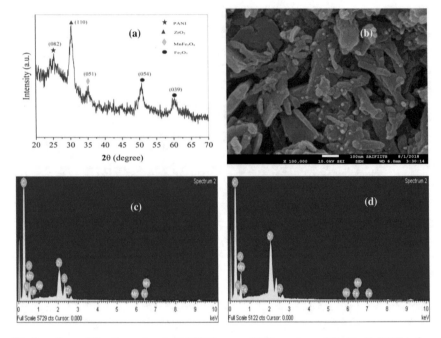

Fig. 1 a X-ray diffraction pattern **b** FESEM image **c** EDX spectrum of Fe–Mn-Zr/PANI before adsorption and **d** EDX spectrum of Fe–Mn-Zr/PANI after adsorption of CR dye

Fe–Mn-Zr/PANI before adsorption and Fig. 1d which shows the EDX spectrum of Fe–Mn-Zr/PANI after adsorption of CR dye. The peaks obtained for Na, S and C in the adsorbent's post-adsorption EDX spectra also reveal traces of the loaded CR dye onto the nanocomposite [11].

3.2   Effect of Solution pH and Initial Dye Concentration on CR Dye Adsorption Efficiency

The solution pH has a significant impact on the ionization of adsorbate molecules. It may also have an impact on the adsorbent surface's surface charge and active functional group. Therefore, the influence of solution pH on CR dye adsorption onto Fe–Mn-Zr/PANI has been investigated in the pH range of 3.0–10.0 with a CR dye concentration of 20 mg/L. The absorbent dose was fixed at 0.2 g/L for this experiment. The solution was stirred in an ultrasonicator bath for about 15 min. The variation of removal efficiency of the adsorbent with solution pH for CR dye has been plotted in Fig. 2a. From the plot, it can be clearly seen that the maximum removal efficiency was obtained at pH 3–5 with 89.26% removal efficiency at pH 5.0 in the acidic condition of the solution. The removal efficiency of CR dye reduced from 80.74 to 35.74% as

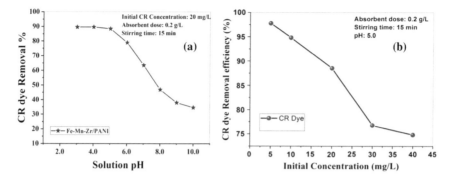

Fig. 2 a Variation of CR dye removal (%) with pH of the solution **b** Variation of CR dye removal (%) with the initial concentration of CR dye

the solution pH increased from 6.0 to 10.0. Hence optimum solution pH was fixed at 5.0 for CR dye as the working pH for this adsorption experiment because a neutral value of pH is more economical than in acidic and basic conditions.

The initial concentration of dye is one of the most important aspects influencing dye adsorption onto Fe–Mn-Zr/PANI adsorbent. The capacity of Fe–Mn-Zr/PANI to adsorb dyes is, of course, dependent on the initial concentration of dye. Figure 2b depicts the variation of colour removal efficiency of the Fe–Mn-Zr/PANI trimetal oxide nanocomposite affected by initial dye concentration. It can be observed that CR dye removal efficiency increases as the initial concentration of dye decreases. Thereafter, it is decreasing with a rise in initial concentration. Specifically, at an initial CR dye concentration of 16 mg/L, the CR removal efficiency was 92.12%, while at an initial dye concentration of 20 mg/L, the removal efficiency reduced to 88.52% after 15 min of reaction time under ultrasonication. Hence, it can be stated that the adsorption of dye was dependent on initial dye concentration.

3.3 Effect of Dose of Adsorbent and Sonication Time on Dye Adsorption Efficiency

Adsorption experiments were performed to investigate the effect of adsorbent dose by varying the adsorbent dose as 0.2, 0.3, 0.4 and 0.5 g/L with 20 mg/L of initial CR dye concentration maintaining the solution pH at 5.0. As depicted by Fig. 3a, the plot has been clearly defined that the adsorption of CR dye onto the adsorbent surface increased with the rise in the dose of Fe–Mn-Zr/PANI nanocomposites. It also depicts that the removal efficiency for CR dye reaches equilibrium at 96.30% corresponding to an adsorbent dosage of 0.5 g/L. This could be attributed to an increase in the number of active sites available for dye adsorption as the amount of adsorbent doses increases.

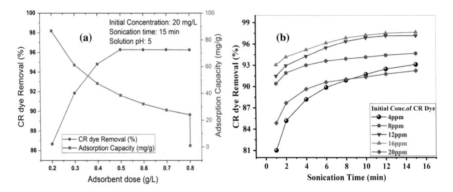

Fig. 3 **a** Effect of adsorbent doses, **b** effect of initial CR dye concentration (ppm) and contact time (min) on CR dye removal efficiency

At the initial dye concentration of 4–20 mg/L, the effect of sonication time on the amount of dye adsorbed was studied with the Fe–Mn-Zr/PANI dose of 0.25 g/L. Figure 3b clearly presents that the extent of removal efficiency of CR dye by Fe–Mn-Zr/PANI nanocomposite was found to increase uniformly with the reaction time under ultrasonication process at an interval of 15 min for 0.25 g/L adsorbent dose and then it almost became equilibrium after a certain period of time.

3.4 Sono-Assisted Kinetic Analysis of Adsorption of CR Dye onto Fe–Mn-Zr/PANI

The kinetic studies help to indicate the mechanism of adsorption reaction and enumerate the valuable information to explore the rate of dye removal. The mechanism of CR dye adsorption was studied by the fitting of the adsorption experimental data into the linear form of three typical kinetic models, namely pseudo-first-order, pseudo-second-order and intra-particle diffusion model [22]. The experimental data was acquired by varying the initial concentrations of 4, 8, 12, 16 and 20 mg/L for CR dye at 0.25 g/L of adsorbent dosage. Table 2 summarizes the numerical values of all significant kinetic parameters determined from the plot for three kinetic models. The slopes and intercepts of log Q_e against t plots were used to calculate the values of pseudo-first-order rate constants (k_f and Q_e). The slopes and intercepts of straight lines obtained by plotting $\frac{t}{Q_t}$ versus t were used to derive the second-order rate constants (k_s and Q_e). The slope and intercept of the plot Q_t versus $t^{1/2}$ give the intra-particle diffusion rate (k_{id} and c). Figure 4a, b show the fitting of adsorption data of CR dye with the linear form of pseudo-first-order and pseudo-second-order kinetic model.

From the calculated adsorption capacity, it can be easily stated that the pseudo-first-order kinetic model does not fit well as compared to the pseudo-second-order

Table 2 Kinetic parameters of CR dye adsorption onto Fe–Mn-Zr trimetal oxide (adsorbent dose of 0.25 g/L and initial concentration of 4–20 ppm)

Models and equation	Parameters	Initial CR concentration (ppm)				
		4	8	12	16	20
Pseudo-first-order kinetic $\log(Q_e - Q_t) = \log Q_e - \frac{k_f}{2.303}t$	$K_f \times 10^2$ (min^{-1})	2.74	2.35	3.06	2.53	3.09
	Q_e (mg g^{-1})	2.46	1.50	4.30	24.49	7.73
	R^2	0.945	0.947	0.981	0.981	0.878
Pseudo-second-order kinetic $\frac{t}{Q_t} = \frac{1}{k_s Q_e^2} + \frac{1}{Q_e}t$	$K_s \times 10^3$ (min^{-1})	385.00	512.00	220.50	225.00	169.00
	Q_e (mg g^{-1})	12.99	31.25	47.62	66.67	76.92
	R^2	0.999	0.992	0.995	0.999	0.999
	H × 105	0.65	5	5	10	10
Intra-particle diffusion $Q_t = k_{id}t^{\frac{1}{2}} + c$	K_{id} (mg g^{-1} min$^{1/2}$)	0.95	0.68	1.32	1.41	3.16
	R^2	0.888	0.030	0.993	0.952	0.931
	C (mg g^{-1})	12.15	28.34	42.62	58.13	65.16
Experimental value	Q_e	14.88	30.29	46.67	62.56	73.70

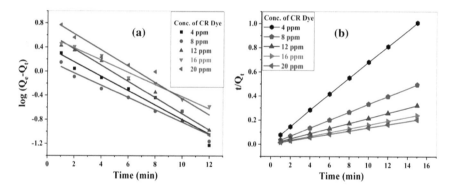

Fig. 4 **a** Pseudo-first-order and **b** pseudo-second-order plot of CR dye adsorption onto Fe–Mn-Zr/PANI

kinetic model. Hence it clearly depicts that the pseudo-second-order kinetic model is better to describe the kinetic model of adsorption of CR dye onto Fe–Mn-Zr/PANI nanocomposites and the rate of reaction plays a major role in the chemisorption process. The values of different parameters such as Q_e, k_f, k_s and R^2 obtained for pseudo-first-order and second-order were given in Table 2, and from the table, it has been found that the values of R^2 for the pseudo-second-order model are greater than the pseudo-first-order model. The obtained theoretical value of Q_e is much closer to the $Q_{e(exp)}$ value, which means that for kinetic adsorption the pseudo-second-order model describes the best-fitted curve [48].

3.5 Isotherm Study of Sono-Assisted Adsorption of CR Dye Onto Fe–Mn-Zr/PANI

The adsorption isotherm is the equilibrium correlation between the concentration of adsorbate on the solid phase and in the liquid phase [41]. Adsorption isotherm studies were performed at room temperature with an initial concentration of 5–70 mg/L for CR dye and adsorbent dose of 0.2 g/L and solution pH 5.0. Figure 5a–c demonstrates the linear fitting of standard isotherm models such as the Langmuir, Freundlich and Temkin, respectively, to which the experimental equilibrium data of adsorption CR dye onto Fe–MN-Zr/PANI were fitted.

The isotherm parameters for CR dye and the R^2 value were obtained from the linear fitting of experimental data, which are summarized in Table 3. After getting the R^2 value it is clear that the Langmuir isotherm model was the simplest explanation of this adsorption reaction with uniform distribution of active sites associated with the interaction between adsorbed molecules. Though the applicability of Langmuir isotherm separation factor R_L was also calculated and the obtained R_L value was found in between 0.1, which indicated that the process of adsorption isotherm is favourable. The intercept and slope of the graph $\frac{C_e}{Q_e}$ versus C_e were used to calculate the Langmuir isotherm constant 'b' and Q_m (theoretical) maximum adsorption

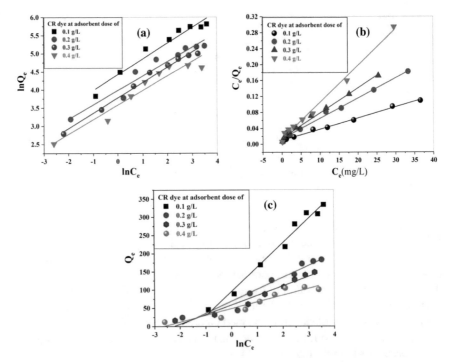

Fig. 5 **a** Langmuir isotherm, **b** Freundlich isotherm, **c** Temkin isotherm model for CR dyes adsorption onto Fe–Mn-Zr/PANI nanocomposite

Table 3 Equilibrium constants for adsorption obtained from Langmuir, Freundlich and Temkin isotherm fittings for adsorption of CR dye onto Fe–Mn-Zr/PANI trimetal oxide

Models	Equation	Parameters	Adsorbent dose 0.2 g/L
Langmuir	$\frac{C_e}{Q_e} =$ $\frac{1}{Q_m b} + \frac{C_e}{Q_m}$	Q_m (mg g^{-1})	196.08
		b (L mg^{-1})	0.35
		R_L	0.36–0.04
		R^2	0.991
Freundlich	$\ln Q_e =$ $\ln K_F +$ $\frac{1}{n} \ln C_e$	n	2.52
		K_F	1.38
		R^2	0.939
Temkin	$Q_e =$ $\beta_1 \ln K_T +$ $\beta_1 \ln C_e$	β_1	32.79
		K_t (L mg^{-1})	8.16
		R^2	0.937

capacity. The important factor R_L is determined based on the following equation:

$$R_L = \frac{1}{(1 + bC_0)} \tag{5}$$

where C_0 is the initial concentration of CR dye and b is the constant. In Langmuir isotherm, the maximum adsorption capacity (Q_m) is found to be 357.14–108.70 mg g^{-1} for CR dye.

Similarly, for the case of the Freundlich isotherm, the intercept and slope of the $\log C_e$ versus Q_e plot were used to calculate the K_f and n values. In the case of the Temkin isotherm model, the constants were determined from the slope and intercept, and the graph was plotted using Q_e versus $\ln C_e$. The R^2 value for Langmuir ($0.991 < R^2 < 0.995$), Freundlich ($0.929 < R^2 < 0.987$) and Temkin ($0.887 < R^2 < 0.982$) shows the removal efficiency CR dye. Thus, among the three standard isotherm models, the Langmuir isotherm has been considered the most appropriate, yielding the best correlation factor for the best experimental results.

3.6 Intra-Molecular Interaction Between CR Dye and Fe–Mn-Zr/PANI

Since the pH of the solution can substantially affect the surface charge of metal oxide-based adsorbents, the pH of the solution has a major impact on the adsorption process at the solid/liquid interface. The optimum solution pH was set at 5.0, which can be regarded as an acidic condition. PANI's surface becomes positively charged in this condition. In the structure of CR dye, there is a sulfonated group ($-SO_3Na$) that disintegrates in aqueous environments and appears in complete anionic form [42,

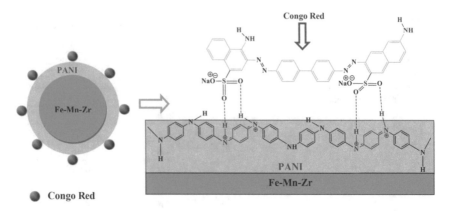

Fig. 6 Schematic diagram of the interaction between anionic CR dye molecules with a positively charged PANI surface

52]. As a result, the primary adsorption mechanism is the electrostatic interaction between the negatively charged CR dye molecules [50] and the positively charged PANI surface [3] and also π–π interaction between CR dye molecules and Fe–Mn-Zr/PANI nanocomposites [20]. The oxygen molecule present in the sulfonated group ($-SO_3Na$) of CR dye interacts with the hydrogen molecule present in amide ($-NH_2$) of the PANI surface. The schematic diagram of the interaction of negatively charged CR molecules onto positively charged PANI surface is shown in Fig. 6.

3.7 Statistical Analysis of Central Composite Design (CCD) and 2D/3D Response Plots

To optimize process variables on response, the proposed design of experiments (DOE) was used. RSM's core concept is to use a series of designed experiments to produce the best possible response of operational factors. RSM can be used to optimize operational factors in order to maximize the production of a specific substance. The primary goal of RSM is to determine the optimal dynamic adsorption conditions for maximum dye removal. To study the interaction between the experimental variables, initial CR dye concentration, adsorbent dose and contact time must be analysed by an experimental design called central composite design for CR dye. CCD helps in optimizing the experimental parameters with minimum possible experimental runs. Design-Expert 11.0 was used to apply the CCD with three factors at five levels. The independent variables, i.e., for CR dye conc. (X_1): 5–25$_{mg}$/L, adsorbent dose (X_2): 0.1–0.5$_g$/L, contact time (X_3): 2–15 min, each was coded at five levels, i.e., $-\alpha$ to $+\alpha$ at the selected ranges obtained from the preliminary experiment. The level of independent variables and the experimental data used by CCD are described in Table 1.

The response surface quadratic model was used to express all of the analytical processes of the experimental results. ANOVA technique, which was obtained from a response surface quadratic model, was used to evaluate the regression coefficient. It was also used to know the extent up to which the model is statistically significant. Thereby, it was used for fitting the mathematical models of the experimental data to optimize the total region for response variables, as shown in Table 4. For the CR dye, a second-order or quadratic polynomial model was employed to estimate the response variables, as given in Eq. (6).

The ANOVA findings are summarized to anticipate the model's soundness. This is a statistical approach that divides overall variation in a collection of data into subgroups linked to specific sources of variation. By dividing the sum of the squares of each variation source by their degree of freedom, the mean square values were determined. The results were evaluated using a variety of descriptive coefficients (R^2) for each coefficient in the equation, which were computed using the Fisher's F test and the probability value. With a lowering in the P-value, the model becomes more significant, and it could predict more correctly the response function. The projected correlation coefficient $(R^2$ Pred.) and adjusted correlation coefficient $(R^2$ Adj.) were also used to assess the model's fit [45].

Table 4 Analysis of variance (ANOVA), quadratic summary statistics and coefficient of regression for the RSM quadratic model developed

Source of variation	Sum of squares	df	Mean squares	F-value	P-value	Status	Regression coefficient	
							Factor	Estimate
Model	605.65	9	67.29	7.84	0.0017	Significant	Intercept	89.51
X_1	330.42	1	330.42	38.47	0.0001		X_1	−4.54
X_2	106.14	1	106.14	12.36	0.0056		X_2	2.58
X_3	22.25	1	22.25	2.59	0.1385		X_3	1.18
X_1X_2	1	1.72	0.20	0.6640	0.6640		X_1X_2	−0.46
X_1X_3	1	0.86	0.10	0.7576	0.7576		X_1X_3	0.33
X_2X_3	1	0.14	0.016	0.9017	0.9017		X_2X_3	−0.13
X_1^2	1	56.73	6.61	0.0279	0.0279		X_1^2	−1.50
X_2^2	1	3.30	0.38	0.5493	0.5493		X_2^2	−0.36
X_3^2	1	53.80	6.26	0.0313	0.0313		X_3^2	1.46
Residual	85.89	10	8.59					
Pure Error	1.58	5	0.32					
Cor Total	691.53	19						
Quadratic summary statistics response (CR removal, %)	R^2	Adj.R^2	Std. Dev	CV%	Press	Adeq Precision		
	0.8758	0.7640	2.93	3.29	701.63	11.247		

df, degree of freedom; CV, Coefficient of variation

The lack of fit cannot be considered significant because the F-value of CR dye is 53.21. Therefore, the proposed quadratic or second-order response surface model in this analysis for estimating CR removal % is considered to be quite satisfactory. The final equation in terms of the actual factor is

$$Y_{CR\%} = + 85.87 + 0.975X_1 - 65.34X_2 - 4.095X_3 - 0.928X_1X_2 + 0.033X_1X_3$$
$$- 0.656X_2X_3 - 0.060X_1^2 - 36.216X_2^2 + 0.366X_3^2 \qquad (6)$$

Figure 7 shows the 2D contour plots as well as the 3D response surface plots of CR dye for a graphical depiction of the independent variables such as contact time, adsorbent dose and initial concentration of CR dye, as well as the simultaneous impact of the aforesaid variables on the decolourization of the dyes, using RSM. In such plots, the response functions of two factors are presented while all other factors are at fixed levels. Figures 2b and 3a depict the effect of initial dye (CR) concentration and adsorbent dose, respectively, on CR dye removal efficiency. It can be clearly depicted that the removal % increases with the decrease in the initial concentration of dye and increase when there is a rise in adsorbent dose. Simultaneous interaction of contact time on the CR dye removal % can be represented in Fig. 3b. With an increase in adsorbent dose surface area increases and multiple active sites which gives a highly efficient adsorption process. The 2D and 3D plots represented in the above figure indicate the negative impact of dye concentration and positive impact of contact time on dye removal percentage.

Figure 8a indicates an excellent contract between experimental and predicted model data for CR dye. From Fig. 8b, it has been clearly defined that the RSM model has residual ranging from + 3.00 to − 3.00 for CR dye, from which it has been assured that the RSM model has a good experimental value with a lower deviation.

3.8   Performance Evaluation Study of Fe–Mn-Zr/PANI Adsorbent

In this study, the effectiveness of ultrasound-assisted CR dye adsorption onto Fe–Mn-Zr/PANI as compared to other reported adsorbents was assessed on the basis of maximum adsorption capacity, equilibrium time, solution pH and mixing method used. Table 5 shows that Q_m of 0.2 g/L Fe–Mn-Zr/PANI for CR dye is 357.143 mg/g, which is reasonably better compared to many other reported adsorbents. For example, adsorbents like bentonite, maghemite nanoparticle, bael shell carbon, nanocrystalline MFe_3O_4 spinel ferrites, bagasse fly-ash, raw pine and hematite nanocomposite have shown Q_m of 158.7, 208.33, 98.03, 244.5, 9.638, 32.65 and 203.60 mg/g, respectively. In comparison to all previously published adsorption methods, ultrasound-aided adsorption of CR dye over Fe–Mn-Zr/PANI reached equilibrium extremely rapidly (within 15 min), indicating the improved performance of ultrasonic-assisted

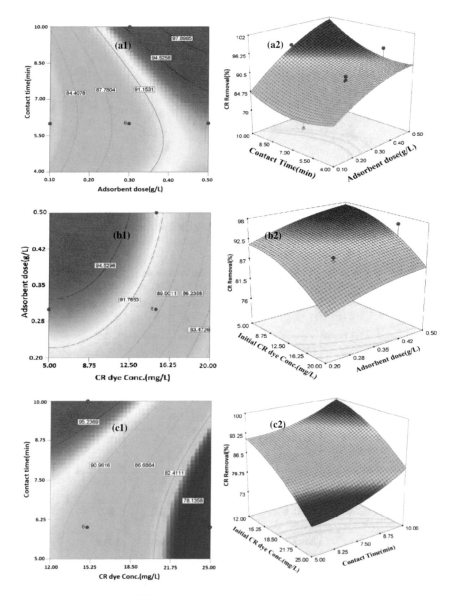

Fig. 7 2D contour plots and 3D response surface plots for simultaneous interaction of **a** contact time and adsorbent dose **b** Fe–Mn-Zr/PANI dose and initial conc. of CR dye **c** initial concentration of CR dye and contact time

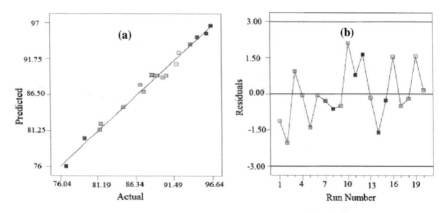

Fig. 8 **a** Experimental versus predicted removal (%) of CR dye **b** residual versus the number of runs for CR dye

Table 5 Evaluation of the performance of Fe–Mn-Zr/PANI adsorbent in terms of adsorption capacity and equilibrium time with some other adsorbents for CR dye adsorption

Adsorbent	Method used	Equilibrium time (min)	Solution pH	Adsorption capacity (mg/g)	References
ZnO@Ze composite	Batch stirring mixing	20	3.0	161.3	[34]
Fe_3O_4-TSPED-tryptophan (FTT)	Ultrasound-assisted mixing	10	3.0	183.15	[40]
Cornulaca monacantha stem (CS)	Batch stirring mixing	120	2.0	43.42	[43]
Cornulaca monacantha stem-based activated carbon (CS^{AC})	Batch stirring mixing	90	2.0	78.19	[43]
Amberite IRA-400	Batch stirring mixing	180	4.5	18.32	[46]
Activated carbon from coffee waste (CW)	Batch stirring mixing @220 rpm	180	3.0	90.90	[32]
Fe_2O_3 nanoparticle	Batch stirring mixing	200	4.0	203.66	[23]
Modified commercial zeolite	Batch stirring mixing	90	7.0	21.11	[30]
Fe–Mn-Zr/PANI trimetal nanocomposite	Ultrasound-assisted mixing	15	5.0	357.14	Present study

adsorption of CR dye. The enhanced performance is due to the fact that the ultrasonic-assisted adsorption process greatly improves the transfer of mass and rate of diffusion, which results in faster attainment of equilibrium. Dye molecules are forced into accessible pores on the Fe–Mn–Zr/ PANI surface by ultrasound waves, resulting in more adsorption sites. Additionally, ultrasound can improve the porous structure of Fe–Mn–Zr/PANI and minimize mesopore clogging, resulting in better mass transfer and, eventually, improved adsorption capacity in a brief period of time [19].

4 Conclusions

In batch mode studies, a novel nano-adsorbent (Fe–Mn–Zr/PANI) for ultrasonic-aided dye adsorption was developed and successfully employed to remove CR dye from an aqueous solution. The optimal pH of the solution was observed to be 5.0. Using kinetic adsorption models, the effects of sonication period (15 min), initial concentration of dye (4–20 mg/L) and dose of adsorbent (0.25 g/L) were assessed. The produced adsorbent was shown to be very efficient in this study, with a CR dye removal of 89.27%. The experimental data have been analysed for pseudo-first-order, pseudo-second-order and intra-particle diffusion kinetic models. But in this study, the pseudo-second-order kinetic model was proved as the best fit curve. Isotherm modelling was used to determine the adsorbent's maximum adsorption capacity and to comprehend the nature of the adsorption process at various adsorbent dosages, and RSM was successfully used to analyse and optimize the adsorption process. Isotherm modelling revealed that Langmuir isotherm was best fitted for equilibrium data in which a high regression coefficient value of 0.991 for CR dye shows a good fitting model. RSM modelling with three independent variables at five levels is employed, which reveals that the RSM model gives more applicability in optimization. Under ideal conditions, the corresponding experimental values of dye adsorption were found to be 97.78% for CR dye using the RSM optimization technique, which is very near to the optimized value.

References

1. Abdi J, Vossoughi M, Mahmoodi NM, Alemzadeh I (2017) Synthesis of metal-organic framework hybrid nanocomposites based on GO and CNT with high adsorption capacity for dye removal. Chem Eng J 326:1145–1158. https://doi.org/10.1016/j.cej.2017.06.054
2. Agarwal S, Tyagi I, Gupta VK, Dastkhoon M, Ghaedi M, Yousefi F, Asfaram A (2016) Ultrasound-assisted adsorption of Sunset Yellow CFC dye onto Cu doped ZnS nanoparticles loaded on activated carbon using response surface methodology based on central composite design. J Mol Liq 219:332–340. https://doi.org/10.1016/j.molliq.2016.02.100
3. Ahmadi K, Ghaedi M, Ansari A (2015) Comparison of nickel doped Zinc Sulfide and/or palladium nanoparticle loaded on activated carbon as efficient adsorbents for kinetic and equilibrium study of removal of Congo Red dye. Spectrochim Acta Part A Mol Biomol Spectrosc 136:1441–1449. https://doi.org/10.1016/j.saa.2014.10.034

4. Angelova R, Baldikova E, Pospiskova K, Maderova Z, Safarikova M, Safarik I (2016) Magnetically modified Sargassum horneri biomass as an adsorbent for organic dye removal. J Clean Prod 137:189–194. https://doi.org/10.1016/j.jclepro.2016.07.068
5. Ansari R, Mosayebzadeh Z (2011) Application of polyaniline as an efficient and novel adsorbent for azo dyes removal from textile wastewaters. Chem Pap 65(1):1–8. https://doi.org/10.2478/s11696-010-0083-x
6. Arabi M, Ghaedi M, Ostovan A (2017) Water compatible molecularly imprinted nanoparticles as a restricted access material for extraction of hippuric acid, a biological indicator of toluene exposure, from human urine. Microchim Acta 184(3):879–887. https://doi.org/10.1007/s00604-016-2063-5
7. Arabi M, Ghaedi M, Ostovan A, Tashkhourian J, Asadallahzadeh H (2016) Synthesis and application of molecularly imprinted nanoparticles combined ultrasonic assisted for highly selective solid phase extraction trace amount of celecoxib from human plasma samples using design expert (DXB) software. Ultrason Sonochem 33:67–76. https://doi.org/10.1016/j.ultsonch.2016.04.022
8. Bhaumik M, McCrindle RI, Maity A (2015) Enhanced adsorptive degradation of Congo red in aqueous solutions using polyaniline/Fe^0 composite nanofibers. Chem Eng J 260:716–729. https://doi.org/10.1016/j.cej.2014.09.014
9. Bhowmik KL, Deb K, Bera A, Debnath A, Saha B (2018) Interaction of anionic dyes with polyaniline implanted cellulose: organic π-conjugated macromolecules in environmental applications. J Mol Liq 261:189–198. https://doi.org/10.1016/j.molliq.2018.03.128
10. Bhowmik KL, Debnath A, Nath RK, Das S, Chattopadhyay KK, Saha B (2016) Synthesis and characterization of mixed phase manganese ferrite and hausmannite magnetic nanoparticle as potential adsorbent for methyl orange from aqueous media: artificial neural network modeling. J Mol Liq 219:1010–1022. https://doi.org/10.1016/j.molliq.2016.04.009
11. Bhowmik M, Deb K, Debnath A, Saha B (2017) Mixed phase Fe_2O_3 /Mn_3O_4 magnetic nanocomposite for enhanced adsorption of methyl orange dye: Neural network modeling and response surface methodology optimization. Appl Organomet Chem 32(3):e4186. https://doi.org/10.1002/aoc.4186
12. Bhowmik M, Debnath A, Saha B (2018) Fabrication of mixed phase calcium ferrite and zirconia nanocomposite for abatement of methyl orange dye from aqua matrix: optimization of process parameters. Appl Organomet Chem e4607. https://doi.org/10.1002/aoc.4607
13. Bhowmik M, Debnath A, Saha B (2019) Fabrication of mixed phase $CaFe_2O_4$ and $MnFe_2O_4$ magnetic nanocomposite for enhanced and rapid adsorption of methyl orange dye: statistical modeling by neural network and response surface methodology. J Dispersion Sci Technol 1–12. https://doi.org/10.1080/01932691.2019.1642209
14. Bhowmik M, Debnath A, Saha B (2020) Effective remediation of an antibacterial drug from aqua matrix using $CaFe_2O_4$/ZrO_2 nanocomposite derived via inorganic chemical pathway: statistical modelling by response surface methodology. Arab J Sci Eng. https://doi.org/10.1007/s13369-020-04465-y
15. Chávez-Guajardo AE, Medina-Llamas JC, Maqueira L, Andrade CAS, Alves KGB, de Melo CP (2015) Efficient removal of Cr (VI) and Cu (II) ions from aqueous media by use of polypyrrole/maghemite and polyaniline/maghemite magnetic nanocomposites. Chem Eng J 281:826–836. https://doi.org/10.1016/j.cej.2015.07.008
16. Chebli D, Bouguettoucha A, Mekhalef T, Nacef S, Amrane A (2014) Valorization of an agricultural waste, Stipa tenassicimafibers, by biosorption of an anionic azo dye congo red. Desal Water Treat 54(1):245–254. https://doi.org/10.1080/19443994.2014.880154
17. Dai J, Sun J, Xie A, He J, Li C, Yan Y (2016) Designed preparation of 3D hierarchically porous carbon material via solvothermal route and in situ activation for ultrahigh-efficiency dye removal: adsorption isotherm, kinetics and thermodynamics characteristics. RSC Adv 6(5):3446–3457. https://doi.org/10.1039/c5ra24774h
18. Das R, Bhaumik M, Giri S, Maity A (2017) Sonocatalytic rapid degradation of Congo red dye from aqueous solution using magnetic Fe^0 /polyaniline nanofibers. Ultrason Sonochem 37:600–613. https://doi.org/10.1016/j.ultsonch.2017.02.022

19. Deb A, Debnath A, Saha B (2019) Ultrasound-aided rapid and enhanced adsorption of anionic dyes from binary dye matrix onto novel hematite/polyaniline nanocomposite: response surface methodology optimization. Appl Organomet Chem. https://doi.org/10.1002/aoc.5353

20. Deb A, Debnath A, Bhowmik KL, Paul SR, Saha B (2021) Application of polyaniline impregnated mixed phase Fe_2O_3, $MnFe_2O_4$ and ZrO_2 nanocomposite for rapid abatement of binary dyes from aqua matrix: response surface optimisation. Int J Environ Anal Chem 1946683. https://doi.org/10.1080/03067319.2021.1946683

21. Deb A, Kamani M, Debnath A, Bhowmik KL, Saha B (2019) Ultrasonic assisted enhanced adsorption of methyl orange dye onto polyaniline impregnated zinc oxide nanoparticles: kinetic, isotherm and optimization of process parameters. Ultrason Sonochem 54:290–301. https://doi.org/10.1016/j.ultsonch.2019.01.028

22. Debnath A, Bera A, Chattopadhyay KK, Saha B (2017) Facile additive-free synthesis of hematite nanoparticles for enhanced adsorption of hexavalent chromium from aqueous media: Kinetic, isotherm, and thermodynamic study. Inorganic Nano Metal Chemistry 47(12):1605–1613. https://doi.org/10.1080/24701556.2017.1357581

23. Debnath A, Deb K, Das NS, Chattopadhyay KK, Saha B (2015) Simple Chemical Route Synthesis of Fe_2O_3 Nanoparticles and its application for adsorptive removal of congo red from aqueous media: artificial neural network modeling. J Dispersion Sci Technol 37(6):775–785. https://doi.org/10.1080/01932691.2015.1062772

24. Debnath B, Majumdar M, Bhowmik M, Bhowmik KL, Debnath A, Roy DN (2020) The effective adsorption of tetracycline onto zirconia nanoparticles synthesized by novel microbial green technology. J Environ Manage 261:110253. https://doi.org/10.1016/j.jenvman.2020.110235

25. Feng J, Hou Y, Wang X, Quan W, Zhang J, Wang Y, Li L (2016) In-depth study on adsorption and photocatalytic performance of novel reduced graphene oxide-$ZnFe_2O_4$ -polyaniline composites. J Alloy Compd 681:157–166. https://doi.org/10.1016/j.jallcom.2016.04.146

26. Ghaedi M, Rahimi MR, Ghaedi AM, Tyagi I, Agarwal S, Gupta VK (2016) Application of least squares support vector regression and linear multiple regression for modeling removal of methyl orange onto tin oxide nanoparticles loaded on activated carbon and activated carbon prepared from Pistacia atlantica wood. J Colloid Int Sci 461:425–434. https://doi.org/10.1016/j.jcis.2015.09.024

27. Gupta VK, Pathania D, Kothiyal NC, Sharma G (2014) Polyaniline zirconium (IV) silicophosphate nanocomposite for remediation of methylene blue dye from waste water. J Mol Liq 190:139–145. https://doi.org/10.1016/j.molliq.2013.10.027

28. Hao Q, Liu S, Yin X, Du Z, Zhang M, Li L, Wang Y, Wang T, Li Q (2011) Flexible morphology-controlled synthesis of mesoporous hierarchical α-Fe_2O_3 architectures and their gas-sensing properties. Cryst Eng Comm 13(3):806–812. https://doi.org/10.1039/c0ce00194e

29. Karpuraranjith M, Thambidurai S (2017) Design and synthesis of graphene-SnO_2 particles architecture with polyaniline and their better photodegration performance. Synth Met 229:100–111. https://doi.org/10.1016/j.synthmet.2017.02.017

30. Khalaf IH, Al-Sudani FT, Razak AAA, Aldahri T, Shorab S (2021) Optimization of Congo red dye adsorption from wastewater by a modified commercial zeolite catalyst using response surface modeling approach. Water Sci Technol 83(6):1369–1383. https://doi.org/10.2166/wst.2021.078

31. Kumar M, Sridhari TR, Bhavani K, Dutta P (1998) Trends in color removal from textile mill effluents. Biores Technol 77(100214844):25–34

32. Lafi R, Montasser I, Hafiane A (2018) Adsorption of congo red dye from aqueous solutions by prepared activated carbon with oxygen-containing functional groups and its regeneration. Adsorpt Sci Technol 026361741881922. https://doi.org/10.1177/0263617418819227

33. Liu R, Fu H, Yin H, Wang P, Lu L, Tao Y (2015) A facile sol combustion and calcination process for the preparation of magnetic $Ni0.5Zn0.5Fe_2O_4$ nanopowders and their adsorption behaviors of Congo red. Powder Technol 274:418–425. https://doi.org/10.1016/j.powtec.2015.01.045

34. Madan S, Shaw R, Tiwari S, Tiwari SK (2019) Adsorption dynamics of Congo red dye removal using ZnO functionalized high silica zeolitic particles. Appl Surf Sci. https://doi.org/10.1016/j.apsusc.2019.04.273

35. Mittal A, Thakur V, Mittal J, Vardhan H (2013) Process development for the removal of hazardous anionic azo dye Congo red from wastewater by using hen feather as potential adsorbent. Desalin Water Treat 52(1–3):227–237. https://doi.org/10.1080/19443994.2013.785030

36. Nayunigari MK, Das R, Maity A, Agarwal S, Gupta VK (2017) Folic acid modified cross-linked cationic polymer: Synthesis, characterization and application of the removal of Congo red dye from aqueous medium. J Mol Liq 227:87–97. https://doi.org/10.1016/j.molliq.2016.11.129

37. Patel H, Vashi RT (2012) Removal of Congo Red dye from its aqueous solution using natural coagulants. J Saudi Chem Soc 16(2):131–136. https://doi.org/10.1016/j.jscs.2010.12.003

38. Patra BN, Majhi D (2015) Removal of anionic dyes from water by potash alum doped polyaniline: investigation of kinetics and thermodynamic parameters of adsorption. J Phys Chem B 119(25):8154–8164. https://doi.org/10.1021/acs.jpcb.5b00535

39. Raval NP, Shah PU, Shah NK (2016) Adsorptive amputation of hazardous azo dye Congo red from wastewater: a critical review. Environ Sci Pollut Res 23(15):14810–14853. https://doi.org/10.1007/s11356-016-6970-0

40. Sahoo JK, Paikra SK, Mishra M, Sahoo H (2019) Amine functionalized magnetic iron oxide nanoparticles: synthesis, antibacterial activity and rapid removal of Congo red dye. J Mol Liq 282:428–440. https://doi.org/10.1016/j.molliq.2019.03.033

41. Saxena R, Sharma S (2016) Kinetic modeling and isotherm studies on removal of methyl orange from aqueous solutions using guar gum powder. Asian J Chem 28(8):1848–1854. https://doi.org/10.14233/ajchem.2016.19898

42. Shahabuddin S, Sarih NM, Mohamad S, Atika Baharin SN (2016) Synthesis and characterization of Co_3O_4 nanocube-doped polyaniline nanocomposites with enhanced methyl orange adsorption from aqueous solution. RSC Adv 6(49):43388–43400. https://doi.org/10.1039/c6ra04757b

43. Sharma A, Siddiqui ZM, Dhar S, Mehta P, Pathania D (2018) Adsorptive removal of congo red dye (CR) from aqueous solution by Cornulaca monacantha stem and biomass-based activated carbon: isotherm, kinetics and thermodynamics. Sep Sci Technol 1–14. https://doi.org/10.1080/01496395.2018.1524908

44. Shen W, Liao B, Sun W, Su S, Ding S (2013) Adsorption of Congo red from aqueous solution onto pyrolusite reductive leaching residue. Desalin Water Treat 52(19–21):3564–3571. https://doi.org/10.1080/19443994.2013.855680

45. Singh NH, Kezo K, Debnath A, Saha B (2017) Enhanced adsorption performance of a novel Fe-Mn-Zr metal oxide nanocomposite adsorbent for anionic dyes from binary dye mix: response surface optimization and neural network modeling. Appl Organomet Chem 32(3):e4165. https://doi.org/10.1002/aoc.4165

46. Sinha S, Behera SS, Das S, Basu A, Mohapatra RK, Murmu BM, Parhi PK (2018) Removal of Congo Red dye from aqueous solution using Amberlite IRA-400 in batch and fixed bed reactors. Chem Eng Commun 205(4):432–444. https://doi.org/10.1080/00986445.2017.1399366

47. Slokar YM, Majcen Le Marechal A (1998) Methods of decoloration of textile wastewaters. Dyes Pigm 37(4):335–356. https://doi.org/10.1016/s0143-7208(97)00075-2

48. Tanhaei B, Ayati A, Lahtinen M, Sillanpää M (2015) Preparation and characterization of a novel chitosan/Al_2O_3/magnetite nanoparticles composite adsorbent for kinetic, thermodynamic and isotherm studies of Methyl Orange adsorption. Chem Eng J 259:1–10. https://doi.org/10.1016/j.cej.2014.07.109

49. Tanzifi M, Yaraki MT, Kiadehi AD, Hosseini SH, Olazar M, Bharti AK, Kazemi A (2018) Adsorption of Amido Black 10B from aqueous solution using polyaniline/SiO_2 nanocomposite: experimental investigation and artificial neural network modeling. J Colloid Interface Sci 510:246–261. https://doi.org/10.1016/j.jcis.2017.09.055

50. Wang L, Li J, Wang Y, Zhao L, Jiang Q (2012) Adsorption capability for Congo red on nanocrystalline MFe_2O_4 (M = Mn, Fe Co, Ni) spinel ferrites. Chem Eng J 181–182:72–79. https://doi.org/10.1016/j.cej.2011.10.088

51. Wang R, Yan X, Ge B, Zhou J, Wang M, Zhang L, Jiao T (2020) Facile preparation of self-assembled black phosphorus-dye composite films for chemical gas sensors and surface-enhanced raman scattering performances. ACS Sustain Chem Eng. https://doi.org/10.1021/acssuschemeng.9b07840

52. Yokwana K, Kuvarega AT, Mhlanga SD, Nxumalo EN (2018) Mechanistic aspects for the removal of Congo red dye from aqueous media through adsorption over N-doped graphene oxide nanoadsorbents prepared from graphite flakes and powders. Phys Chem Earth Parts A/B/C. https://doi.org/10.1016/j.pce.2018.08.001

Silica-Polymer Composite for Dyes Removal

Bouhadjar Boukoussa⦿ **and Adel Mokhtar**

Abstract Water pollution is still increasing as a result of industrial discharges, today the scientific community is looking for efficient materials for the treatment of polluted water. In order to obtain the best results in terms of efficiency in the removal of pollutants, several materials have been tested in the literature. Among these solids there are composite materials based on polymer and silica which have known wide application, in particular in this field. Porous silica is known by its higher specific surface area having a negative charge and polymers by their functional groups. The combination of both matrices results in the formation of a new polymer-silica composite which has new properties especially for the removal of organic and inorganic pollutants. In this chapter we present some strategies for the preparation of composites polymer-silica and their applications in adsorption and catalysis field, thus very detailed explanations on the interactions between the composite and the pollutants were provided.

Keywords Composites · Polymer-silica · Adsorption · Catalysis · Dye degradation

1 Introduction

With the significant increase in the growth of the population and the development of the industry, plenty of discharges of organic and inorganic contaminants have been evacuated toward the aquatic sources [1, 2]. According to a new report by UNICEF and the WHO, about 2.2 billion people around the world do not have safely managed drinking water services. Over the past decades, several water treatment techniques

B. Boukoussa (✉) · A. Mokhtar (✉)
Laboratory of Materials Chemistry, University of Oran1, Ahmed Ben Bella, El- Mnaouer, BP 1524, 31000 Oran, Algeria

B. Boukoussa
Department of Materials Engineering, Faculty of Chemistry, University of Science and Technology Mohamed Boudiaf, El-Mnaouer, BP 1505, Oran, Algeria

A. Mokhtar
Department of Process Engineering, Faculty of Science and Technology, University of Relizane, 48000 Relizane, Algeria

© The Author(s), under exclusive license to Springer Nature Singapore Pte Ltd. 2022
A. Khadir and S. S. Muthu (eds.), *Polymer Technology in Dye-containing Wastewater*,
Sustainable Textiles: Production, Processing, Manufacturing & Chemistry,
https://doi.org/10.1007/978-981-19-1516-1_3

have been exploited to reduce or eliminate permanently these hazardous species. But they have some drawbacks such as slow treatment, and the problem of disposing of waste after the treatment, as well as many emerging anthropogenic contaminants could not be effectively removed, especially those present in trace amounts.

To date several treatment methods have been carried out with the aim of reducing the pollution of wastewater, among the most used techniques there are coagulation/flocculation [3], filtration membrane [3, 4], adsorption [3], advanced oxidation [5], electrochemical methods [6], biodegradation, detoxification [7] and others. Among these methods, adsorption and catalysis have been considered as simple, inexpensive and efficient methods for the removal of organic and inorganic pollutants, moreover, adsorption has always been considered an important step in catalysis. Zeolites, clays, activated carbon, polymers, metal oxides are the materials most used in water treatment due to their remarkable properties. But in order to have a significant elimination of pollutants, researchers have always tried to combine between different synthetic or natural materials [8–13]. Particular attention has been paid to polymers due to the presence of several functional groups which are considered to be active sites [8–11]. Among these polymers there are some which are unstable in the reaction medium; certain are thermally unstable, which limits their applications. To solve this problem it has been shown in several studies that the combination of polymers with other fillers such as silica, clay, zeolites or other charges causes the formation of a new stable and efficient material which combines the advantages of both materials [8, 9, 14, 15, 11, 12].

Recently, polymers-silica composites have known a particular interest due to their remarkable properties (Polymers containing various functional groups and silica has a specific surface area and a higher porosity) which make them potential materials for wastewater treatment [16–19]. It is known that the nature of the polymer, the nature of silica and the method of preparation play a very important role in the performance of the prepared composite for the removal of organic pollutants [16–19]. Due to the presence of functional groups on the surface of these composites, it is possible to remove a wide variety of organic and inorganic pollutants. These functional groups (chelating groups) make it possible to obtain complexes during the adsorption of transition metals and under the action of a reducing agent a good dispersion of nanoparticles having ultra-fine sizes can be obtained [20–23]. This opens a new way for the preparation of new effective Metal NPs-Polymer-Silica catalysts for wastewater treatment. The application of polymer-silica composites does not lie only in the treatment of wastewater depending on the nature of the polymers other properties can be improved (see Fig. 1). In this chapter a general study on the different strategies used for the preparation and characterization of composites based on polymer and silica will be discussed. Thus we present the application of these solids in water treatment.

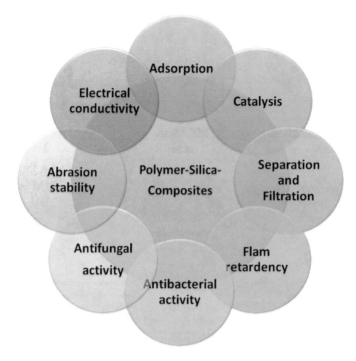

Fig. 1 Schematic structures of the various applications of composites polymer-silica

2 Polymer-Silica Composite

Polymers have known a wide application, particularly in the industrial field, due to their remarkable properties. Many attempts have been made on this kind of material to obtain new properties. The dispersion of charges on polymers was considered to be a more interesting method which subsequently generates new properties. The properties of these composites differ depending on the nature of the polymer used, nature of the filler, their content and their structures.

In order to obtain a composite with improved properties several nanofillers have been used such as clays, silica, metal oxides, activated carbon, zeolites, MOFs and others. It was shown in several studies that the combination between nanofiller and polymers improves thermal stability, strength and heat resistance, biodegradability of biodegradable polymers, the mechanical stability of the resulting composite and also decreases gas permeability and flammability [24, 25, 22, 26]. The presence of functional groups in polymers and active sites in nanofillers also can improve the adsorption affinity via organic and inorganic pollutants [16–19, 15]. Among these nanofillers, porous silica has been of particular interest due to its higher surface area and its greater porosity than the other materials cited above, which significantly influences the adsorption and catalytic properties. In the mesoporous silica family there are several types of materials having different structures and which can be obtained

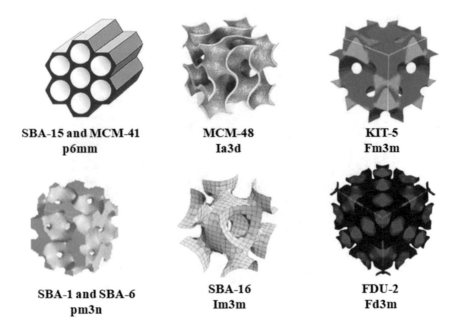

| SBA-15 and MCM-41
p6mm | MCM-48
Ia3d | KIT-5
Fm3m |
| SBA-1 and SBA-6
pm3n | SBA-16
Im3m | FDU-2
Fd3m |

Fig. 2 Structure of some mesoporous silica

from different types of surfactants using hydrothermal process (see Fig. 2). These kinds of materials are characterized by a large pore size, high specific surface area, large wall thickness and high hydrothermal/thermal stability. All these properties are the key factors which have widened the application field of these solids, notably in catalysis and adsorption. Mesoporous silica is known by the presence of high density of silanol groups which make them effective for functionalization by organosilanes or by impregnation.

3   Preparation Ways of Polymer-Silica Composite

3.1   *Impregnation*

This preparation method was considered to be a simple, rapid and efficient method for the dispersion of polymer on the surface of the silica. To have a good dispersion of polymer on the surface of the silica, it is very important to use a silica having a large surface area such as mesoporous silica (M41S, SBA, …) [16]. To have such a property, particularly in the adsorption of organic and inorganic pollutants, it is very important to control the polymer content because at higher contents it is possible to reduce the specific surface area of the resulting composite due to the blocking of its porosity by polymer [16]. In general, the procedure for preparing the polymer-silica

Fig. 3 Impregnation of polymer on the surface of the mesoporous silica

composite consists on the dispersion of polymer in a solvent following the addition of the porous silica (see Fig. 3). In general, the possible interactions between the polymer and the silica are of the hydrogen or van der waals type.

3.2 In-Situ Polymerization

The preparation of polymer-silica composites by in-situ polymerization of monomer inside the pores has been of particular interest because this method makes it possible to immobilize a large variety of polymers [27–31]. Among the advantages of this method is that the formation of the polymer will be inside the pores of the silica which is difficult to achieve by the impregnation method because it is known that the size of the polymers is important compared to the pore size of mesoporous silica [16–19]. This is why different mesoporous materials having different pore sizes were tested for the immobilization of polymers inside their pores. Among the disadvantages of this method, the molecular weight of the resulting polymer is limited.

The method of preparing polymer-silica composites firstly consists of the adsorption of monomer by the mesoporous silica following the addition of an initiating agent to initiate the polymerization of monomer within the pores of the mesoporous silica [27–31] (see Fig. 4).

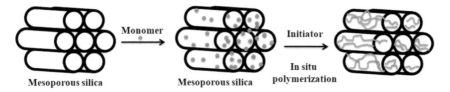

Fig. 4 In-situ polymerization of a monomer inside the pores of mesoporous silica

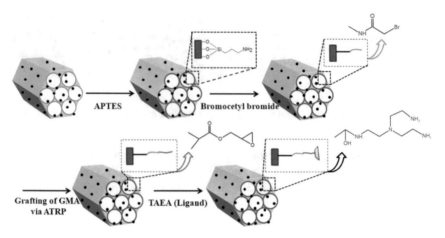

Fig. 5 Preparation of magnetic mesoporous silica grafted with poly(glycidylmethacrylate) brush, protocol adopted from the literature [35]

3.3 Functionalization

The functionalization of silica has known wide application because it is possible to chemically bond the polymer with the silica. The presence of silanol groups on the surface of the porous silica is the key factor which influences the functionalization of the surface. The functionalization of the porous silica can be carried out by two methods either by post-synthesis route, in which the organosilane groups are linked with the surface silanol groups [32]. Either by the co-condensation of organosilanes with the silica precursors $Si(OR')_4$ during the synthesis of porous silica, the functional groups are incorporated directly into the silica framework [33]. Among these two methods the first is the most used due to its higher thermal stability and a more ordered structure than co-condensation [34]. From the functionalized form of silica, it is possible to graft polymers. For example, it is easy to prepare a composite containing the porous silica functionalized by amine and grafted with poly(glycidyl methacrylate) (see Fig. 5). The polymer poly(glycidyl methacrylate) has reactive epoxy groups which can be used for the coupling of amino groups of functionalized silica [35]. The use of functionalized porous silica makes it possible to graft a large variety of polymers either by chemical bonds between the functionalized silica and the polymer or by an in-situ polymerization process.

3.4 Encapsulation

The preparation of polymer-silica composite by the encapsulation method has known wide application due to the simplicity of the method [36, 18, 37, 38]. The most used polymers for this method are biopolymers such as alginate, cellulose and chitosan.

Among the advantages of this method, it is possible to obtain different forms of polymer-silica composites such as films, beads, micro-beads, aerogels, hydrogels [36, 18, 37, 38]. These materials can be easily recovered after their use. Example of the preparation of alginate-silica composite: the synthesis way involves the preparation of a solution containing sodium alginate and porous silica. The modification of the reaction mixture by a solution containing Ca^{++} makes it possible to obtain the hydrogel form of the polymer-silica composite; in this case the Ca^{++} plays a role of a crosslinking agent [18]. The content of porous silica in the composite plays a very important role in the textural and structural properties and also in their adsorption behavior for organic pollutants [18].

4 Characterizations

4.1 XRD

X-ray diffraction is one of the most widely used analytical methods to characterize polymer-silica composites, several information can be provided with the help of this characterization such as the type of the structure of the mesoporous silica (hexagonal, cubic, and lamellar) so it is possible to calculate the lattice parameter of the mesoporous silica and also the degree of the mesoporous silica organization. Among the key information that can be provided is that the mesh parameter can be swollen upon impregnation of the polymer within the pores of the mesoporous silica [10, 11]. This technique also makes it possible to observe the structure of the conductive or semi-conductor polymers in the region of the higher angles.

4.2 FTIR

Infrared spectroscopy has been known for its efficiency to determine the functional groups existing in the composite polymer-silica. It allows the determination of the groups of the two matrices, it is often observed that the absorption bands of the functional groups shifts toward the large wavenumbers as a result of the interaction between the two matrixes. The absorption bands located around ~3300 cm^{-1} characterize the hydrophilic character of the composite which can be observed very well by TGA [19]. This analysis also confirms the functionalization of the porous silica during the preparation of the composite. For example, the functionalization of mesoporous silica by aminosilane leads to the formation of new bands at 2800–2900 cm^{-1} attributed to the bonds of –CH_2 and –CH_3 and also bands at 1500 cm^{-1} attributed to the amine originating from the aminosilane [14, 18, 39, 40].

4.3   SEM

The use of SEM scanning electron microscopy provides general information on the shape and homogeneity of the particles and on the nature of the surface of the composites. For example in the work of Hachemaoui et al. [41] they have shown that the lyophilized composite beads based on alginate and mesoporous silica have external surface rougher due to the presence of Fe_3O_4@MCM -41. But the SEM images of the same composite obtained inside the beads showed a porous structure of the composite due to the freeze-drying treatment [41]. A porous structure facilitates the diffusion of the reagents inside the beads, which makes this composite a very good catalyst or adsorbent for treating polluted water.

4.4   TEM

Transmission electron microscopy (TEM) is an important technique to determine the spatial distribution of particles, structural defects and internal structure and also the detailed topography of a composite, it also allows to assess the size of the nanoparticles dispersed in the surface of the composite. For example TEM images of PMMA/AIL16/Ag(24%) and PS/AIL16/Ag (24%) composites showed black spots which correspond to silver nanoparticles. The size of AgNPs was approximately 4.60–4.44 nm for the PMMA/AIL16/Ag(24%) composite and for the PS/AIL16/Ag(24%) composite was 6.88–4.51 nm. When the Ag content exceeds 24%, certain agglomeration phenomena of Ag NPs in a certain local position on the surface of the polymer/AIL16 spheres can be observed [42].

4.5   TGA

TGA thermogravimetric analysis is a technique that allows to take the necessary information on the stability and degradation of polymer-silica composites. Additional information on the hydrophilic and hydrophobic character of composite can be given in the temperature range of 35–150 °C. In this temperature range, the presence of a strong loss of mass is mainly due to the physisorbed water molecules, confirming the hydrophilic character of the composite, which is reflected in the multiple interactions between the functional groups of the composites and water molecules [17, 19]. This analysis also makes it possible to understand the degradation behavior and the hydrophilic/hydrophobic character of the composite during the variation of the polymer or silica content [17, 19]. At temperatures higher than 200 °C information on the degradation processes of the polymer can be given and it is possible to know the nature of the products formed during the degradation using TGA coupled with GC-Mass.

5 Adsorption

Water pollution has become the topical issue following industrial discharges. Dyes, toxic organic and inorganic products and pharmaceutical discharges are the most well-known pollutants due to their toxicities. Several attempts have been applied to reduce and degrade these pollutants, the most known methods are adsorption, oxidation, photocatalysis, separation, etc.

Among all these methods adsorption has experienced wide application due to its simplicity and low energy consumption. To improve the retention of these pollutants, several adsorbents have been tested; it has been shown in several studies that the polymers have an interesting adsorption capacity via organic and inorganic pollutants due to the presence of several functional groups on their surfaces, which are considered to be active sites for the adsorption of organic and inorganic pollutants. But these polymers are sometimes unstable in the reaction medium and they can be degraded. The combination of polymers with mesoporous silica has shown interesting results and the adsorption capacity of pollutants has been greatly improved compared to pure polymer or pure silica.

Alginate is one of the best known examples due to its availability, biocompatibility and its interesting adsorption capacity. Alginate is a biopolymer having a negative charge, possessing carbonyl and hydroxyl groups which also allow the improvement of the adsorption of organic pollutants much more cationic dyes due to the several electrostatic interactions, hydrogen and van der waals interactions. The studies conducted in the literature have shown that the combination of alginate with mesoporous silica greatly increases the adsorption capacity of cationic dye compared to pure alginate or pure mesoporous silica. The adsorption capacity increases with increasing the mesoporous silica content. It is known that mesoporous silica also exhibits a negatively charged surface and the combination of both materials increases the adsorption capacity of cationic dyes due to the increase in number of sites and the surface charge, also the thermal properties can be improved. The most important thing about the preparation of polymer-silica composites, it is easy to control their shapes (beads, films, …) which makes them easy to recover after the adsorption process.

The affinity of the polymer-silica composite via organic pollutants differs depending on the nature of the polymer used. For example, the encapsulation of mesoporous silica with chitosan gives the formation of a positively charged composite which has affinity via anionic dyes. It is possible to prepare bi-functional composites having an affinity with cationic and anionic dyes due to the presence of different sites in their structures.

Table 1 shows some examples of the adsorption of dyes and metals by polymer-silica composites. From this table it is clear that the adsorption capacity is important. The adsorption capacity can be influenced by the mass of composite, the concentration of pollutant, the pH of the medium and the temperature. The adsorption of metals (such as Cu, Pb, Ni, …) on polymer-silica composites has shown encouraging results following the presence of chelation sites subsequently allowing chemical adsorption

Table 1 Application of polymer-silica composite in adsorption of pollutants

Adsorbent	Silica	Polymer	Pollutant	Qmax (mg/g)	Refs
PVA/SBA-15	SBA-15	Polyvinylalcohol	Methylenebluedye	77	[16]
PPy/SBA-15	SBA-15	Polypyrrole	Methylene blue dye	58.82	[17]
PPy/SBA-15	SBA-15	Polypyrrole	Methyl orange dye	41.66	[17]
poly(GDMA)/MCM-41	MCM-41	Poly(GDMA)	Methylene blue dye	111.11	[19]
SBA-15/ CPAA	SBA-15	Poly acrylic acid	Reactive Orange 16	~280	[30]
Magnetic-SBA-15/ CPAA	Magnetic-SBA-15	Poly acrylic acid	Acid Blue 25	236.68	[29]
PPy/MCM-41	MCM-41	Polypyrrole	Acid blue 62	55.55	[28]
Magnetic-NH2-MCM-41/p(GMA)[a]	Magnetic-NH_2-MCM-41	Poly(glycidyl methacrylate) P(GMA)	Direct Blue-6	11.8	[35]
ALG-SBA-15	SBA-15	Ca-Alginate	Methylene blue dye	333.33	[18]
Chitosan/SBA-15	SBA-15	Chitosan	Brilliant Red dye	19.23	[36]
SBA-15/calcium alginate	SBA-15	Ca-Alginate	Pb(II)	1029.58	[38]
PVP-SBA-15	NH_2-SBA-15	Polyvinylpyrrolidone	Cu (II)	128	[43]
PVP-SBA-15	NH_2-SBA-15	Polyvinylpyrrolidone	Pb(II)	175	[43]
PVP-SBA-15	NH_2-SBA-15	Polyvinylpyrrolidone	Ni(II)	72	[43]

[a] Magnetic MCM-41 composite modified by 3-aminopropyl triethoxysilane (APTES) and grafted with poly(glycidyl methacrylate) p(GMA)

and leading to the formation of Metal@polymer-silica complexes which can be used in various applications.

5.1 Mechanism of Adsorption of Organic and Inorganic Pollutants

The mechanism of adsorption of organic or inorganic pollutants on the polymer-silica composite differs depending on the nature of the pollutant (cationic, anionic or nonionic), the pH of the reaction medium because it can influence the nature of the surface of the adsorbent and also the nature of the polymer used for the retention of the pollutant. For example the adsorption of cationic dye (MB) on composite ALG-SBA-15 at pH 6.5, it is known that Ca-alginate and SBA-15 have negatively charged surfaces at pHs higher than 3, the combination between both matrices generates a composite with improved adsorption properties. The dominant interactions are electrostatic and hydrogen interactions between the adsorbent and the adsorbate (see Fig. 6). It was shown that the content of silica in the alginate matrix significantly influences the adsorption capacity of MB dye due to the increase of sites number in the resulting material [18]. Thus the modification of the silica by the amines results in a composite ALG@NH$_2$-SBA-15 having weaker properties compared to the material ALG-SBA-15 due to the presence of repulsive forces between the amines and MB dyes. This type of material (composite based on chitosan or alginate and silica) can be used as good candidate for the capture of transition metals due to the presence of

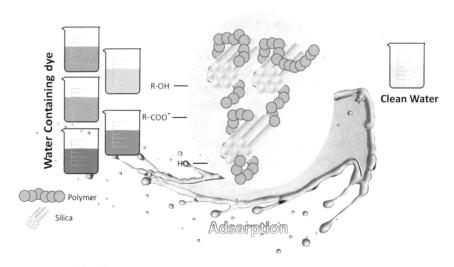

Fig. 6 Mechanism of MB dye adsorption on Alginate@SBA-15 composite

chelating groups. The dominant interactions in this case are electrostatic interactions subsequently forming metal-binder complexes.

6 Polymer-Silica Composite in Catalysis Field

Among the advantages of a polymer-silica composite it is easy to prepare a novel porous catalyst. The work of Shen et al. has shown that the modification of spherical polymers such as polymethylmethacrylate (PMMA) or polystyrene (PS) by amphiphilic ionic liquid 1-hexadecyl-3-methylimidazolium chloride (AIL16) and Ag NPs (PMMA/AIL16/Ag ou PS/AIL16/Ag) were used as templates for the preparation of bimodal porous silica decorated with Ag (macropore-mesoporore) (see Fig. 7) [42]. The calcination of these composites released the porosity of the resulting material by removing the polymer matrix. This technique allowed the good dispersion of AgNPs in the pores of the porous silica. The application of these solids in the reaction of reduction of 4-nitrophenol to 4-aminophenol under the presence of a reducing agent NaBH$_4$ has given interesting results with a lower reaction time [42].

The preparation of a membrane based on polyaniline, aminosilane containing bimitallic nanoparticles (Ag-Metal, M: Cu, Fe, Al and Zn) has shown interesting results during the filtration and catalytic reduction of 4-nitrophenol and Methyl Orange MO [1]. The nature of the metal plays a very important role via the reduction of these pollutants in which it has been shown that the PAN-Si-Cu-Ag membrane has shown both good stability and high catalytic activity for the reduction of 4-NP and MO, this activity is linked to the synergistic effect between Cu, Ag and NaBH$_4$ [1].

Recently a more efficient catalyst based on mesoporous silica SBA-15 and poly (methacrylic acid) (PMAA) decorated with silver nanoparticles has shown its efficiency via the degradation of a variety of dyes such as congo red (CR) methyl orange (MO), metanil yellow (MY), rhodamine B (RhB), erichrome black T (EBT) and methylene blue (MB) [44]. This composite has made it possible to obtain a large amount of Ag NPs on its surface and to stabilize them. Among the advantages of

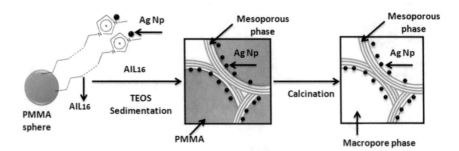

Fig. 7 Schematic representation of different procedure for the preparation of Ag-decorated bimodal porous silica, figure adopted from [42]

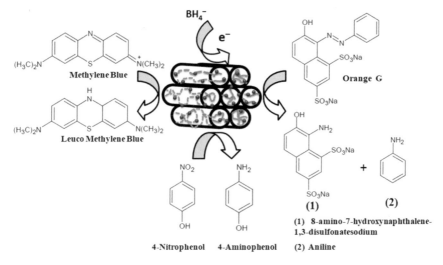

Fig. 8 Catalytic reduction of some organic pollutants on the polymer-silica composite containing metal nanoparticles

these composite, nanoparticles with ultra-fine sizes can be obtained. Their dispersion on a matrix having a larger specific surface area increases their activity via the degradation of the dyes.

The mechanism for reducing organic pollutants on the surface of polymer-silica composites containing nanoparticles can follow the following steps: In the first step, the organic pollutant and the reducing agent NaBH$_4$ diffuse inside the pores of the composites. In the second step, the transfer of electrons will take place on the surface of the metallic nanoparticles which transfers the electrons from the donor (NaBH$_4$) to the acceptor (organic pollutant) [45, 46, 39, 47, 48]. In the third step, several complex reactions between the surface of the nanoparticles, the chemisorbed pollutant and the hydrides formed during the dissociation of NaBH$_4$ can take place [45, 46, 39, 47, 48]. And finally in the last step, the product formed is desorbed on the surface of the catalyst (see Fig. 8).

7   Conclusion and Future Perspectives

Polymer-silica composite materials are promoter materials that can be used in wastewater pollution treatment. The choice of polymer and also the nature of the silica (nanoporous, mesoporous, etc.) can significantly influence their behavior toward the elimination of organic and inorganic pollutants. These solids can form complexes during the adsorption of transition metals (such as Cu, Zn, etc.) which is a key factor for the stabilization of these nanoparticles. It is possible to use them in the catalysis field for the degradation of organic pollutants (such as reduction reactions in the presence of NaBH$_4$, or oxidation reactions, etc.). Polymer-silica composites can be easily modified by heat treatment for the preparation of new porous materials effective for the removal of organic and inorganic pollutants. It is really desirable to study the cytotoxicity of these composites and also their analogs modified by nanoparticles to broaden their application field. Polymer-silica composites containing metal nanoparticles can be used as antibacterial or antifungal agents, so it is very interesting to test them in this field.

References

1. Li P, Wang Y, Huang H, Ma S, Yang H, Xu Z (2021) High efficient reduction of 4-nitrophenol and dye by filtration through Ag NPs coated PAN-Si catalytic membrane. Chemosphere 263:127995. https://doi.org/10.1016/J.CHEMOSPHERE.2020.127995
2. Wang J, Fan J, Li J, Wu X, Zhang G (2018) Ultrasound assisted synthesis of Bi2NbO5F/rectorite composite and its photocatalytic mechanism insights. Ultrason Sonochem 48:404–411. https://doi.org/10.1016/j.ultsonch.2018.06.008
3. Badawi AK, Zaher K (2021) Hybrid treatment system for real textile wastewater remediation based on coagulation/flocculation, adsorption and filtration processes: Performance and economic evaluation. J Water Proc Eng 40:101963. https://doi.org/10.1016/J.JWPE.2021.101963
4. Mokhtar A, Abdelkrim S, Zaoui F, Sassi M, Boukoussa B (2020) Improved stability of starch layered-materials composite films for methylene blue dye adsorption in aqueous solution. J Inorg Organomet Polym Mater 30(9):3826–3831. https://doi.org/10.1007/s10904-020-01536-3
5. Bengotni L, Trari B, Lebeau B, Michelin L, Josien L, Bengueddach A, Hamacha R (2021) Effect of diatomite addition on crystalline phase formation of TiO$_2$ and photocatalytic degradation of MDMA. New J Chem 45(30):13463–13474. https://doi.org/10.1039/D1NJ01529J
6. Hamous H, Khenifi A, Orts F, Bonastre J, Cases F (2021) Carbon textiles electrodes modified with RGO and Pt nanoparticles used for electrochemical treatment of azo dye. J Electroanal Chem 887:115154. https://doi.org/10.1016/J.JELECHEM.2021.115154
7. Hameed BB, Ismail ZZ (2018) Decolorization, biodegradation and detoxification of reactive red azo dye using non-adapted immobilized mixed cells. Biochem Eng J 137:71–77. https://doi.org/10.1016/J.BEJ.2018.05.018
8. Bouhadjar L, Boukoussa B, Kherroub DE, Hakiki A, Elaziouti A, Laouedj N, Benhadria N, Chikh K (2018) Adsorption behavior of carbon dioxide on new nanocomposite CuO/PPB: effect of CuO Content. J Inorganic Organomet Polymers Mater 29(2):326–331. https://doi.org/10.1007/S10904-018-1002-9
9. Boukoussa B, Abidallah F, Abid Z, Talha Z, Taybi N, Sid El Hadj H, Ghezini R, Hamacha R, Bengueddach A (2017) Synthesis of polypyrrole/Fe-kanemite nanocomposite through in situ polymerization: effect of iron exchange, acid treatment, and CO$_2$ adsorption properties. J Mater Sci 52(5):2460–2472. https://doi.org/10.1007/s10853-016-0541-0
10. Boukoussa B, Hakiki A, Nunes-Beltrao AP, Hamacha R, Azzouz A (2018) Assessment of the intrinsic interactions of nanocomposite polyaniline/SBA-15 with carbon dioxide: correlation between the hydrophilic character and surface basicity. J CO$_2$ Utilization 26:171–178. https://doi.org/10.1016/j.jcou.2018.05.006
11. Ghomari K, Boukoussa B, Hamacha R, Bengueddach A, Roy R, Azzouz A (2017) Preparation of dendrimer polyol/mesoporous silica nanocomposite for reversible CO$_2$ adsorption: effect of pore size and polyol content 52(15):2421–2428. https://doi.org/10.1080/01496395.2017.1367810

12. Mokhtar A, Abdelkrim S, Hachemaoui M, Adjdir M, Zahraoui M, Boukoussa B (2020) Layered silicate magadiite and its composites for pollutants removal and antimicrobial properties: a review. Appl Clay Sci 198:105823. https://doi.org/10.1016/j.clay.2020.105823

13. Mokhtar A, Abdelkrim S, Sardi A, Benyoub A, Besnaci H, Cherrak R, Hadjel M, Boukoussa B (2020) Preparation and characterization of anionic composite hydrogel for dyes adsorption and filtration: non-linear isotherm and kinetics modeling. J Polym Environ 28(6):1710–1723. https://doi.org/10.1007/s10924-020-01719-6

14. Boukoussa B, Hakiki A, Bouazizi N, Beltrao-Nunes AP, Launay F, Pailleret A, Pillier F, Bengueddach A, Hamacha R, Azzouz A (2019) Mesoporous silica supported amine and amine-copper complex for CO_2 adsorption: detailed reaction mechanism of hydrophilic character and CO_2 retention. J Mol Struct 1191:175–182. https://doi.org/10.1016/j.molstruc.2019.04.035

15. Cherifi Z, Boukoussa B, Zaoui A, Belbachir M, Meghabar R (2018) Structural, morphological and thermal properties of nanocomposites poly(GMA)/clay prepared by ultrasound and in-situ polymerization. Ultrason Sonochem 48:188–198. https://doi.org/10.1016/j.ultsonch.2018.05.027

16. Abid Z, Hakiki A, Boukoussa B, Launay F, Hamaizi H, Bengueddach A, Hamacha R (2019) Preparation of highly hydrophilic PVA/SBA-15 composite materials and their adsorption behavior toward cationic dye: effect of PVA content. J Mater Sci 54(10):7679–7691. https://doi.org/10.1007/s10853-019-03415-w

17. Boukoussa B, Hakiki A, Moulai S, Chikh K, Kherroub DE, Bouhadjar L, Guedal D, Messaoudi K, Mokhtar F, Hamacha R (2018) Adsorption behaviors of cationic and anionic dyes from aqueous solution on nanocomposite polypyrrole/SBA-15. J Mater Sci 53(10):7372–7386. https://doi.org/10.1007/s10853-018-2060-7

18. Boukoussa B, Mokhtar A, El Guerdaoui A, Hachemoui M, Ouachtak H, Abdelkrim S, Ait Addi A, Babou S, Boudina B, Bengueddach A, Hamacha R (2021) Adsorption behavior of cationic dye on mesoporous silica SBA-15 carried by calcium alginate beads: experimental and molecular dynamics study. J Mol Liq 115976. https://doi.org/10.1016/j.molliq.2021.115976

19. Cherifi Z, Boukoussa B, Mokhtar A, Hachemaoui M, Zeggai FZ, Zaoui A, Bachari K, Meghabar R (2020) Preparation of new nanocomposite poly(GDMA)/mesoporous silica and its adsorption behavior towards cationic dye. React Funct Polym 153:104611. https://doi.org/10.1016/j.reactfunctpolym.2020.104611

20. Tang D, Zhang W, Qiao Z, Liu Y, Huo Q (2016) Functionalized mesoporous silica nanoparticles as a catalyst to synthesize a luminescent polymer/silica nanocomposite. RSC Adv 6(20):16461–16466. https://doi.org/10.1039/C5RA25135D

21. Long W, Brunelli NA, Didas SA, Ping EW, Jones CW (2013) Aminopolymer-silica composite-supported Pd catalysts for selective hydrogenation of alkynes. ACS Catal 3(8):1700–1708. https://doi.org/10.1021/CS3007395

22. Meer S, Kausar A, Iqbal T (2016) Attributes of polymer and silica nanoparticle composites: a review. 55(8):826–861. https://doi.org/10.1080/03602559.2015.1103267

23. Sodhi RK, Paul S (2019) Palladium(0) nanoparticles immobilized onto silica/starch composite: sustainable catalyst for hydrogenations and suzuki coupling. Bull Chem Reaction Eng Catal 14(3):586–603. https://doi.org/10.9767/BCREC.14.3.4395.586-603

24. Krasucka P, Stefaniak W, Kierys A, Goworek J (2015) Polymer–silica composites and silicas produced by high-temperature degradation of organic component. Thermochim Acta 615:43–50. https://doi.org/10.1016/J.TCA.2015.07.004

25. Lee DW, Yoo BR (2016) Advanced silica/polymer composites: materials and applications. J Ind Eng Chem 38:1–12. https://doi.org/10.1016/J.JIEC.2016.04.016

26. Zhang J, Liu N, Wang M, Ge X, Wu M, Yang J, Wu Q, Jin Z (2010) Preparation and characterization of polymer/silica nanocomposites via double in situ miniemulsion polymerization. J Polym Sci Part A Polym Chem 48(14):3128–3134. https://doi.org/10.1002/POLA.24094

27. Aghajani K, Tayebi AH (2017) Synthesis of SBA-15/PAni mesoporous composite for adsorption of reactive dye from aqueous media: RBF and MLP networks predicting models. Fibers Poly 18(3):465–475. https://doi.org/10.1007/s12221-017-6610-4

28. Binaeian E, Tayebi H-A, Rad AS, Afrashteh S (2018) Adsorption of acid blue on synthesized polymeric nanocomposites, PPy/MCM-41 and PAni/MCM-41: isotherm, thermodynamic and kinetic studies. 55(3):269–279. https://doi.org/10.1080/10601325.2018.1424554

29. Ghanei M, Rashidi A, Tayebi H-A, Yazdanshenas ME (2018) Removal of acid blue 25 from aqueous media by magnetic-SBA-15/CPAA super adsorbent: adsorption isotherm, kinetic, and thermodynamic studies. J Chem Eng Data 63(9):3592–3605. https://doi.org/10.1021/ACS.JCED.8B00474

30. Tayebi HA, Ghanei M, Aghajani K, Zohrevandi M (2019) Modeling of reactive orange 16 dye removal from aqueous media by mesoporous silica/ crosslinked polymer hybrid using RBF, MLP and GMDH neural network models. J Mol Struct 1178:514–523. https://doi.org/10.1016/J.MOLSTRUC.2018.10.040

31. Torabinejad A, Nasirizadeh N, Yazdanshenas ME, Tayebi H-A (2017) Synthesis of conductive polymer-coated mesoporous MCM-41 for textile dye removal from aqueous media. J Nanostr Chem 7(3):217–229. https://doi.org/10.1007/S40097-017-0232-7

32. Knowles GP, Delaney SW, Chaffee AL (2006) Diethylenetriamine[propyl(silyl)]-Functionalized (DT) Mesoporous silicas as CO_2 adsorbents. Ind Eng Chem Res 45(8):2626–2633. https://doi.org/10.1021/IE050589G

33. Stein A, Melde BJ, Schroden RC (2000) Hybrid inorganic±organic mesoporous silicatesð-nanoscopic reactors coming of age.https://doi.org/10.1002/1521-4095

34. Lim MH, Stein A (1999) Comparative studies of grafting and direct syntheses of inorganic−organic hybrid mesoporous. Mater Chem Mater 11(11):3285–3295. https://doi.org/10.1021/CM990369R

35. Arica TA, Ayas E, Arica MY (2017) Magnetic MCM-41 silica particles grafted with poly(glycidylmethacrylate) brush: modification and application for removal of direct dyes. Microporous Mesoporous Mater 243:164–175. https://doi.org/10.1016/J.MICROMESO.2017.02.011

36. Bahalkeh F, Juybari MH, Mehrabian RZ, Ebadi M (2020) Removal of brilliant red dye (Brilliant Red E-4BA) from wastewater using novel Chitosan/SBA-15 nanofiber. Int J Biol Macromol 164:818–825. https://doi.org/10.1016/J.IJBIOMAC.2020.07.035

37. Jose Varghese R, Parani S, Remya VR, Maluleke R, Thomas S, Oluwafemi OS (2020) Sodium alginate passivated $CuInS_2/ZnS$ QDs encapsulated in the mesoporous channels of amine modified SBA 15 with excellent photostability and biocompatibility. Int J Biol Macromol 161:1470–1476. https://doi.org/10.1016/J.IJBIOMAC.2020.07.240

38. Song Y, Yang LY, Wang YG, Yu D, Shen J, Ouyang XK (2019) Highly efficient adsorption of Pb(II) from aqueous solution using amino-functionalized SBA-15/calcium alginate microspheres as adsorbent. Int J Biol Macromol 125:808–819. https://doi.org/10.1016/J.IJBIOMAC.2018.12.112

39. Hachemaoui M, Boukoussa B, Ismail I, Mokhtar A, Taha I, Iqbal J, Hacini S, Bengueddach A, Hamacha R (2021) CuNPs-loaded amines-functionalized-SBA-15 as effective catalysts for catalytic reduction of cationic and anionic dyes. Colloids Surf A 623:126729. https://doi.org/10.1016/j.colsurfa.2021.126729

40. Hakiki A, Boukoussa B, Habib Zahmani H, Hamacha R, Hadj Abdelkader N, Bekkar F, Bettahar F, Nunes-Beltrao AP, Hacini S, Bengueddach A, Azzouz A (2018) Synthesis and characterization of mesoporous silica SBA-15 functionalized by mono-, di-, and tri-amine and its catalytic behavior towards Michael addition. Mater Chem Phys 212:415–425. https://doi.org/10.1016/j.matchemphys.2017.12.039

41. Hachemaoui M, Mokhtar A, Abdelkrim S, Ouargli-Saker R, Zaoui F, Hamacha R, Habib Zahmani H, Hacini S, Bengueddach A, Boukoussa B (2021) Improved catalytic activity of composite beads calcium alginate@MIL-101@Fe3O4 towards reduction toxic organic dyes. J Polym Environ 1–14. https://doi.org/10.1007/s10924-021-02177-4

42. Shen J, Zuo L, Meng Y, Fu T, Chi L, Wang T, Liu J (2021) Amphiphilic ionic liquid assembly route for the synthesis of polymer/Ag spheres and Ag-decorated bimodal porous silica. J Mol Liq 337:116477. https://doi.org/10.1016/J.MOLLIQ.2021.116477

43. Betiha MA, Moustafa YM, El-Shahat MF, Rafik E (2020) Polyvinylpyrrolidone-Aminopropyl-SBA-15 schiff Base hybrid for efficient removal of divalent heavy metal cations from wastewater. J Hazard Mater 397:122675. https://doi.org/10.1016/J.JHAZMAT.2020.122675

44. Mohan A, Rout L, Thomas AM, Nagappan S, Parambadath S, Park SS, Ha CS (2020) Silver nanoparticles impregnated pH-responsive nanohybrid system for the catalytic reduction of dyes. Microporous Mesoporous Mater 303. https://doi.org/10.1016/j.micromeso.2020.110260

45. Benali F, Boukoussa B, Ismail I, Hachemaoui M, Iqbal J, Taha I, Cherifi Z, Mokhtar A (2021) One pot preparation of CeO_2 Alginate composite beads for the catalytic reduction of MB dye: effect of Cerium percentage. Surfaces Interfaces 101306. https://doi.org/10.1016/j.surfin.2021.101306

46. Benhadria N, Hachemaoui M, Zaoui F, Mokhtar A, Boukreris S, Attar T, Belarbi L, Boukoussa B (2021) Catalytic reduction of methylene blue dye by copper oxide nanoparticles. J Cluster Sci 1–12. https://doi.org/10.1007/s10876-020-01950-0

47. Mekki A, Mokhtar A, Hachemaoui M, Beldjilali M, Meliani Fethia M, Zahmani HH, Hacini S, Boukoussa B (2021) Fe and Ni nanoparticles-loaded zeolites as effective catalysts for catalytic reduction of organic pollutants. Microporous Mesoporous Mater 310:110597. https://doi.org/10.1016/j.micromeso.2020.110597

48. Zaoui F, Sebba FZ, Liras M, Sebti H, Hachemaoui M, Mokhtar A, Beldjilali M, Bounaceur B, Boukoussa B (2021) Ultrasonic preparation of a new composite poly(GMA)@Ru/TiO2@Fe3O4: application in the catalytic reduction of organic pollutants. Mater Chem Phys 260(2020):124146. https://doi.org/10.1016/j.matchemphys.2020.124146

Polymer-Based Photocatalysis for Remediation of Wastewater Contaminated with Organic Dyes

Doaa M. EL-Mekkawi

Abstract Photocatalysis is an eco-friendly strategy to clean polluted waters under ambient conditions using sunlight or artificial light sources. Photocatalysis can degrade different classes of organic dyes into simple bio-assimilable species or less toxic components, and even mineralize them into harmless CO_2 and H_2O without leaving undesired byproducts. The utilization of polymer-based photocatalysts in the remediation of wastewater is initially motivated by several economic advantages such as facile separation and reuse, flexible designs, effectiveness, low cost and availability. Polymeric photocatalysts include photo-inactive polymers loaded with conventional photoactive species such as transition metal oxides, plasmonic metals and metal complexes. Some organic polymers are photoactive, possess outstanding semiconductive features and are notably capable of photocatalytically degrading organic dyes pollutants in water. The photocatalytic performance of the conventional photoactive species can be drastically improved when combined with the photoactive polymers. This chapter summarizes the latest representative advances of polymers in the photocatalytic treatment of wastewater contaminated with organic dyes. It begins with an introduction that includes a short general background on the importance of photocatalysis in the remediation of organic dyes, particularly with respect to polymeric photocatalysis. Then, a comprehensive overview of the conventional and up-to-date polymer-based photocatalysts is introduced, exploring their different types and the proposed mechanistic pathways. The commonly used techniques for the characterization of polymer-based photocatalysts are then concisely explored. The gained advantages, as well as the activity evaluation approaches, are also addressed. Finally, in conclusion, the future perspectives, recommendations and existing challenges are outlined.

Keywords Polymer composites · Metal nanoparticles · Photocatalytic efficiency · UV/visible light · Sunlight · Conjugated polymer · Conducting polymer ·

D. M. EL-Mekkawi (✉)
Physical Chemistry Department, National Research Centre, NRC, 33 EL Bohouth St., Dokki, P.O. 12622, Giza, Egypt
e-mail: doaa_egypt@yahoo.com; dm.mohammed@nrc.sci.eg

© The Author(s), under exclusive license to Springer Nature Singapore Pte Ltd. 2022   57
A. Khadir and S. S. Muthu (eds.), *Polymer Technology in Dye-containing Wastewater*,
Sustainable Textiles: Production, Processing, Manufacturing & Chemistry,
https://doi.org/10.1007/978-981-19-1516-1_4

Coordination polymer · Porous organic polymer · Organic dyes · Textile wastewater · Water treatment · Photocatalysis

1 Introduction

Dyeing is one of the most crucial sources of water pollution worldwide. Wastewater polluted with organic dyes, particularly that emerged from textile and printing industries, are predominantly loaded with high levels of biochemical oxygen demand (BOD) and chemical oxygen demand (COD) [1]. Thousands of types of synthetic dyes have been extensively utilized in various industries possessing various functional groups, physicochemical behavior and structural configurations. For example, azo, thiazine and fluorescein dyes are the most commonly used type of dyes in the textile and printing industries. These dyes are highly soluble in water due to their ionic configuration. Further, the existence of the unsaturated chromophore groups with high bond energy makes them highly stable. For instance, the bond energies of N=N in methyl orange and C=N in rhodamine B (RhB) are 518 and 407 kJ mol^{-1}, respectively [2]. These high values of bond energies make these dyes persistent pollutants which significantly resist the conventional biodegradation treatment processes. The BOD and COD ratio between 1:2 and 1:3 is ordinarily accepted and implies facile biodegradability of the organic pollutants [3]. Additionally, these wastewaters are characterized by their heavy color due to large amount of dissolved organic dyes. Efficient removal of color is a substantial issue since even low concentrations of dye are obviously visible [4, 5]. Direct discharge of these polluted waters will cause potentially deleterious effects on human beings and other organisms. These dyes may lead to several diseases such as respiratory illness, cancer, allergies and mutagenesis, etc. Therefore, implementing appropriate treatment of these polluted waters before discharge is highly required so that the concentration of the organic species (COD) can be reduced to the allowable limits.

Conventional techniques of wastewater treatment such as sedimentation, ultrafiltration, chemical methods and reverse osmosis have disadvantages of the elevated operating cost and the potential generation of poisonous byproducts. For instance, adsorption can effectively eliminate pollutants from wastewater. However, the adsorbed species needs further treatment for the final disposal. The advanced oxidation process is a promising alternative. This process refers to a range of techniques that rely on the generation of highly reactive species such as singlet oxygen ($^1O_2$), superoxide ($O_2^{\bullet-}$) and hydroxyl ($OH^{\bullet}$) radicals for the degradation of organic pollutants. Photocatalysis is the most widespread advanced oxidation process. Over the last few decades, photocatalysis acquired significant attention to remove organic dyes from polluted waters. Photocatalysis is an eco-friendly strategy to clean polluted waters under ambient conditions using sunlight or artificial light sources. More importantly, photocatalysis can degrade different classes of organic dyes into biodegradable species or less toxic components, and even eventually mineralize them into harmless CO_2 and H_2O without leaving undesired byproducts [6]. Photocatalysis mostly involves the generation of electron–hole pairs upon the irradiation of a semiconductor with light energy larger than or equal to its bandgap. These charge carriers diffuse to the photocatalyst surface to react with the adsorbed species to form active radicals. More than 150 photocatalysts have been introduced for the photodegradation of organic dyes under light illumination. The utilized catalysts were mostly based on metal oxides, sulfides, nitrides, hydroxides, complexes, carbonaceous and polymeric materials [7–11].

One major obstacle hindering the utilization of slurry photocatalysts is their post separation after wastewater treatment. If the photocatalysts nanoparticles are not separated, they may cause severe toxicity to aquatic and human systems. As an alternative, the photocatalyst can be immobilized on appropriate support, which inhibits the necessity of the separation processes. Several methodologies were explored for the immobilization of photocatalysts on solid supports. These supports include glass, quartz, silica, activated carbon, zeolites, cordierite, etc. [12].

Polymers have been extensively recognized for photocatalytic wastewater treatment due to their facile preparation, flexible design options, low expense, minimal influence on both environment and energy consumption. Therefore, attention has been paid to boost photocatalyst nanomaterials by polymers for workable and secure application purposes [13]. In the photocatalytic degradation processes, only a small amount of catalysts is immobilized on polymers, making the treatment operation more eco-friendly. Additionally, potential byproducts are inhibited and the separation and recycling of polymeric-based catalysts become easily achieved. Polymers such as polyurethane, polypropylene, polyvinyl chloride, polyvinyl alcohol, polyaniline etc. have been explored for immobilization of photocatalysts [4].

Further, many photocatalysts are only photoactive in ultraviolet-light irradiation. Thus, they are not sensitive to the full range of sunlight irradiation. For example, ZnO, MnO_2, TiO_2, and ZrO_2 are investigated for catalytic degradation [1, 14–16]. They displayed slightly poor photocatalytic performance toward the photodegradation of large organic molecules because of their low surface area. When these metal oxides are supported on porous materials and polymers, they exhibited reasonable photocatalytic performance under light irradiation. Therefore, organic/inorganic composites have been the goal of recent studies. They have been extensively investigated for wastewater remediation, particularly for the decomposition of organic dye molecules from water.

Polymeric photocatalysts include photo-inactive polymers loaded with conventional photoactive species such as transition metal oxides, plasmonic metals and metal complexes. Some organic polymers are photoactive, possess outstanding semiconductive features and notably capable of photocatalytically degrading organic dyes pollutants in water. Conjugated, conducting, porous and coordination polymers represent the most commonly investigated photoactive polymer classes in various photocatalytic applications. The photocatalytic performance of the conventional photoactive species is drastically improved when combined with these photoactive polymers. Combining inorganic semiconductor photocatalysts with the photoactive organic polymers creates synergistic and complementary features which reinforce the semi-conductor performance toward the removal of dye species from water. Benefiting from their extraordinary electron delocalization, effective light harvesting and facile charge carrier movement, photoactive polymers can act as both sorbents and photo-catalysts for the decontamination of organic dyes from water. The most popular types of polymer-based photocatalysts as well as the investigated classes of photoactive polymers and polymeric hybrids in the photdegradation of dyes are illustrated in Fig. 1.

Many published reviews cover the photocatalytic potential applications of polymer-based photocatalysts [17–26]. Advancement research on polymer-based photocatalysis is going on. Recent decades have witnessed the release of dozen of published research articles in the relevant field. Figure 2 illustrates the progress in the research field starting from 2003 until now. The data has been developed using the Web of Science search engine. Upon diverse improvement of photocatalysts using polymeric hosts at the molecular level, many spectacular accomplishments in photo-catalysis have recently been introduced, which clearly rationalizes a comprehensive review of this research field.

In this chapter, the latest representative advances of polymers in the photocat-alytic treatment of wastewaters contaminated with organic dyes are highlighted. An inclusive overview of the various types of polymer-based catalysts is exhaustively

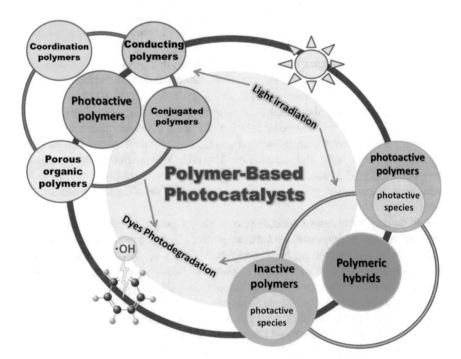

Fig. 1 Schematic view of the most popular types of polymer-based photocatalysts as well as the investigated classes of photoactive polymers and polymeric hybrids in the photodegradation of dyes

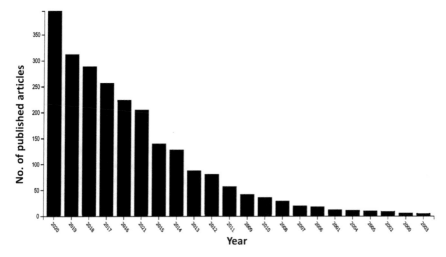

Fig. 2 The progress in the research field in polymer-based photocatalysis. The data has been developed using the Web of Science search engine

explored. A focus on an in-depth understanding of how polymers could influence the optical features, generate charge separation and transportation, and photocatalytic performance is demonstrated. The chapter begins with a brief discussion of conventional photocatalysts and their possible photocatalytic degradation mechanistic pathways. This is followed by exploring the developed types of photoactive polymers and polymeric hybrids in the photocatalytic decomposition of organic dyes. Concise descriptions of the utilized synthesis methodologies are also referred. After that, the gained merits, as well as the activity evaluation approaches, are addressed. Finally, the future perspectives, recommendations and existing challenges are outlined.

2 Conventional Inorganic Photocatalysts

IUPAC defined photocatalysis as a "change in the rate of a chemical reaction or its initiation under the action of ultraviolet, visible, or infrared radiation in the presence of a substance—the photocatalyst—that absorbs light and is engaged in the chemical transformation of the reaction partners" [27]. On these bases, the photocatalytic decomposition of organic species was conventionally interpreted as follows: The photocatalyst is first activated by the absorption of an incident photon having energy equal to or higher than its bandgap energy (E_g). As a consequence of the light absorption, negative electrons (e^-) transfer from the valence band (VB) to the conduction band at the catalytic surface leading to the formation of positive holes (h^+) at the VB. The generated electrons may then reduce the existing molecules to form $^\bullet O_2$, which further transforms into hydroxyl radicals ($^\bullet OH$). Simultaneously, the generated holes (h^+) may interact with hydroxyl groups (OH^-) to form hydroxyl radicals ($^\bullet OH$). The formed radical species are highly reactive oxidizing agents which are able to oxidize organic dyes into elementary species such as water, carbon dioxide and other less harmful species than the original pollutants [6].

Photocatalysis can be utilized for various remediation stages, from partial degradation of organic species to complete mineralization reactions. The substantial advantage of photocatalysis is the direct conversion of light energy into chemical energy. This causes a reduction in energy exhaustion and provides a green cost-effective pathway to sustainably combat environmental pollution.

Semiconductor materials are most utilized in photocatalysis. TiO_2 is one of the most extensively utilized metal oxide photocatalysts due to its comparatively high activity, abrasion resistance and low cost. However, its relatively wide bandgap is often a major disadvantage. For example, the bandgap values of the anatase and rutile phases of TiO_2 are 3.2 and 3.0 eV, respectively. This indicates that TiO_2 can only be photoactivated in the UV light region, which accounts for only 5% of the overall sunlight intensity. As a consequence, the photocatalytic efficiency is low under visible light irradiation. ZnO has been also widely used in the photodegradation of organic pollutants including dyes. It was originally proposed as an alternative photocatalyst to TiO_2. ZnO is an n-type semiconductor that absorbs also in the UV region due to its wide bandgap energy (3.37 eV). Attention has been also paid to ZnO-based materials because of their significant optical response to the ultraviolet region. ZnO-based materials induce remarkable photocatalytic activities, sometimes over TiO_2 [28, 29]. ZnO is a low price semiconductor compared to TiO_2 [30]. Thus, it is much considered and studied for large-scale applications. However, its major drawback originated from its low spectral response (sensitive only to the ultraviolet range). Cerium oxide (CeO_2) attracts special attention. It also displayed a significant photocatalytic activity. Ceria possesses a bandgap similar to TiO_2 (3.19 eV). In contrast with titania, the *4f* electronic configuration of ceria assists in charge transfer between oxygen and the surface-adsorbed dye pollutant [28]. Facile switches between the Ce^{3+} and Ce^{4+} oxidation states generate plenty of oxygen vacancy defects which are accountable for the efficient photocatalytic capability of nano-sized ceria particles [31].

Further, ceria nanoparticles illustrated outstanding features such as high oxidizing power, long-term stability, low price and enhanced electron transfer ability. These features make them effective photocatalysts for environmental operations [32].

Other semiconductors such as zinc and cadmium sulfides, tungsten, tin and zirconium oxides have also been used as photocatalysts for the decomposition of dyes. However, they are not quite stable and possess low corrosion resistance. Further, utilization of noble metal nano-sized particles, such as gold, silver, palladium or platinum has been extensively investigated as outstanding plasmonic photocatalysts. Noble metals enhance the absorption of materials in both visible light and UV light regions. This is attributed to the localized surface plasmon resonance (LSPR) effect [33, 34] and UV light due to interband electron transitions [33, 35], making them convenient catalysts for natural solar irradiation.

Similar to the majority of nano-sized inorganic particles, nano-sized inorganic photocatalysts tend to agglomerate/aggregate due to their high surface energy. This results in a remarkable reduction in photocatalytic performance due to the decrease in surface area. In addition, the separation and reutilization of the photocatalyst nanoparticles are quite complicated [36]. Immobilization of photocatalysts nanoparticles on polymeric substrates is an adequate approach to eliminate these drawbacks. This will result in acquiring hybrid composites with synergistically enhanced physicochemical characteristics.

Mainly, two possible mechanistic pathways for the generation of the reactive species have been often considered to photodegrade organic dyes by inorganic photocatalysts under visible light illumination. The first is a direct approach that requires the exploitation of a narrow bandgap semiconductor having an adequate optical response to the visible light region. The second is an indirect approach which involves the photosensitization of a wide bandgap semiconductor. In the second approach, a photosensitizer (co-catalyst) adsorbed on the photocatalyst surfaces is first excited by visible light illumination followed by an electron transfer from its excited state to the conduction band of the semiconductor to initiate the photodegradation processes. The majority of metal-based semiconductors have been reported to exhibit the first type of activity. Meanwhile, the second type mostly includes some surface-modified inorganic semiconductors and some specific substances such as bismuth-based materials [2].

Several studies have been made to extend the photoresponse of catalysts into the visible region and enhance their photocatalytic performance. These studies included modification of the photocatalysts by metal [37] and non-metal dopants [38] noble metal deposits [39] and formation of composites with narrow semiconductors.

The performance of the photocatalytic reactions relies on the structural morphology of the photocatalyst. Thus, precise control of the preparation conditions is essential for developing novel and highly active photocatalysts showing good separation between photogenerated holes and electrons, high surface-to-volume ratio and light absorptivity with excellent durability characteristics. Thus, the embedding and loading of photocatalyst nanostructured materials into polymeric substrates could demonstrate a good strategy to build up flexible, agglomeration-resistant and easily separated photocatalytic systems. Overall, currently, it is possible to modulate the bandgap energy and extend the spectral response of the photocatalytic reaction into the visible domain. The recombination of the photogenerated charge carriers can also be suppressed, while the photocatalytic performance is consistently improved.

3 Polymer-Supported Inorganic Photocatalysts

As previously illustrated, the nano-sized dimensions of the inorganic photocatalysts create some drawbacks such as difficult separation and recycling, as well as their potential threats to humans and the environment. An effective strategy to get rid of these obstacles is to immobilize the photocatalysts nanoparticles onto host substrates of larger dimensions. Consequently, a new type of material, namely nanocomposites, is produced. The polymeric host is a promising choice to attain these goals, particularly due to its practically unlimited architectural varieties. Polymeric substrates could possess multifunctional surface features and adequate robustness for long-term usage. The fast advancement of polymer/inorganic composites originates from their capability to integrate the photo-electrochemical features of nano-sized photo-catalysts with the elasticity of the transparent polymeric matrix [40, 41]. Thus, they can be suitable for different applications (optical, electrical, catalytical or medical applications).

The adaptation of polymers to support nano-sized photocatalysts has been extensively studied. Besides the previously mentioned advantages, polymers can enhance the resistance of the photocatalytic systems against UV radiation. They also can improve usability and stability. Many polymers are also characterized by their chemical inertness and availability [18]. Further, the polymers improve the surficial adsorption of the organic pollutants, increasing the activity of the photodecomposition processes [22].

3.1 Immobilized Inorganic Photocatalysts on the Surfaces of Photo-Inactive Polymer Substrates

3.1.1 TiO$_2$ Supported on Polymers

The first reported immobilized photocatalyst on a polymer host was introduced in 1995 [42]. The authors explored the utilization of TiO$_2$ immobilized on polythene films for the photodecomposition of phenol. 50% of phenol was successfully removed under solar irradiation. Since then, several polymeric structures have been introduced as substrates for the different inorganic photocatalysts. Recently, several techniques have been introduced to immobilize different photocatalysts on polymer surfaces for photocatalytic dye removal from water.

A comparative study was carried out on the photodegradation of methylene blue (MB) dye by TiO$_2$ immobilized on polyethersulfone using three different methods of dip-coating. The first two methods included immobilization of titania from suspension at 210 °C in the absence or presence of ultrasonication. The third method involved pre-grafting of polyethersulfones with the surface carboxylic group using electron beam radiation in the presence of polyacrylic acid. The anchored carboxylic on the modified polyethersulfone surface facilitated immobilization through its interaction with the surface hydroxyl group on titania. The MB degradation was found to be approximately the same for all samples. Moreover, TiO$_2$/polyethersulfone showed excellent durability (up to nine successive cycles) against MB degradation [43]. Recently, immobilization of TiO$_2$ on polymeric disposals including cotton, polyamide, paper, polyethylene terephthalate, polypropylene and polyurethane was studied by our research group. The TiO$_2$/polyethylene terephthalate immobilized via dip-coating followed by thermal and pressure treatment was chosen.

The TiO$_2$/polyethylene terephthalate displayed the best recyclability, robust polymer/TiO$_2$ linkage and catalytic performance. Analyses confirmed the strong binding and homogenous distribution of TiO$_2$ nanoparticles on the polyethylene terephthalate surfaces. Immobilized TiO$_2$/polymers were utilized to treat both synthetic and heavily contaminated real textile wastewaters. The solar catalytic degradation of wastewater resulted in remarkable photobleaching. A significant reduction in COD and mineralization percentages of up to 42% were obtained [15].

Recently, researchers began to make their studies on the utilization of eco-friendly polymers such as biopolymers as alternatives to those derived from non-renewable sources. For example, Liu et al. introduced xylan as a biopolymer for the synthesis of a xylan/PVA/TiO$_2$ hybrid. Polyvinyl alcohol (PVA) was used as a crosslinker and xylan as a template. TiO$_2$ nanoparticles were found to be homogeneously adhered to the composite surfaces [44]. The prepared composite showed acceptable catalytic performance under the visible light domain in the photodecomposition of ethyl violet and Astrazon Brilliant Red 4G dyes. In a single-step polymerization reaction, Yu and coworkers covered the surface of plasma pretreated cotton fabrics with polyaniline/TiO$_2$ composite [45]. The oxygen plasma treatment resulted in the generation of extra hydroxyl groups on the cotton surface. In their preparation, sulfosalicylic and sodium dodecylbenzene sulfonate were added as dopants to improve the catalytic performance. The composite showed good catalytic activity under simulated solar illumination for Rhodamine B decomposition. Shoueir et al. introduced a novel approach by coating chitosan fibers with a thin layer of nanostructured Au@TiO$_2$ [46]. The plasmon nature of Au promoted the photocatalytic reaction under visible irradiation. The plasmonic fiber was examined in the decomposition of various pollutants, including MB dye. The removal percentage of MB was found to be 98.8% within only 12 min of light illumination. The composite displayed a reasonable degradation rate and significant stability even after eight cycles (loss ~4.1%).

3.1.2 ZnO Supported on Polymers

Several studies have been introduced to explore the catalytic activity of ZnO/polymethyl methacrylate (PMMA) composites in wastewater treatment. For instance, a thin coat of ZnO was loaded on polymethyl methacrylate powders by atomic layer deposition technique at 80 °C followed by sonication and solution casting [47]. The catalytic performance was examined on the decomposition of phenol and MB in water, using ultraviolet light. The prepared powders and films of ZnO/polymethyl methacrylate displayed significant photocatalytic activity toward the removal of MB under the UV light domain. The powder form of ZnO composite showed a slightly higher degradation rate due to its higher exposed surface area. Furthermore, ZnO/polymethyl methacrylate nanocomposite displayed good stability and durability for seven degradation cycles. Arslan et al. studied improving the catalytic performance and the charge separation effectiveness of ZnO/polyacrylonitrile (PAN) under visible irradiation by incorporating a plasmonic

metal [48]. Electrospun polyacrylonitrile (PAN) fibers were loaded with a decorated nanolayer of ZnO with Pd nanocubes (Pd@ZnO@PAN-NF). The dimensions
of nanocubes range from 7 to 22 nm. The catalytic performance of the prepared
nanofibers was measured and compared with the undecorated fibers (ZnO@PAN-
NF) in the decomposition of MB dye under visible illumination. Pd enhanced the
catalytic reaction through the energy transfer process between palladium and zinc
oxide. Palladium nanocubes acted as an electron trap leading to the reduction of
the electron–hole recombination process. On the other side, ZnO@PAN-NF had no
activity against the photodegradation of MB under the same experimental conditions.

3.2 Inorganic Photocatalysts Embedded Within the Photo-Inactive Polymer Substrates

3.2.1 TiO$_2$ Embedded Within the Polymer Matrix

Several studies explored the utilization of embedded inorganic photocatalysts inside
polymeric matrices. The inclusion of photocatalyst species within a polymer matrix
reduces the catalyst leaching from the polymeric surfaces, eliminates surface aggre-
gations, retains the nanostructured configuration of the photocatalysts and also may
promote some synergistic processes and create new features.

For example, the steep reduction of the penetrated stream across the polymeric
membrane is one of the major drawbacks during the treatment of wastewater. Accumulation of different substances on the membrane surface results in a decrease in its
efficiency. Many solutions were determined to eliminate this problem. The combination of photocatalysis and membrane operations is an outstanding approach. This
integration practically enhanced both the decomposition of organic pollutants and
the permeability of the polymeric membrane. Benhabiles and co-workers demonstrated a study on combining membrane separation and photocatalysis "photocatalytic membranes reactor". They incorporated different dosages of TiO$_2$ (0.12,
0.25, and 0.5 wt%) within the porous sponge-like polyvinylidene fluoride/PMMA
membranes using the phase inversion technique [49]. The hybrid nanocomposite
showed good catalytic performance in MB decomposition under ultraviolet illumination. The photodecomposition of MB increased upon raising the percentage of
TiO$_2$ nanoparticles within the membrane.

Nakatani and co-workers designed an unfamiliar reverse micelle pattern composed
of the hydrophilic part (acrylic acid; PAA) and hydrophobic part (polystyrene; PS)
for the incorporation of TiO$_2$ gel [50]. The catalytic performance of the poly (PS-
block-PAA) was examined in MB photodegradation using visible illumination. The
degradation of MB was mostly affected by the polymeric chain length and the
molar ratio of acrylic acid/polystyrene. First, MB molecules were adsorbed on the

polymer chains. Only the adsorbed MB species surrounding the gel was photodecomposed. The photocatalytic performance was greatly enhanced by the addition of Cu-phthalocyanine within the micelle. In that case, the degradation rate increased three-fold. The authors suggested the occurrence of effective e–h separation at the photo-generated interfacial region between titania and Cu-phthalocyanine using visible illumination. Further, the composite demonstrated significant catalytic durability, so it can be reutilized.

3.2.2 ZnO Embedded Within the Polymer Matrix

Ding et al. synthesized hierarchical flexible nanofiber membranes composed of zinc oxide embedded within the polyimide. The nanocomposite was prepared via both electrospinning and direct ion-exchange methods [51]. Initially, poly amic acid nanofibers were prepared by electrospinning followed by their immersion $ZnCl_2$ solutions. The prepared nanofibers were then thermally treated. Consequently, imidization of poly (amic acid) into polyimide and the oxidation of $ZnCl_2$ into nano-sized ZnO particles occurred. Different shapes of ZnO such as nanoplatelets or nanorods were prepared by varying the initial concentration of $ZnCl_2$ in solution. These various nanostructures provided a broad range of nanomaterials having tunable catalytic performance. The catalytic features of polyimide/ZnO membranes were assessed using methylene blue photodegradation. The optimal photocatalytic performance of the membrane was also identified.

The utilization of biodegradable polymeric supports is a promising eco-friendly approach for a sustainable solution of water remediation. Lefatshe et al. synthesized nanostructured ZnO particles within the cellulose via in situ solution-casting method [52]. The catalytic performance of cellulosic nanocomposite was studied in the degradation MB under ultraviolet illumination. The degradation results were compared with those obtained by pure ZnO nanostructures. ZnO dispersed in nanocellulose matrix displayed better activity than pure ZnO. This behavior was due to the high exposed surface area provided by cellulose. This high surface area resulted in raising the number of active sites and produced improved activity of the nanocomposite. Another green approach was introduced by Rajeswari et al. They reported the synthesis of cellulose acetate-polystyrene membrane incorporated with ZnO nanoparticles by solution dispersion method [53]. The utilized polymeric catalyst is characterized by its high thermal stability and improved photocatalytic activity. The performance of the catalyst was tested against the photodegradation of congo red (CR) and reactive yellow-105 (RY-105) under sunlight illumination. The complete photodegradation of the dyes was rapidly achieved at neutral pH. The photodegradation efficiency was up to 98%. The recyclability tests also revealed the reasonable durability of the hybrid membrane under solar illumination.

3.2.3 CeO_2 Embedded Within the Polymer Matrix

In situ crystallization of CeO_2 incorporated within polystyrene was demonstrated by Fischer and co-workers. They utilized surface-modified CeO_2/polystyrene with phosphonate and phosphate groups to nucleate and stabilize the crystallization of the nanocrystalline CeO_2 [54]. The catalytic performance of CeO_2/polymer hybrid was examined in the photodecomposition of RhB under ultraviolet radiation. The authors revealed that the oxidation of RhB occurred at a higher rate upon using CeO_2/polystyrene nanoparticles than without photocatalyst or by using CeO_2 powders. This confirmed the effectiveness of the hybrid catalyst on RhB B degradation. Morselli et al. prepared poly (vinylidene fluoride-*co*-hexafluoropropylene)/CeO_2 (PVDF-HFP/CeO_2) nanocomposite fibrous membranes [55]. The nanocomposite was prepared by electrospinning followed by in situ preparation of CeO_2 by thermal treatment of the precursor salt of cerium. The catalytic decomposition performance of the fibers was enhanced by doping Au. Au particles were whether in situ or ex situ introduced to form (PVDF/CeO_2/Au_in) or (PVDF/CeO_2/Au_ex), respectively. The catalytic performance was examined on MB photodegradation by visible illumination. PVDF-HFP/CeO_2 membrane displayed similar degradation performance as the pristine PVDF-HFP membranes. This indicated that CeO_2 has no observable catalytic performance in the visible domain. However, the doped nanocomposites with gold, particularly the ex situ prepared membrane, determined a significant enhancement of the catalytic performance. These observations were due to the narrower bandgap values of the Au-doped samples. This allowed the enhancement of photodecomposition reactions under visible illumination. Ni and co-workers investigated the catalytic photodegradation of MB using synthesized PS/RGO@CeO_2 (RGO is reduced graphene oxide) under visible illumination [32]. The catalytic activity of ceria was synergistically improved by introducing RGO. Besides, these composite nanoparticles were simply separated and reused for many photocatalytic runs.

3.2.4   Plasmonic Metals Embedded Within the Polymer Matrix

Plasmonic metal photocatalysts are presently widely explored because of their distinctive quantum effect and strong absorptivity in visible and UV domains. Plasmonic metal strongly interacts with incident photons by their outer-shell electrons leading to the generation of energetically high electrons on its surfaces. These electrons can further activate dye species in the vicinity of its surface and will initiate the decomposition reactions [56]. The photocatalytic features of noble metal mainly rely on their allocation, geometry, dimension and surficial interactions with supports. Below, some recent studies on the polymeric systems embedding noble metal nanoparticles for the catalytic photodegradation of organic dyes will be demonstrated.

Hareesh et al. synthesized polycarbonate incorporated with silver and gold Ag/PC and Au/PC with the aid of synchrotron X-ray radiation [57]. The authors investigated the catalytic photodecomposition of MB dye using Ag/PC and Au/PC composites under ultraviolet illumination. MB displayed higher removal rates using the polymeric composites than that obtained by the unbounded metals. Aggregation of unsupported metal nanoparticles drastically reduced the photocatalytic reaction rate.

Chibac et al. introduced a facile low cost and eco-friendly synthesis method of gold/polymer nanocomposite films. Synthesis was carried out via in situ photogeneration of Au simultaneously with the photo-assisted polymerization of methacrylated glycomonomers [58]. The prepared films were flexible free-standing. Composites showed reasonable decomposition rates against MB and methyl orange (MO) using visible illumination when compared with the previously published systems. The catalytic activity of the films was slightly reduced for the first photocatalytic run using MB due to the initial low adsorption of MB by the nanocomposite.

Overall, the most important benefit relies on the possibility of polymers to host conventional UV active photocatalysts, such as titanium and zinc oxides, and more modern UV/visible photoactive species such as CeO_2 and noble metals. As illustrated, polymer supports eliminated aggregation and leaching of various photoactive species during the degradation processes. They also facilitated their recovery, and mostly, preserved or even improved their catalytic efficiency compared to the slurry systems.

4 Organic Polymers-Based Photocatalysts

As previously mentioned, in a photocatalytic reaction, the energy necessary for excitation is controlled by the bandgap of the photocatalyst. The ultraviolet domain encompasses approximately 5% of sunlight. However, the visible domain (400–800 nm) comprises around 53% of the sunlight radiation. Thus, it is essential to develop catalysts with reduced bandgap (below 3.0 eV) to enhance the exploitation of sunlight, particularly the visible light domain.

Some organic polymers themselves display outstanding photocatalytic behaviors. They showed the superior capability of driving several photocatalytic processes including the catalytic degradation of organic dyes. Conjugated, conducting, porous and coordination polymers represent the most commonly investigated polymer classes in various photocatalytic applications. These types of polymers illustrated promising semiconductive behavior and were extensively investigated in the remediation of organic dyes in water as illustrated below.

Inorganic and organic photocatalysts display wide variations in the nature of their charge carriers. These variations directly affect charge separation and transportation. Therefore, a thorough investigation of the photophysical and physicochemical fundamentals that govern organic photocatalysis with respect to specified features of organic substance is required. Organic polymer photocatalysts are specifically outstanding due to their distinctive chemical versatility, which permits tuning of building units to achieve various promising photo-, physical- and chemical criteria [20].

In this section, a comprehensive interpretation of these polymeric photocatalysts will be introduced and examples for the most relevant recent studies will be explored, particularly the studies on the utilization in the contaminated waters with dye species.

Similar to the conventional inorganic photocatalysts, when these polymers absorb the energetically appropriate photons (mostly visible light), electrons are excited and transferred from VB to CB. As a consequence, a simultaneous formation of holes at VB and electrons at CB occurred. These charge carriers then move to the polymer surface for redox reaction. A stream of highly active radicals, including singlet oxygen ($^1O_2$) and superoxide ($O_2^{\bullet-}$), is produced via singlet electron transfer and energy transfer from the excited polymer to molecular O_2. The formed superoxide radical $O_2^{\bullet-}$ then interacts with H_2O molecules to produce hydroxyl radical ($^{\bullet}OH$). Simultaneously, the holes disperse over the polymer surfaces and attract further dye molecules. This facilitates the separation of the charge carriers. Finally, the dyes species get decomposed by one or more of the produced active oxidative radicals.

4.1 Conjugated Polymers (CPs)

Conjugated polymers (CPs) represent a modern platform in the field of photocatalysis. In the past few decades, research has been devoted to finding more convenient catalysts for practical operations. Among them, the π-conjugated polymers have been recently significantly considered as the outstanding host for various photocatalytic processes. CP is distinguished by a backbone chain of alternating double and single bonds. The overlapping p-orbitals in CP generate a range of mobile π-electrons that can produce important and beneficial optical and electronic features.

Several promising properties enable conjugated polymers to compete with inorganic semiconductors in photocatalytic operations. CPs are eco-friendly semiconductive substances, composed of abundant elements in the earth and own adjustable energy levels for redox reactions. CPs can be easily synthesized under moderate conditions. They possess significant chemical resistance to photobleaching. Their molecular configuration can be finely adjusted to achieve optimal use of the visible domain, while most conventional inorganic photocatalysts only have ultraviolet activity. Further, the extended π-conjugation over the polymer backbone gives them the unique transport and separation characteristics of photo-generated charge carriers, which are essential for triggering photo-redox reactions [23].

The major photodegradation mechanism using semiconductive polymers can be summarized as follows. Under the irradiation of light, the polymer absorbs incident photons and forms e–h pairs. The photo-generated charges then transfer to the surface of the semiconductor where the redox reaction occurs. However, the majority of the photogenerated charges can exhibit bulk or surface recombination, which is a competitive process with the surface redox reaction.

CP encompasses several classes of polymers, including carbon nitrides g-C_3N_4, linear conjugated polymers, water-soluble conjugated polyelectrolytes (CPEs), conjugated porous polymers (CPPs), conjugated microporous polymers (CMP), covalent organic frameworks (COFs) and covalent triazine frameworks. Structurally, linear conjugated polymers are composed of alternative conjugated units. Meanwhile, conjugated porous polymers have a three-dimensional porous framework, which facilitates charge transportation in various directions. The statistical copolymerization for the preparation of CPPs grants opportunities to tune the bandgaps and porous structure of polymeric photocatalyst [23]. Further, crystalline COFs possess a piled two-dimensional configuration that is useful for charge transfer, thus improving the photocatalytic performance.

Effective CP photocatalysts should have the following characteristics: (a) Wide spectral response and strong absorptivity in the visible domain in order to efficiently exploit sunlight; (b) Efficient separation and transportation of charges in order to facilitate the surficial redox reactions; (c) Adequately arranged energy levels to allow effective charge separation and redox reactions.

In order to successfully achieve these criteria, reasonable design and preparation of CPs for an efficient photocatalyst are required.

Conjugated polyelectrolytes (CPEs) polymers are a family of polymeric photocatalysts enriched with charged sidechains and conjugated backbones. This unique structure enabled the enhancement of the polymer wettability in water, thus improving the degradation performance. Ghasimi et al. demonstrated (CPE) for the photodegradation of MB and RhB dyes [59]. They synthesized their CPE polymer by linking 1-alkyl-3-vinylimidazolium bromide to the sidechain of the poly(benzothiadiazolyl fluorene) main chain. The polymerization process was self-initiated by using visible light. The bandgap of the polymer was 2.11 eV. The prepared CPE successfully photodecomposed the targeted dyes. Further, the polymer illustrated excellent durability where RhB degradation was successfully carried out till 10 cycles, RhB.

Linear conjugated polymers having distinctive nanostructures also showed good performance in organic dye photodegradation. Ghosh et al. synthesized poly(diphenylbutadyne) nano-sized fibers via photo-assisted polymerization using a soft templating strategy [26]. The length of prepared fibers was a few micrometers, and the diameter was nearly 19 nm. MO exhibited 75% decomposition after 240 min of illumination. However, only 17% decomposition was accomplished using Ag-modified TiO_2. The catalytic activity of the nanofibers was almost constant even after 15 repeated runs. Using the same synthetic strategy, Ghosh et al. further prepared poly(3,4-ethylenedioxythiophene) with vesicle and spindle structures [24, 25]. The thickness of their spindles was found to be 40 nm, while the length was several hundred nanometers. However, the average diameter of the vesicles was about 1 mm and the average thickness of the vesicle's wall was nearly 40 nm. The catalytic performance was found to be shape-dependent. The spindles showed 100% removal of methyl orange dye under visible illumination. On the contrary, the vesicles had no catalytic activity under the same conditions. Further, the photocatalytic performance of the spindles was found to be more active than that of the previously synthesized nano-sized fibers by [26].

4.2 Conducting Polymers (CNPs)

Conducting polymers (CNPs) are composed of conjugated organic configuration. CNP characterizes by the existence of alternating double and single bonds and mobile π-electrons. Electrons in the conducting polymers are highly delocalized, have high mobility and are able to easily transfer across the polymer matrix [60]. Therefore, CNPs possess distinctive electrochemical and optical features [19]. Polyanilines, polypyrroles, polyfurans, polyacetylenes, poly(3,4-ethylenedioxythiophene), poly(p-phenylenevinylene) and polythiophenes are the most utilized CNPs. CNPs participated in a broad range of applications. For example, they can be used in sensors, flexible displays, solar cells, energy devices, biomedical and environmental applications [19].

It was previously revealed that the inclusion of conductive substances, such as nickel foam, metallic wire mesh and graphene, can efficiently overcome the drawbacks related to the fast recombination of charges within the photocatalysts [61]. In a similar way, by incorporating CNP, the generated electrons will migrate to conductive substances, which accept electrons in order to reduce the undesired rebound of the generated charges. Furthermore, besides these outstanding advantages, the introduction of CNPs will provide processability, particularly through the solution operation. CNPs have been extensively examined in different applications, such as power generation, solar and electronic devices and light-emitting diodes. In photocatalytic applications, CNP may be introduced as individual species in the form of composites with other photocatalysts. The composite photocatalysts can be synthesized by the inclusion of CNPs (particularly polyaniline, pyrrole and thiophenes) with semiconductor nanoparticles. CNP provides compatible band configurations with inorganic semiconductors, thus suppressing the rebound of the generated charges [61].

As previously illustrated, the catalytic photodegradation reaction is composed of three major processes: light absorption by a catalyst, separation of the photogenerated e–h pairs, and the interfacial photocatalytic oxidation and reduction reactions. When UV/visible light illuminates CNP, excitation of a photon having energy higher than bandgap VB to CB spontaneously occurs. This process is denominated as $\pi-\pi*$ electronic transition. The photoexcited electrons transfer to CNP surfaces and promote the adsorbed molecular O_2 to produce superoxide radicals ($O_2^{\cdot-}$). The formed superoxide radicals decompose organic pollutants through oxidation reactions in water. Additionally, the generated holes may also directly oxidize organic species [60].

Ghosh et al. [24], Yuan et al. [62], and Floresyona et al. [63] introduced nano-sized structures of poly(3,4-ethylenedioxythiophene), polypyrrole and poly(3-hexylthiophene) hexagonal mesophases for the catalytic photodecomposition of various organics including dyes. They revealed that mineralization of pollutants and polymer durability enhanced using ultraviolet and visible irradiation compared to their bulk powders (see Table 1). Although CNPs displayed good photocatalytic capability, their utilization in real operations may face several challenges, such as scalable preparation of CNPs, reutilization and durability under real environmental and treatment conditions [60].

Table 1 Different polymer-based photocatalysts utilized for the decomposition of organic dyes under light irradiation

Polymer-based photocatalyst	Preparation method	Targeted dyes	Light domain/source	Performance indices	Degradation performance (Optimal)	Techniques and Methods	Durability	References
Inorganic photocatalysts supported on polymer surfaces								
TiO$_2$ supported on polymer								
TiO$_2$/polyethersulfone	Three different methods of dip-coating	MB	UV/sunlamp; 11.3 ± 1.3 mW cm^{-2}	Pseudo-first-order rate constant (k)	k=0.11 min^{-1}	UV–Vis spectrophotometry	The same activity up to 9 cycles	[43]
TiO$_2$/polymeric textile disposals	Dip-coating followed by thermal treatment under pressure	MB, AGY*, textile wastewaters	UVA/UVA light bulbs; 2.74 mW/cm^2	Time for complete decolorization	100% color removal for MB and AGY	UV–Vis Spectrophotometry		[15]
			Sunlight	Mineralization: %COD removal	Up to 42% COD removal for textile dyes	COD		
				BOD removal		BOD		
TiO$_2$ polyethylene terephthalate	Dip-coating followed by thermal treatment under pressure	MB	Sunlight	Decolorization rate	100% color removal	UV–Vis Spectrophotometry	Active up to 7 cycles	[16, 64]
				Mineralization: %COD removal	100% COD removal	COD		
Xylan/polyvinyl alcohol/TiO$_2$	Green template	Ethyl violet / Astrazon Brilliant Red 4G	Visible light	Decolorization rate	~94% color removal for both dyes	UV–Vis Spectrophotometry	The same activity after 3 cycles	[44]
Polyaniline/TiO$_2$/cotton fabrics	Plasma-treated cotton followed by one-step in situ polymerization	RhB	Simulated sunlight/300W Xenon lamp	Decolorization rate	Up to 87.67% color removal	UV–Vis–NIR spectrophotometry	UV protection was enhanced even after 10 washing cycles	[45]

(continued)

Table 1 (continued)

Polymer-based photocatalyst	Preparation method	Targeted dyes	Light domain/source	Performance indices	Degradation performance (Optimal)	Techniques and Methods	Durability	References
Au@TiO$_2$/chitosan	Photo-assisted deposition followed by soaking in chitosan fiber	MB	Visible light/ 25 W equivalent halogen lamp	Decolorization rate	98.8% color removal	UV–VIS spectrophotometry	8 cycles (activity loss ~ 4.1%)	[46]
				Pseudo-first- order rate constant (k)	0.85×10^{-2} min^{-1}			
ZnO supported on polymer								
ZnO/polymethyl methacrylate	Atomic layer deposition followed by sonication and solution casting	MB	UV/ UV lamp~2mW/cm^2	Decolorization rate	60% (powder)	UV–Vis spectrophotometry	Stable up to 7 cycles	[47]
				Pseudo-first-order rate constant (k)	40% (film)			
					0.41×10^{-2} min^{-1} (powder)			
					0.26×10^{-2} min^{-1} (film)			
Pd@ ZnO@polyacrylonitrile core-shell nanofiber	Atomic layer deposition on electrospun PAN fiber	MB	Visible light/ solar simulator equipped with 300 W xenon arc	Pseudo-first-order rate constant	N/A	UV–VIS spectrophotometry	N/A	[48]
Inorganic photocatalysts embedded within polymer substrates								
TiO$_2$ embedded within the polymer								
TiO$_2$/polyvinylidene fluoride/polymethyl methacrylate membrane	Phase inversion method	MB	UV/ Zφ type 500 W	Decolorization rate	99% color removal	UV–VIS spectrophotometry	N/A	[49]
				Rate constant k	0.0117 min^{-1}			
				Halflife time t$_{1/2}$	59 min			
TiO$_2$/polystyrene-block-acrylic acid (PS-*b*-PAA)	Reverse micelle	MB	Visible/(LED) lamp of 25 W	k (TiO$_2$/PS-*b*-PAA)	0.17×10^{-2} min^{-1}	UV–VIS spectrophotometry	Reactivity loss (20%) after 1 cycle	[50]
Cu phthalocyanine@ TiO$_2$/PS-*b*-PAA				k (Cu phthalocyanine@ TiO$_2$/PS-*b*-PAA)	0.6×10^{-2} min^{-1}			

(continued)

Table 1 (continued)

Polymer-based photocatalyst	Preparation method	Targeted dyes	Light domain/source	Performance indices	Degradation performance (Optimal)	Techniques and Methods	Durability	References
ZnO embedded within the polymer								
Polyamide/ZnO nanofiber membrane	Electrospinning and direct ion-exchange methods	MB	UV	Decolorization rate k	up to 98% $1.7\ h^{-1}$	UV–VIS spectrophotometry	N/A	[51]
ZnO/cellulose	Via in situ solution casting	MB	UV	Decolorization rate k	79% $0.1174\ h^{-1}$	UV/Vis/NIR spectrophotometry	N/A	[52]
ZnO/cellulose acetate-polystyrene membrane	Solution dispersion blending	Congo red (CR)	Sunlight	Decolorization rate	95% for CR	UV/Vis/NIR spectrophotometry	No remarkable changes were noticed till 4 cycles	[53]
		Reactive yellow-105 (RY)			98% for RY			
CeO₂ embedded within the polymer								
Phosphonate and phosphate-modified CeO₂/polystyrene	Surfactant-free miniemulsion polymerization followed by chemical precipitation	RhB	UV/(mercury vapor lamp)	k	$2.86 \times 10^{-2}\ min^{-1}$ $4.36 \times 10^{-2}\ min^{-1}$	Spectrofluorometry (decrease of the fluorescence emission intensity of RhB)	N/A	[54]
Au/CeO₂/polyvinylidene fluoride-*co*-hexafluoropropylene) fibrous membranes	Electrospinning and thermal treatment salt of cerium	MB	Visible/LEDs array	k	$1.4 \times 10^{-3}\ min^{-1}$	UV/Vis/NIR spectrophotometry	N/A	[55]
Polystyrene/reduced graphene oxide@CeO₂	dispersion polymerization followed by hydrolysis of Ce salt.	MB	UV/45 W fluorescent lamp	k	$0.015\ min^{-1}$	UV/Visible spectrophotometry	<10% performance decay after 3 cycles	[32]

(continued)

Table 1 (continued)

Polymer-based photocatalyst	Preparation method	Targeted dyes	Light domain/source	Performance indices	Degradation performance (Optimal)	Techniques and Methods	Durability	References
Noble metal embedded within the polymer								
Ag or Au/polycarbonate	Synchrotron X-ray radiation assisted method	MB	UV/250 W mercury lamp	Decolorization rate	Ag/PC > Au/PC	UV/Visible spectrophotometry	N/A	[57]
Hybrid Au/methacrylates	UV curing process	MB	Visible/	Decolorization rate and k	MB (98%)	UV/Visible spectrophotometry	~25% performance decay after 1 cycle	[58]
		MO			4.7×10^{-2} min^{-1}			
					MO (94%)			
					3.18×10^{-2} min^{-1}			
Organic polymers-based photocatalysts								
Metal-free polymer photocatalysts								
Conjugated Polymers (CPs)								
Porous imidazolium-based ionic-liquid	Self-initiated intramolecular radical cross-linking process under visible-light irradiation	MB	Visible/white LED lamp	Decolorization rate	> 90% for both dyes	UV/Visible spectrophotometery	Stable up to 10 cycles	[59]
		RhB						
Polydiphenylbutadiyne nanofiber	Photopolymerization utilizing a soft templating process	MO	Visible (using cut-off filter at λ > 450 nm)/(xenon lamp	Decolorization rate	75%	UV/Visible spectrophotometry	Same activity up to 15 cycles	[26]
Poly(3,4-ethylenedioxythiophene) with vesicle and spindle structures	Soft templates via chemical oxidative polymerization	MO	Visible/300W xenon lamp, UV cut-off filter (λ > 450nm)	Decolorization rate	100% (spindle)	UV/Visible spectrophotometry	Spindles retained >95% of their activity after 6 cycles	[24, 25]
Conductive Polymers (CNPs)								
Polypyrrole	Chemical polymerization and radiolysis	MO	UV and visible/ 300 W Xenon lamp	Decolorization rate	80% (UV)	UV/Visible spectrophotometery	Stable up to 4 cycles	[62]
				TOC %removal	10% (visible)	TOC		
					~79% (UV)			

(continued)

Table 1 (continued)

Polymer-based photocatalyst	Preparation method	Targeted dyes	Light domain/source	Performance indices	Degradation performance (Optimal)	Techniques and Methods	Durability	References
Poly(3-hexylthiophene)	Soft templates	RhB	Simulated solar light/ 300W Xenon	Time of Decolorization	120 min UV; 140 min (visible)	UV/Visible spectrophotometery	Stable up to 4 cycles	[63]
Porous Organic Polymers (POPs)								
BT3-a	Suzuki-Miyaura cross coupling reaction via miniemulsion polymerization (BT3-a)	RhB	Visible/23 W household energy-saving light bulb	Decolorization rate	80% (BT3-a)	UV/Visible spectrophotometery	Stable up to 5 cycles	[65]
BT3-b	Sonogashira–Hagihara cross-coupling reaction via miniemulsion polymerization (BT3-a).				50% (BT3-b)			
Core-shell MOF/COF hybrid composite (NH$_2$-MIL-68@TPA-COF)	Solvothermal method	RhB	Visible/ 300-W xenon lamp coupled with a UV cutoff filter (>420 nm)	k	0.077 min^{-1}	UV/Visible spectrophotometery	N/A	[66]
Imine-based COF with a triazine core	Sonically assisted copolymerization	MO	Visible/300 W xenon lamp equipped with an optical cutoff filter (λ≥420 nm)	k; Mineralization: %TOC	0.09 min^{-1}; 25.6%	UV/Visible spectrophotometery; TOC	Retained 80% of their original efficiency up to 4 cycles	[67]
Metal-modified POPs								
(Cu-POP): coupling between 1,4-diethynylbenzene and 1,3,5-bromobenzene or	Sonogashira-Hagihara cross-coupling	RhB, MB, CR, MO	Visible/350 W Xe lamp attached with a long-pass UV cut-off filter (λ > 420 nm)	Decolorization rate	(Cu-POP) up to 99%	UV/Visible spectrophotometery	Slight decrease in degradation efficiency over 5 cycles	[2]
(Pd-POP): coupling between 1,4-diethynylbenzene and 1,2-bis(3,5-dibromophenyl) diazene				k	(Pd-POP) up to 90%; (Cu-POP) up to 67 × 10^{-3} min^{-1}			

(continued)

Table 1 (continued)

Polymer-based photocatalyst	Preparation method	Targeted dyes	Light domain/source	Performance indices	Degradation performance (Optimal)	Techniques and Methods	Durability	References
					(Pd-POP) up to 39.1×10^{-3} min^{-1}			
Two Anderson-type polyoxometalates coupled with	Sonogashira–Hagihara cross-coupling	RhB, MB	Visible/450 nm optical filter was set in front of the xenon lamp light	Decolorization rate	MB up to 92%, RhB up to 66%	UV/Visible spectrophotometery	Stable up to additional 5 cycles	[68]
Conjugated microporous polymer modified with ferrocene	Palladium catalyzed Sonogashira–Hagihara cross coupling copolymerization	MB	Visible/simulated solar light irradiation (AM1.5G)	Decolorization rate	up to 99%	UV/Visible spectrophotometery	Slight decrease in degradation efficiency over 5 cycles	[69]
Coordination Polymer (COPs)								
Ag(I)-doped COP-based material [{Pb(Tab)$_2$(bpe)}$_2$(PF$_6$)$_4$.1.64AgNO$_3$]n	Ambient temperature solid state reaction	12 different dyes	UV/ 400 W medium pressure mercury lamp	Decolorization rate	up to 95% of MO	UV/Visible spectrophotometery	Stable till 5 cycles	[70]
Fe^{3+}, Cr^{3+}, Ru^{3+}, Co^{2+} and Ni^{2+}, Fe^{3+} doped [Zn(cca)(4,4'bipy)]n	Hydrothermal synthesis followed by ion-exchange reaction	RhB	Visible/300 W medium pressure mercury lamp with a cutoff filter ($\lambda \geq$ 420 nm)	Decolorization rate	94% RhB using Fe^{3+}-COP (the highest activity)	UV/Visible spectrophotometery	Stable up to additional 5 cycles	[71]
Inorganic photocatalyst/semiconductive polymer hybrids								
TiO$_2$/conducting polymers								
TiO$_2$/polyaniline	In-situ chemical oxidative polymerization of aniline with ammonium persulfate in the presence of TiO$_2$	RhB, MB	UV	Decolorization rate, k	80% RhB; 73% MB; RhB: 0.7642×10^{-2} min^{-1}; MB: 0.684×10^{-2} min^{-1}	UV/Visible spectrophotometery	Slight decrease in activity after 3 cycles	[72]

(continued)

Table 1 (continued)

Polymer-based photocatalyst	Preparation method	Targeted dyes	Light domain/source	Performance indices	Degradation performance (Optimal)	Techniques and Methods	Durability	References
TiO_2, $BiVO_4$, graphene oxide (GO) and polyaniline composite	Hydrothermal loading process	MB	Visible/Xe lamp with a cutoff filter ($\lambda \geq 400$ nm)	Decolorization rate / k	85% / 1.06×10^{-2} min^{-1}	UV/Visible spectrophotometery	No change in degradation rates after 5 cycles	[73]
TiO_2/polypyrrole	Chemical oxidative polymerization method	MB	Sunlight illumination	Decolorization rate	93%	UV/Visible spectrophotometery	Efficiency only decreased by 20% after 4 cycles	[74]
TiO_2/poly 3-hexylthiophene	Phase-separated film shattering method followed by surface functionalization then TiO_2 loading	MB	Visible/ 100W solar simulator (UV was excluded)	Decolorization rate / k	93% / 0.992×10^{-2} min^{-1}	UV/Visible spectrophotometery	N/A	[75]
TiO_2/Poly-o-phenylenediamine	Oxidative polymerization in the presence of TiO_2	MB	UV-Visible /1000 W xenon lamp	k	0.33×10^{-2} min^{-1}	UV/Visible spectrophotometery	Stable till 5 cycles	[76]
ZnO/conducting polymers								
polyaniline/ZnO	In situ chemical polymerization process	MB	Visible	k	1.944×10^{-2} min^{-1}	UV/Visible spectrophotometery	N/A	[77]
polyaniline/ZnO sheets	In situ polymerization	MB	Visible/Xenon arc lamp (300 W) attached with UV cutoff filter	Decolorization rate	76%	UV/Visible spectrophotometery	N/A	[14]

AGY: Basic Yellow 28, k is the pseudo-first-order rate constant (min^{-1})

4.3   Porous Organic Polymers (POPs)

Porous organic polymers (POPs) are also an emerging class of multi-dimensional porous network materials. It is constructed from various organic building blocks. These blocks are connected through strong covalent linkages. Attention has been paid to POPs because of their distinctive geometries that give rise to promising features. In general, POPs are split into two categories according to the arrangement pattern of their building units. The first one is the amorphous POPs. This category includes hyper-crosslinked polymers, polymers of intrinsic microporosity, porous aromatic frameworks and conjugated microporous polymers. The second is the crystalline POPs which include covalent organic frameworks (COFs) [78]. It is worthy to mention here that conjugated microporous polymers and COF are also considered as special types of conjugated polymers (conjugated porous polymers) as mentioned previously.

The morphologies of POPs are ordinarily demonstrated as accumulations of particles. Polymers having particular structures, such as the nano-sized spheres, sheets and tubes, can be attained by precise controlling of the experimental conditions, monomer type and preparation methodology. The surface area values of POPs extend from hundreds to thousands of m^2/g. These values approach that of other porous substances, such as zeolites, porous carbon and metal–organic frameworks [79]. The different surface features of POPs, such as the surface area, structure and size of pores, and surface functional groups can be controlled by inserting specific functional building blocks [78].

The majority of POPs are unsolvable and display high stability in various solvents, which facilitates their reusability and separation. This is attributed to the high degree of polymerization and robust chemical linkages. Further, POPs have high surface area, well-organized building blocks (particularly for COFs) and relatively symmetrical pores. These features enable organic pollutants to easily diffuse and react inside the cages and networks of POPs in photocatalytic reactions. POPs can act as heterogeneous supports for molecular catalysts and enhance their photocatalytic performance [79].

For example, conjugated microporous polymers are characterized by their outspread π-conjugated skeletons when compared with other amorphous POPs. This distinctive structure provides conjugated microporous polymers with outstanding absorptivity of photons, which is useful for photocatalytic reactions. Additionally, COFs having repeated arranged and expanded skeletons with symmetrical nano-sized pores facilitate the accurate combination of different functional building blocks [78].

Similar to what was mentioned previously, the major fundamentals of POPs in photocatalysis are summarized as follows. When semi-conductive POP is illuminated with visible light with an energy similar to its bandgap, photoexcited charges will be generated near CB and VB, respectively. Simultaneously, oxidation and reduction surface reactions occur between POP and dye by effective charge separation. However, in reality, the major generated electrons and holes may recombine. This charge recombination will decrease the performance of POP as a photocatalyst.

To eliminate this recombination and enhance the separation between electrons and holes, different approaches have been interpreted. This includes the structural design with electron donor–acceptor systems, asymmetrical structures, highly crystalline polymers, incorporation of noble metals and hybridization [21].

Several studies have been introduced to explore the utilization of POPs in the photocatalytic remediation of wastewaters contaminated with organic dyes. Wang et al. introduced a comprehensive review on the utilization of POP-based photocatalysts in wastewater remediation [79]. They focused on the latest trends of POPs in photocatalytic applications including the photodegradation of organic pollutants, particularly organic dyes.

Although the incorporation of metal within the organic/inorganic hybrid porous systems may enhance the catalytic performance, it may also result in the existence of secondary pollutants. Therefore, a set of metal-free porous polymers have been introduced. For example, Ma et al. introduced a set of conjugated microporous polymers having various bandgaps and morphologies. They prepared various shapes of conjugated microporous polymers including nano-sized spheres, rods, and rings by altering the types of the donors and acceptors constituents [65]. B-BT3-b was, for example, prepared by the combination of 1,3,5-triethynylbenzene and 4,7-dibromobenzo[c]-1,2,5-thiadiazole. B-BT3-b displayed the best performance in the photodecomposition of RhB using a light-saving household lamp as a source of illumination. The semiconductive and optical features can be controlled by varying compositions. Imine-based COF having a triazine core showed an enhanced degradation rate of methyl orange when compared with the original imine COF [67]. Hybrid compounds mostly acquire the structural and functional characteristics of original components in addition to the newly gained advantages. Peng et al. combined NH_2-MIL-68 and phenylamine-modified (TPA) COF to prepare a core–shell hybrid product (NH_2-MIL-68@TPA-COF) [66]. The rate of the photocatalytic reaction using the hybrid increased 1.4-fold. The higher activity was attributed to the lower bandgaps and enhanced specific surface area after hybridization.

Nath et al. introduced a tailored synthesis procedure of energy-state tuned polymers that bridged several photodegradation mechanistic pathways. They synthesized and studied copper and palladium incorporated in porous organic polymers (POP) for the mineralization of different azo, thiazine and fluorescein dyes. Semiconductive POPs prepared via coupling between 1,4-diethynylbenzene and 1,3,5-bromobenzene (POP-1) or 1,2-bis(3,5-dibromophenyl) diazene (POP-2) showed photodegradation capabilities against different dyes such as RhB, MB, Congo red (CR), methyl orange (MO) and a mixture of MB and MO [2]. The authors illustrated the capability of the synthesized polymers to directly generate the reactive oxygen species under both natural sunlight and visible light illumination. POP-1 produces both singlet oxygen and superoxide radicals that provide POP-1 with outstanding performance. Under solar illumination, POP-1 also effectively photocatalyzed the decomposition of RhB, MB, CR and MO dyes. Moreover, in a multi-component mixture of cationic and anionic dyes, the supported photocatalysts selectively degraded dyes without the use of any additional reagent. The studies revealed that the degradation mechanism occurred simultaneously by the conventional photosensitization and substrate-sensitized secondary pathway.

On the other side, organic dyes and metal complexes are extensively used as photocatalysts in various applications. Combining organic dyes and metal complexes with functional monomers is a good approach to endow POPs catalytic behavior. Li et al. integrated two Anderson-type polyoxometalates with conjugated microporous polymers to photodegrade RhB and MB [68]. The catalytic performance of polyoxometalates-conjugated microporous polymers hybrid was promoted when compared to the ordinary polyoxometalates. Ma et al. investigated the photocatalytic degradation behavior of a series of conjugated microporous polymers modified with ferrocene [69]. Ferrocene was chosen because it can provide an outstanding sandwich structure with two cyclopentadienyl rings linked to a Fe^{2+} center. This structure served in the formation of cavities with large apertures to encapsulate dye pollutants. Besides, the distinctive configuration and electron distribution of ferrocene improve the porous polymers with specific photocatalytic features. The combination of conjugated microporous polymers, ferrocene and tetrakis(4-ethynylphenyl) ethene-conjugated microporous polymer hybrid displayed outstanding capability in the decomposition of MB dye.

4.4 Coordination Polymers (COPs)

Coordination polymer (COP) is a rising class of crystalline inorganic or organometallic polymeric networks. In COP, metal cations or clusters are connected to organic ligands. The conduction band (CB) in COP is basically composed of unoccupied d orbitals of the metal cation and the LUMOs of the organic ligands [6]. Several research attempts have been conducted to utilize COPs as photocatalysts in the treatment of polluted wastewater in order to extend the absorption of light to the visible region. Another advantage of COPs is the wide variety of organic and organometallic ligands and the tunable preparation. These features enable adjusting the absorption band in order to utilize visible light for the photodegradation of organic contaminants. Therefore, COPs provide a promising domain for emerging effective catalysts for the photodegradation of organic contaminants in wastewater.

Many studies were carried out on the synthesis and application of COP-based photocatalysts, giving rise to several substantial outcomes [80–82]. Wang et al. introduced a comprehensive review on photocatalytic COPs in wastewater remediation [83]. Later, some similar studies on the COP-based catalysts for photodegradation of organic contaminants in wastewater were introduced [70, 71].

Scientific literature addressed many COP-based photocatalysts which are sensitive to the ultraviolet region. The design and fabrication of COPs having a good photoresponse to the visible region with convenient catalytic performance is the current challenge. Broad bandgap and rapid rebound of the generated charged during the catalytic reaction are the essential obstacles. Thus, the improvement of COP-based catalysts seems essential. Lately, some preparation methodologies to improve COP-based materials have been introduced. These methodologies included doping COPs with metal cations [70, 71] as well as loading COPs on carbonaceous nanomaterials [84–86]. Incorporation of metal cations within COP's framework could be a valuable approach to reduce the bandgap and consequently enhance the catalytic performance. Loading COP onto carbonaceous substrates such as graphene, graphene oxide or carbon fiber may also enhance the sensitivity to the visible light and lower the electron–hole recombination. These multi-component systems may enhance the photocatalytic decomposition of dyes under visible illumination.

Lately, Dai et al. highlighted these two strategies and their applications in the catalytic photodegradation of dyes in a brief overview [6]. They presented a survey on the recent studies on COP-based materials doped with metal cations as well as COPs loaded on carbonaceous supports for the degradation of different classes of organic dyes. For example, they illustrated the photoreactivity studies conducted by Wang et al. on Ag^+-doped COP-based composite $[\{Pb(Tab)_2(bpe)\}_2(PF_6)_4.1.64AgNO_3]_n$ for the photodecomposition of several azo dyes [70], where TabH is 4-(trimethylammonio-benzenethiol) and bpe is 1,2-bis(4-pyridyl ethylene). The energy bandgaps of the ligands (Tab and bpe) and Ag-COP were estimated to be 2.51, 2.52, and 2.46 eV, respectively. This implied the semiconducting behavior of the components and also revealed the generation of narrow bandgap energies, thanks to the existence of Ag^+ dopants. The doped Ag-COP displayed high catalytic activity for photodegradation of different azo dyes under visible light illumination. Ag-COP displayed significantly higher activity than the individual components ($AgNO_3$ and $Pb(OAc)_2$) under the same catalytic conditions. The activity of Ag-COP was also compared with that of many previously reported inorganic catalysts against the degradation of different 12 azo dyes. The dyes included methyl orange, acid orange 7, orange I, orange IV, orange G, congo red, acid red 27, sunset yellow, amido black 10B, nigrosin, acid chrome blue K and eriochrome black T. Comparison indicated the significantly high activity of Ag-COP with respect to that of the reported catalysts. Wang et al. proposed the generation of a new energy level between the 4d orbital of silver cation and the CB of the bpe ligand. This resulted in the formation of a new CB of Ag-COP with lower energy than the bpe. Consequently, the bandgap energy decreased resulting in facile migration of electrons from the valence band to the new conduction band and thus rapid production of active radicals.

In another study, Xu et al. revealed the influence of the type of transition metal cation dopant on the photocatalytic activity of $[Zn(cca)(4,4'bipy)]_n$ against the photodecomposition of rhodamine b dye, where (H_2cca) referred to 4-carboxycinnamic acid and ($4,4'bipy$) referred to 4,4'-bipyridine. Among different doped cations (including Fe^{3+}, Cr^{3+}, Ru^{3+}, Co^{2+} and Ni^{2+}), Fe^{3+} displayed the most efficient catalytic activity under visible light illumination [71].

5 Inorganic Photocatalyst/Photoactive Polymer Hybrids

5.1 TiO$_2$/Photoactive Polymer Hybrids

As previously discussed, the conducting polymers (e.g., polyaniline, polypyrrole, polyacetylene and polythiophene) are characterized by their extended π-conjugated electron system over numerous monomer units. Therefore, they possess high charge carriers' mobility which is notable in their large light absorptivity that changed from visible to near-infrared domains [87]. Therefore, they can be utilized as photosensitizers for semiconductors, improving their catalytic performance under both UV and solar irradiation. In this section, some examples from recent studies on the utilization of conducting polymers in enhancing the catalytic activities of inorganic photocatalysts for the remediation of wastewaters from organic dyes will be discussed. Riaz et al. introduced a review on the outstanding role of conducting polymers in enhancing the catalytic performance of titania [17]. CNP possesses a lower bandgap compared to metal oxide. Thus, in CNP/metal oxide hybrid, CNP serves as a photosensitizer to absorb a broad range of visible domain [88]. The photoexcited electrons in LUMO of CNP are transferred to the conduction band of metal oxide (for instance, TiO$_2$). These electrons then react with adsorbed H$_2$O molecules to produce superoxide radicals (O$_2$$^{\bullet-}$). Meanwhile, the photogenerated holes may interact with H$_2$O to form •OH [60].

Polyaniline (PANI) is the most substantial conducting polymer that possesses distinctive physicochemical features. Under light illumination, PANI easily donates electrons and participates in different catalytic and electrical reactions. Later, PANI composites gained great attention as outstanding photocatalysts for the decomposition of organics under light irradiation.

Reddy et al. synthesized TiO$_2$/polyaniline hybrid material by in situ oxidative polymerization of aniline in the presence of titanium dioxide nanoparticles [72]. TiO$_2$/PANI was investigated as a photocatalyst under ultraviolet illumination for photodegradation of RhB, MB and phenol. TiO$_2$/polyaniline showed better performance than the unmodified TiO$_2$ nanoparticles. This indicated the sensitizing effect of polyaniline. Further, the photocatalytic activity was significantly enhanced by raising the TiO$_2$ dosage in the hybrid materials. The best activity was observed for polyaniline/TiO$_2$ hybrid with 20 wt% TiO$_2$, with an efficiency of over 80% after 3 h of illumination. The rate constants were estimated to be: 0.7642 × 10^{-2}, 0.684 × 10^{-2} and 0.38 × 10^{-2} min^{-1} for RhB, MB, and phenol, respectively. Zhao et al. demonstrated a novel non-toxic hybrid composite composed of TiO$_2$, BiVO$_4$, graphene oxide (GO) and polyaniline (BVGTA) prepared through a one-pot hydrothermal reaction [73]. BiVO$_4$ and GO were introduced in order to drive a catalytic reaction using visible illumination. The catalytic activity of BVGTA was investigated in the photodegradation of phenol and MB under visible illumination. The results were compared with the performance of BiVO$_4$-GO-TiO$_2$ (BVGT). The photocatalytic performance of the hybrid composite, including PANI, BVGTA, was higher than that of BVGT, after 3 h of illumination. The degradation percentage of MB was found to be 85% and the rate constant value was estimated to be 1.06×10^{-2} min^{-1}.

Polypyrrole was also introduced as a photosensitizer to TiO$_2$ by Sangareswari and Meenakshi Sundaram [74]. Synthesis of polypyrrole-TiO$_2$ nanocomposite was conducted via chemical oxidative polymerization of pyrrole at different dosages of TiO$_2$. The composite showed good photocatalytic performance under sunlight illumination against MB dye. The percentage of MB removal was up to 93% after 90 min of irradiation. Further, the hybrid composites showed acceptable durability as, after the fourth run, the catalytic decomposition of MB is only reduced to 20%.

Che et al. synthesized poly 3-hexylthiophene (P3HT) using a phase-separated film shattering method [75]. A cationic amphiphile and a phospholipid were introduced to modify P3HT with positive charges. Poly 3-hexylthiophene was then hybridized with negatively charged TiO$_2$ nanoparticles via adsorption through electrostatic attraction. Analyses revealed that poly 3-hexylthiophene was coated with TiO$_2$ nanoparticles in a hybrid core–shell structure with a conjugated polymer core. The catalytic performance of the core–shell was assessed in MB photodegradation using white and visible light sources. The highest performance was observed under white illumination for the hybrid poly 3-hexylthiophene/TiO$_2$ with a molar ratio of 1:0.2.

Yang et al. utilized the conducting polymer, poly-o-phenylenediamine, to synthesize poly-o-phenylenediamine/TiO$_2$ composites by in situ oxidative polymerization in the presence of TiO$_2$ nanoparticles [76]. Different molar ratios of poly-o-phenylenediamine/TiO$_2$ (from 1:6 to 4:1) were investigated. The catalytic performance of the nanocomposites was assessed in the photodecomposition of MB using visible illumination. The optimal molar ratio for the hybrid nanocomposite 4:1 illustrated catalytic rate constant equal to 0.33×10^{-2} min^{-1} and good durability (reused 5 cycles) at 15 h of operation.

5.2   ZnO/Photoactive Polymer Hybrids

Qin et al. introduced a synthesized polyaniline-zinc oxide as a highly active photocatalyst [77]. The rate constant of the reaction is found to be 1.944×10^{-2} min^{-1} using highly concentrated synthetic wastewater of MB. Polyaniline/ZnO was synthesized via in situ chemical polymerization process. Another study conducted by Ameen et al. indicated the achieved catalytic performance against photodegradation of MB dye using polyaniline/ZnO nanocomposites was nearly 76% under visible illumination [14]. The polyaniline/ZnO composite displayed a catalytic performance threefold higher than that obtained by pure PANI. This was due to the enhanced charge separation in polyaniline/ZnO nanocomposites.

Usman et al. introduced polyaniline/NiO composite (polyaniline @NiO) for the photocatalytic removal of RhB dye. On one side, they selected polyaniline due to its cheap method of preparation, great response toward visible light and its efficient separation ability for the generated charge carriers. Polyaniline itself showed the photocatalytic degradation ability toward organic pollutants. However, its major disadvantage was the significant recombination between the generated charges [8]. On the other side, nickel oxide is a wide bandgap semiconductor and mainly absorbs in the UV light region. It possesses a large number of charge carriers with high diffusion rates. This would facilitate the charge transportation process. NiO is predominately utilized in various catalytic processes due to its high thermal and chemical stability. Therefore, Usman et al. investigated the binary composite of photocatalysts (PANI and NiO) under visible light illumination. First, polyaniline-absorbed visible light results in the generation of electrons and holes. Second, NiO facilitated the transportation of the generated charges to the dye species leading to better charge separation. The study revealed that polyaniline@NiO possesses rich catalytic sites to degrade RhB from water.

6 Characterization of Polymer-Based Photocatalysts

X-ray powder diffraction (XRD), Fourier-transform infrared spectroscopy (FTIR), Raman spectroscopy, electron microscopy, nitrogen adsorption, UV/visible spectroscopy and elemental analyses are examples of commonly used characterization tools for polymer-based photocatalysts [60].

Raman and FTIR are utilized to examine the existence of the components of the polymeric composites. For instance, EL-Mekkawi et al. collected FTIR spectra of titanium dioxide immobilized on a polyethylene terephthalate fabric (TiO_2/PET) before and after photocatalytic treatment of textile wastewater under sunlight [15]. The bond between the C atom in the polyethylene terephthalate chain and the O atom of titanium dioxide (C–O) was confirmed by peaks at 1040–1332 cm^{-1}. This indicated the good binding between titanium dioxide and polyethylene terephthalate surfaces. Further, Usman et al. confirmed the existence of polyaniline and NiO in their conductive hybrid photocatalyst by using FTIR [8]. The characteristic wavenumbers and the intensity values of FTIR peaks of the investigated polymer host probably shift due to the interaction with other components such as dopants of metal oxide nanoparticles [60].

The XRD diffraction peaks reflect the regular structures in the investigated polymeric composites (e.g., COF composites). For example, diffraction peaks at $2\theta = 20°$ and $25°$, and broad peak in the region of $2\theta = 20$–$30°$ indicated the regular structures in PANI [89] and PPy [90], respectively. Further, Liu et al. demonstrated the measured X-ray diffractograms of the xylan and xylan/polyvinyl alcohol/TiO_2 composites. A crystalline signal at $19.0°$ appeared in the diffractogram of pure xylan. However, the signal was shifted to $19.5°$ in the composite. This revealed that combining polyvinyl alcohol and titanium dioxide led to a crystallinity change. Additionally, compared with diffractograms of pure xylan, the appearance of specific peaks at $25.4°$, $36.2°$, $47.9°$ and $54.8°$ confirmed the successful synthesis of the composite [44].

The structural and morphological properties of the polymeric catalysts are usually investigated by scanning and transmission electron microscopy (SEM/TEM analysis). For example, Usman et al. visualized the microstructure analysis of polyaniline/NiO, the porosity of PANI fibers and the uniformly nano-sized spherical NiO by using SEM measurements [8]. Li et al. revealed that porphyrin-based porous organic polymer was composed of agglomerated spheres and their average diameter was approximately 200 nm [91].

Energy-dispersive X-ray spectroscopy (EDX) and X-ray photoelectron spectroscopy (XPS) are considered as helpful analytical tools for investigating the surface chemical composition of polymeric composites. Ni et al. investigated the structure and composition of the as-synthesized PS/RGO@CeO$_2$ nanocomposites [32]. The cerium and oxygen elements were detected in the EDX spectrum, suggesting the existence of ceria nanoparticles in the composite. Further, Liu et al. measured the XPS spectra of the xylan and xylan/PVA/TiO$_2$ composites [44]. They found that the XPS spectrum of xylan did not exhibit any signals for Ti 2p, referring to the absence of Ti element in the xylan. On the contrary, two strong signals were observed at 458 and 464 eV in the xylan/PVA/TiO$_2$ composite spectrum. The Ti content in the composite was evaluated to be 0.5%. These findings confirmed the successful grafting of titania on the composite surface.

UV/Vis spectroscopy has been used to study the optical properties of the materials. UV/Visible/Near IR absorption diffuse reflectance spectra (DRS) spectroscopy has been utilized to explore the spectral response of the polymeric photocatalyst toward the UV, visible and IR domains [59]. The doped nano-sized poly(3,4-ethylenedioxythiophene) (PEDOT) structure displayed a broad absorption band in the near IR domain with an intense peak at 390 nm. On the contrary, the doped bulk PEDOT had no remarkable absorption in the visible region [26]. Polyaniline showed intense absorption in the UV and visible domains. After the complex formation between polyaniline and TiO$_2$, the absorption of titanium dioxide was extended to the visible part of the spectrum [76]. Additionally, no absorption peaks of NiO were detected in the visible region by Usman et al. [8]. Only absorption peaks appeared in the ultraviolet region. The bandgaps of polyaniline NiO were estimated using tauc analysis. Their values were found to be 2.6 and 3.1 eV, respectively.

The thermal stability of polymeric photocatalysts has been investigated by thermogravimetric analysis (TGA). Depending on the components, polymeric catalysts displayed different degrees of thermal stabilities [91, 92]. For example, conductive polymers displayed good thermal stability. The remarkable loss of weight of PANI was found at the temperature domains: 100, 190 and 400 °C. The weight losses at 100 and 190 °C were attributed to the loss of residual H$_2$O, contaminants or oligomers. However, the weight loss at 400 °C was attributed to the thermal decomposition of the polymeric backbone [92]. In the same way, poly(3,4-ethylenedioxythiophene) displayed weight loss at 216 °C and 220–430 °C. The weight loss at 220 °C was due to loss of H$_2$O. However, the weight loss at temperatures higher than 220 °C was due to the combustion of oligomers. Finally, the poly(3,4-ethylenedioxythiophene) main chain decomposed at 430 °C [93].

The photoluminescence technique is utilized to investigate the interfacial charge separation and the rate of recombination of generated charges. In some studies, photoluminescence was utilized as an essential tool to assess the catalytic activity of photocatalysts. Yang et al. measured the photoluminescence spectra of pure titania and its hybrid with silver and polypyrrole. The composites illustrated reduced photoluminescence intensities compared to pure titania. This indicated reduced recombination rates of charges under UV illumination [94].

Nitrogen adsorption/desorption isotherms are frequently used to investigate the surface and porosity of polymeric photocatalysts. The permanent porosity of polypyrrole/porous organic polymer (ppy/POP) was proved by N_2 adsorption. In the light of the obtained data, Li et al. proved that ppy/POP was a typical microporous material [91].

In some research studies, an inductive coupled plasma (ICP) or atomic absorption spectroscopy (AAS) was utilized to examine the leaching of the supported metals or metal oxides. Fischer et al. tested the passed water during the photocatalytic decomposition of MB by TiO_2/polyethersulfone to detect the possibly leached titania using ICP measurements [43].

Some studies examined the polymer wettability by measuring the water contact angle (WCA) by using an optical tensiometer. Shoueir et al. measured WCA on the surfaces of chitosan and plasmonic fibers (Au@TiO_2/chitosan) to explore surface hydrophilicity. Chitosan fiber exhibited hydrophilic properties, with a WCA of 42° ± 3°. WCA of the Au@TiO_2/chitosan was enhanced to 63° ± 3°. The result indicated that hydrophilicity was reduced after modification. The difference was attributed to the existence of tiny Au@TiO_2 nanoparticles on the fiber. However, the authors referred that the change was not remarkable, and the Au@TiO_2/chitosan fiber was still sensitive to H_2O diffusion and exhibited a mass transfer mechanism that applies to hydroxyl radicals; thus, pollutants can readily get adsorbed and decomposed to the modified plasmonic fiber [46].

7 Advantages of the Polymer-Based Photocatalysis (Gained Criteria)

Polymers and their nano-sized composites have proven to be promising catalysts for photodecontamination of water from organic dyes. For practical purposes, the photocatalysts should be modulated to acquire some electronic, optical and economic criteria. This section demonstrates the most important gained criteria upon utilizing the polymeric photocatalysts for the removal of organic dyes from wastewater.

7.1 Preservation of the Photoactive Materials

One of the appreciable advantages of utilizing polymers in the fabrication of hybrid photoactive composites is their ability to serve as a host for various photoactive materials. These materials include the conventional UV active photocatalysts, such as TiO_2 and ZnO and more modern photoactive species such as CeO_2 and noble metals. Polymer supports serve in the elimination of aggregation and leaching of these various photoactive species during the decomposition of organic dyes. They facilitate their recovery, and mostly, preserve or even improve their catalytic efficiency compared to the slurry systems.

7.2 Flexibility of Designs

The wide diversity in the nature and behavior of organic dyes has promoted researchers to design compatible catalysts to be used for dye removal. The selection of the catalysts relied on their response to the investigated dyes [8]. As illustrated above, the molecular design has certainly demonstrated a pivotal role in achieving the favored features of the produced polymers. These features included intense and broad visible light absorption, suitable band configuration, effective charge separation and transportation, improved hydrophilicity and surface reaction performance. These synergy enhancements could cause significantly improved photocatalytic activity. For example, copolymerization between the well-designed donor and acceptor components was employed to design CPs for photocatalytic operations. The precise combination could enhance light-harvesting, promote charge mobility and help in the bandgaps and porosity modulation of the polymeric photocatalysts [23].

7.3 Improved Optical and Electronic Properties

The semiconducting properties such as charge separation efficiency, charge transportation kinetics and energy bandgap are essential parameters for light-harvesting capability and catalytic efficiency. As illustrated, polymer-based photocatalysts displayed outstanding activity in the photocatalytic degradation of dyes by visible-light irradiation. Different classes of polymers were successfully utilized as visible active catalysts for the photodecomposition of organic dyes. Additionally, many construction strategies have been introduced to develop electronically and optically improved polymeric photocatalysts with high performance. Several methods were reported to extend the spectral response of the photocatalysts into the visible domain and to enhance charge separation. This included the incorporation of UV active species such as metal oxides and noble metals into the polymer matrices. Further, some promising studies were demonstrated to construct metal-free polymers with outstanding photocatalytic activity under UV/visible illumination by a simple design of monomers and frameworks.

7.4 Economic Features

The polymeric-based photocatalysts are characterized by many favorable economic features such as their high stability, low price, availability and recyclability. These promising features support their utilization for large-scale applications.

8 Photocatalytic Activity Evaluation

The photocatalytic experiments are mostly carried out in laboratories using batch systems. Prior to light irradiation, a small amount of photocatalysts is stirred in the dark to ensure complete dark adsorption of pollutants on the surfaces of photocatalysts. Light sources are selected according to the spectral response of the photocatalyst, polymer and/or composites. The utilization of artificial light setup is most suitable for laboratory experimentations. For instance, Ussia et al. irradiated ZnO/poly(2-hydroxyethylmethacrylate)/graphene oxide cryo-sponge (ZnO/pHEMA-GO) by UV artificial light source, centered at 368 nm for the removal of the adsorbed methylene blue molecules. The supported wide bandgap ZnO semiconductor granted the cryo-sponge adsorbent good photoresponsibility toward the UV light region, allowing its recyclability upon illumination [13]. Usman et al. utilized an artificial visible light source for illumination of polyaniline/NiO. Their studies revealed that the visible light promoted the generation of electron–hole pairs at polyaniline surfaces. Meanwhile, NiO facilitated the charge transportation from the catalyst to the RhB leading to better charge separation and consequently better removal efficiency [8]. Similarly, the photoresponse of the titanium dioxide nanotubes was extended toward visible light when combined with chlorinated polyvinyl chloride. Chlorinated polyvinyl chloride was selected due to its small bandgap [95]. The photocatalytic performance of chlorinated polyvinyl chloride/TiO_2 nanotubes was examined in the presence of visible light illumination toward the degradation of RhB. A simulated solar spectrum light source was utilized for irradiation. UV cut-off glass filter was utilized to eliminate the ultraviolet radiation.

However, the catalytic performance of the photoactive materials under natural sunlight conditions was also explored for real-life field implementations. Nath et al. investigated the absorption spectra of copper and palladium-supported porous organic polymers (POPs). The absorption spectra of the composites interfere with a broad portion of natural sunlight irradiance. This indicates the possible utilization of natural sunlight to initiate the photocatalytic reaction. The supported catalysts illustrated much better activity in natural sunlight irradiation than that obtained when using artificial visible light [2]. Eskizeybek et al. prepared PANI/ZnO nanohybrid through chemical oxidative polymerization of aniline [88]. They studied the photodegradation of MB and malachite green (MG) dyes using solar and artificial UV sources. The study revealed that a small dose of polyaniline/ZnO composite photodegraded both dyes with a removal efficiency of up to 99% under natural sunlight. Hasanat et al. explored the activation of polypyrrole-TiO_2 by both UV and natural sunlight for the decomposition of Reactive Red 45 dye (RR45) Krehula et al. [96]. Polypyrrole/TiO_2 hybrid displayed outstanding performance under both UVA and sunlight illumination. The optimal rate constant under sunlight illumination was estimated to be 7.83×10^{-2} min^{-1}. The rate constant under UVA illumination was calculated to be 6.28×10^{-2} min^{-1}. However, this composite displayed poor stability. The activity of the catalyst was reduced to 80% after two cycles. Low stability was due to the saturation of the photocatalyst surfaces and the agglomeration of the photo-catalyst particles. Further examples of the utilized light sources and the stability of the polymeric composites are given in Table 1.

Scientific literature addressed several methods and techniques for the photocat-alytic performance quantification of the polymeric composites against the removal of dyes. Table 1 illustrates examples of some of the most recent studies on the photo-catalytic decomposition of dyes. The table summarizes the techniques, methods and activity indices that have been commonly utilized to evaluate and follow up the photodecomposition of organic dyes by various polymer-based photocatalysts. As dyes are heavily colored materials, the degradation efficiencies were mostly followed up by monitoring the change in their color spectrophotometrically in the presence of polymer catalysts at different illumination time intervals. Data analyses were mostly interpreted in terms of the percentage of removal, degradation rate constant and the half-life of the reaction.

The photocatalytic performance of the polyaniline/NiO composite was evaluated with respect to the color removal of RhB dye. Change in the absorption spectrum of RhB dye under visible light illumination was followed up spectrophotometrically at 554 nm. The pseudo-first-order kinetic model was chosen to investigate the photo-catalytic activity rate [8]. Nath et al. evaluated their photocatalysts supported on POP spectrophotometrically by monitoring the decrease in the absorbance values at the maximum absorption intensities. They found that the photodegradation reac-tion only followed the pseudo-first-order rate at its initial stage. They attributed this deviation to the low ratios of dyes to dissolved molecular oxygen after a certain illumination period. Therefore, they used the initial reaction rate constant values to interpret their findings. They also used the estimated removal percentage as well as the half-lifetime values to compare the capability of their photocatalysts toward the removal of different types of targeted dyes. Additionally, they used the calculated rate constant values to compare the performance of their photocatalysts with other traditional and novel catalysts in literature. In comparison, it was suggested that their system has exceptional catalytic activities with respect to the degradation of the investigated dyes under visible light illumination [2].

The dye method is the most common technique for the assessment of catalytically photoactive surfaces. As illustrated in Table 1, methylene blue (MB) was extensively reported as a test pollutant model in the assessment of the catalytic performance of semiconductors. As MB is a highly colored organic species, the assessment test using MB can be carried out by following up its rate of decolorization spectrophotomet-rically. Researchers mostly utilized MB to examine the reactivity of their polymeric catalysts (see Table 1). MB has been extensively selected as a model pollutant because of its characteristic color, high absorptivity in UV and visible ranges, well-known structure and molecular size, high stability against light irradiation and its accessibility to photodegrade by the common semiconductors. In 2010, the international standards organization (ISO) published a technical standard (ISO 10678:2010) for the "determination of photocatalytic activity of surfaces in an aqueous medium by degradation of methylene blue". However, the rate of photodecolorization of MB is not equivalent to the rate of photomineralization, which is often a slower reaction. Therefore, following up the mineralization rate is a more accurate method of evaluation (see Table 1) [64]. The mineralization efficiency of photocatalytic systems can be conducted by using several techniques such as COD and total organic carbon analysis (TOC). The estimated reduction percentages of TOC values are used to confirm the efficiency of the investigated photocatalysts toward the real water treatment processes. Nath et al. investigated the mineralization efficiency of their system by using the total organic carbon analysis (TOC). The estimated reduction percentages of TOC values confirmed the high efficiency of the investigated photocatalysts [2].

9 Conclusion, Future Perspectives and Recommendations

In this chapter, the recent advances in polymer-based photocatalysis for the remediation of organic dyes in wastewaters were summarized. Over the last few years, hundreds of polymers have received a great deal of attention and demonstrated a new platform in the field of photocatalysis on account of their inherent merits of abundance, tunable band structure and good stability in photocatalytic operations. All classes of polymers have been discussed in this chapter, including conjugated polymers (CPs), conducting polymers (CNPs), porous polymers (POP) and coordination polymers, as well as their hybrids with inorganic photoactive species.

9.1 Optimal Construction of Photocatalysts

In spite of the pronounced advance made so far, the research in this field is still insufficient. The presented polymers still exhibit low catalytic efficiency, which is away from the practical demands. To achieve better photocatalytic performance, advanced studies should be developed to optimize the photophysical features of polymers to enhance the generation of active radicals. The diverse molecular structures of organic polymers enabled the precise control of their physicochemical characteristics. There are infinite opportunities and challenges to develop efficient polymeric photocatalysts for different photocatalytic applications. The relationship between structure, characteristics and role of polymeric photocatalysts was mostly obtained from empirical correlations [20]. Thus, more fundamental studies are required to enhance the rational design of material properties. Intelligent imitation and computational analyses can anticipate the features of polymeric photocatalysts constructed from various monomers and reduce the heavy workload [79].

9.2 Scalability Studies

The majority of the photocatalytic systems described in this chapter are investigated under laboratory conditions. Most of the polymer-based photocatalysis studies are based on static systems in a batch mode. This experimental setup does not represent real-world wastewater remediation in which photocatalyst should endure a dynamic and continuous flow system, with minimum dissipation of metal species from the composites. Only a few studies are demonstrated for large-scale photocatalytic treatment of wastewater polluted with dyes using polymer-based photocatalysts. For example, our research group published two research studies on the fabrication and implementation of a flow system bench-scale solar prototype designed for industrial wastewater treatment. The results were promising and efficient mineralization of textile dyes was achieved [15, 16]. However, there is still a lack of information on large-scale operations, so intensive investigations are urgently needed. The outstanding economic features of many polymeric-based photocatalysts, such as their high stability, low price, availability and recyclability, reinforce considering them as an essential platform for future large-scale applications.

9.3 Photocatalytic Activity Evaluation

As illustrated in this chapter, researchers mostly utilized MB to examine the reactivity of their polymeric photocatalysts. The assessment of the catalytic performance using one dye pollutant does not always reflect the real performance of the photocatalyst. Some experimental analysis techniques may quantify the conversion rate of a particular pollutant rather than the assessment of the treated wastewater quality. For instance, it was previously confirmed that the photomineralization of MB took place at a longer timescale than the oxidative photodecolorization process. Thus, the rate of photodecolorization of a dye is not equivalent to the rate of photomineralization, which is often a slower reaction. For example, COD has been accepted as an international environmental standard for the assessment of organic pollution. EL-Mekkawi et al. proposed a standardization method based on the COD analyses for the accurate quantification of the treated waters under identical experimental conditions [64]. Despite the great efforts that have been made to enhance the optical response of photocatalysts in the visible domain, there has not been any remarkable advance to date toward examining the possible utilization of polymeric photocatalysts within the present water treatment units, which may employ ultraviolet sources for advance treatment stage. Therefore, here we highly recommend the suggestion made by Kumar et al. to conduct comparative studies on the electrical consumption per order of magnitude (EE/O) of CP-based photocatalysts with the conventional ones [60].

9.4 Stability of the Polymeric Photocatalyst

The stability of the polymer-based photocatalysts against light irradiation should be thoroughly investigated. The polymer itself may undergo severe or partial decomposition during the illumination of light. Based on our experience, COD or TOC values may unexpectedly increase during illumination while the targeted dye pollutants exhibit remarkable decolorization. These findings may be noticed during the degradation of organic dyes in synthetic wastewaters at low initial COD or TOC concentrations of the tested dyes in water. However, this unexpected increase is mostly not recognized during the treatment of industrial real wastewaters using the partially degraded polymers. This partial increase in COD of TOC values probably does not get noticed at high levels of dissolved organic pollutants and as it may be within the permitted experimental error values.

9.5 Sustainable Development

Extending research on cheap and green alternatives is very essential for economic and environmental sustainable development. Seeking alternative green pathways of preparation, low consumption of chemicals and short preparation methodologies should be deeply investigated for the sake of sustainable development. As mentioned in this chapter, some attempts have been made to use alternative eco-friendly biopolymers materials such as chitosan, cellulose and cotton. The preparation and utilization of metal-free photoactive polymers as a green alternative have also been introduced. Expansion of such green research studies should also be considered. Further, exploiting the clean and costless sunlight energy for driving the photodegradation reactions of dyes is a worthy and environmentally responsible alternative to be considered.

References

1. El-Mekkawi D, Abdel-Mottaleb MSA (2005) The interaction and photostability of some xanthenes and selected azo sensitizing dyes with TiO_2 nanoparticles. Int J Photoenergy 7:301527. https://doi.org/10.1155/S1110662X05000140
2. Nath I, Chakraborty J, Heynderickx PM, Verpoort F (2018) Engineered synthesis of hierarchical porous organic polymers for visible light and natural sunlight induced rapid degradation of azo,

thiazine and fluorescein based dyes in a unique mechanistic pathway. Appl Catal B Environ 227:102–113. https://doi.org/10.1016/j.apcatb.2018.01.032

3. Singh S, Sharma R, Khanuja M (2018) A review and recent developments on strategies to improve the photocatalytic elimination of organic dye pollutants by BiOX (X=Cl, Br, I, F) nanostructures. Korean J Chem Eng 35:1955–1968. https://doi.org/10.1007/s11814-018-0112-y

4. Haspulat B, Gulce A, Gulce H (2013) Efficient photocatalytic decolorization of some textile dyes using Fe ions doped polyaniline film on ITO coated glass substrate. J Hazard Mater 260:518–526. https://doi.org/10.1016/j.jhazmat.2013.06.011

5. Abd EL-Wahab RM, EL-Mekkawi DM, El-Dars FM, Farag AB, Selim MM (2010) Utilization of synthetic zeolites for removal of anionic dyes Egypt. J Chem 53:449–464. https://doi.org/10.21608/EJCHEM.2010.1236

6. Dai M, Li H-X, Lang J-P (2015) New approaches to the degradation of organic dyes, and nitro- and chloroaromatics using coordination polymers as photocatalysts. Cryst Eng Comm 17:4741–4753. https://doi.org/10.1039/c5ce00619h

7. Rahangdale PK (2013) Newly developed TiO_2-2-HABT functional material for water treatment. Emerg Mater Res 3:31–36. https://doi.org/10.1680/emr.13.00029

8. Usman M, Adnan M, Ali S, Javed S, Akram MA (2020) Preparation and characterization of PANI@NiO visible light photocatalyst for wastewater treatment. Chem Select 5:12618–12623. https://doi.org/10.1002/slct.202003540

9. El-Mekkawi DM, Labib AA, Mousa HA, Galal HR, Mohamed WAA (2017) Preparation and characterization of nano titanium dioxide photocatalysts via sol gel method over narrow ranges of varying parameters. Orient J Chem 33(1):41–51. https://doi.org/10.13005/ojc/330105

10. El-Mekkawi DM, Galal H (2013) Removal of a synthetic dye "Direct Fast Blue B2RL" via adsorption and photocatalytic degradation using low cost rutile and Degussa P25 titanium dioxide. J Hydro-Environ Res 7:219–226. https://doi.org/10.1016/j.jher.2013.02.003

11. Shebl M, Khalil SME, Kishk MAA, El-Mekkawi DM, Saif M (2019) New less toxic zeolite-encapsulated Cu (II) complex nanomaterial for dual applications in biomedical field and wastewater remediation. Appl Organomet Chem 33(10):e5147. https://doi.org/10.1002/aoc.5147

12. Ibrahim AM, El-Mekkawi DM, Selim MM (2020) Supported TiO_2 on kaolin, cordierite and calcite for photocatalytic removal of dyes from wastewater. Clean-Soil Air Water 48:1900361. https://doi.org/10.1002/clen.201900361

13. Ussia M, Di Mauro A, Mecca T, Consolo F, Nicotra G, Spinella C, Cerruti P, Impellizzeri G, Privitera V, Carroccio S (2018) ZnO-pHEMA nanocomposites: an eco-friendly and reusable material for water remediation. Appl Mater Interfaces 10:40100–40110. https://doi.org/10.1021/acsami.8b13029

14. Ameen S, Akhtar MS, Kim YS, Yang O-B, Shin H–S (2011) An effective nanocomposite of polyaniline and ZnO: preparation, characterizations, and its photocatalytic activity. Colloid Polym Sci 289:415–421. https://doi.org/10.1007/s00396-010-2350-3

15. El-Mekkawi DM, Abdelwahab NA, Mohamed WA, Taha NA, Abdel-Mottaleb MSA (2020) Solar photocatalytic treatment of industrial wastewater utilizing recycled polymeric disposals as TiO2 supports. J Cleaner Prod 249:119430. https://doi.org/10.1016/j.jclepro.2019.119430

16. El-Mekkawi DM, Nady N, Abdelwahab NA, Mohamed WAA, Abdel-Mottaleb MSA (2016a) Flexible bench-scale recirculating flow CPC photoreactor for solar photocatalytic degradation of methylene blue using removable TiO2 immobilized on PET sheets. Int J Photoenergy 9270492 https://doi.org/10.1155/2016/9270492

17. Riaz U, Ashraf SM, Kashyap J (2015) Role of conducting polymers in enhancing TiO_2-based photocatalytic dye degradation: a short review. Polym Plast Technol 54:1850–1870. https://doi.org/10.1080/03602559.2015.1021485

18. Singh S, Mahalingam H, Singh PK (2013) Polymer-supported titanium dioxide photocatalysts for environmental remediation: a review. Appl Catal A Gen 462:178–195. https://doi.org/10.3390/molecules22050790

19. Das TK, Prusty S (2012) Review on conducting polymers and their applications. Polym Plast Technol Eng 51(14):1487–1500. https://doi.org/10.1080/03602559.2012.710697

20. Banerjee T, Podjaski F, Kröger J, Biswal BP, Lotsch V (2021) Polymer photocatalysts for solar-to-chemical energy conversion. Nat Rev Mater 6:168–190. https://doi.org/10.1038/s41578-020-00254-z

21. Byun J, Zhang KAI (2020) Designing conjugated porous polymers for visible light-driven photocatalytic chemical transformations. Mater Horiz 7(1):15–31. https://doi.org/10.1039/C9MH01071H

22. Colmenares JC, Kuna E (2017) Photoactive hybrid catalysts based on natural and synthetic polymers: a comparative overview. Molecules 22:790. https://doi.org/10.3390/molecules22050790

23. Dai C, Liu B (2020) Conjugated polymers for visible-light-driven photocatalysis. Energy Environ Sci 13:24–52. https://doi.org/10.1039/c9ee01935a

24. Ghosh S, Kouame NA, Remita S, Ramos L, Goubard F, Aubert P-H, Dazzi A, Deniset-Besseau A, Remita H (2016a) Visible-light active conducting polymer nanostructures with superior photocatalytic activity. Sci Rep 5:18002. https://doi.org/10.1038/srep18002

25. Ghosh S, Maiyalagan T, Basu RN (2016b) Nanostructured conducting polymers for energy applications: towards a sustainable platform. Nanoscale 8(13):6921–6947. https://doi.org/10.1039/C5NR08803H

26. Ghosh S, Kouamé NA, Ramos L, Remita S, Dazzi A, Deniset-Besseau A, Beaunier P, Goubard F, Aubert P-H, Remita H (2015) Conducting polymer nanostructures for photocatalysis under visible light. Nat Mater 14:505–515. https://doi.org/10.1038/nmat4220

27. Braslavsky SE (2007) Glossary of terms used in photochemistry, 3rd edition (IUPAC Recommendations 2006). Pure Appl Chem 79:293–465. https://doi.org/10.1351/pac200779030293

28. Kandavelu V, Kastien H, Thampi KR (2004) Photocatalytic degradation of isothiazolin-3-ones in water and emulsion paints containing nanocrystalline TiO_2 and ZnO catalysts. Appl Catal B Environ 48:101–111. https://doi.org/10.1016/j.apcatb.2003.09.022

29. Ozgur U, Alivov YI, Liu C, Teke A, Reshchikov M, Dogan S, Avrutin V, Cho S-J, Morkoc HA (2005) Comprehensive review of ZnO materials and devices. J Appl Phys 98:041301. https://doi.org/10.1063/1.1992666

30. Liang S, Xiao K, Mo Y, Huang X (2012) A novel ZnO nanoparticle blended polyvinylidene fluoride membrane for anti-irreversible fouling. J Memb Sci 394:184–192. https://doi.org/10.1016/j.memsci.2011.12.040

31. Sims CM, Maier RA, Johnston-Peck AC, Gorham JM, Hackley VA, Nelson BC (2019) Approaches for the quantitative analysis of oxidation state in cerium oxide nanomaterials. Nanotechnology 30:085703. https://doi.org/10.1088/1361-6528/aae364

32. Ni XJ, Zhang JF, Hong L, Yang C, Li YX (2019) Reduced graphene oxide@ceria nanocomposite-coated polymer microspheres as a highly active photocatalyst. Colloids Surf A 567:161–170. https://doi.org/10.1016/j.colsurfa.2019.01.059

33. Sarina S, Waclawik ER, Zhu H (2013) Photocatalysis on supported gold and silver nanoparticles under ultraviolet and visible light irradiation. Green Chem 15:1814–1833. https://doi.org/10.1039/C3GC40450A

34. Xiao Q, Jaatinen E, Zhu H (2014) Direct photocatalysis for organic synthesis by using plasmonic-metal nanoparticles irradiated with visible light. Chem Asian J 9:3046–3064. https://doi.org/10.1002/asia.201402310

35. Yamada K, Miyajima K, Mafun F (2007) Thermionic emission of electrons from gold nanoparticles by nanosecond pulse-laser excitation of interband. J Phys Chem C 111:11246–11251. https://doi.org/10.1021/jp0730747

36. Oriekhova O, Le Coustumer P, Stoll S (2017) Impact of biopolymer coating on the colloidal stability of manufactured CeO_2 nanoparticles in contrasting water conditions. Colloids Surf A 533:267–274. https://doi.org/10.1016/j.colsurfa.2017.07.069

37. Ghasemi S, Rahimnejad S, Rahman Setayesh S, Rohani S, Gholami MR (2009) Transition metal ions effect on the properties and photocatalytic activity of nanocrystalline TiO_2 prepared in an ionic liquid. J Hazard Mater 172:1573–1578. https://doi.org/10.1016/j.jhazmat.2009.08.029

38. Wang X, Blackford M, Prince K, Caruso RA (2012) Preparation of boron-doped porous titania networks containing gold nanoparticles with enhanced visible-light photocatalytic activity. Appl Mater Interfaces 4:476–482. https://doi.org/10.1021/am201695c
39. Zhang J, Li L, Yan T, Li G (2011) Selective Pt deposition onto the face (110) of TiO_2 assembled microspheres that substantially enhances the photocatalytic properties. J Phys Chem C 115:13820–13828. https://doi.org/10.1021/jp203511z
40. Mahouche-Chergui S, Guerrouache M, Carbonnier B, Chehimi MM (2013) Polymer-immobilized nanoparticles. Colloids Surf A 439:43–68. https://doi.org/10.1016/j.colsurfa.2013.04.013
41. Guerrouache M, Mahouche-Chergui S, Mekhalif T, Dao TTH, Chehimi MM, Carbonnier B (2014) Engineering the surface chemistry of porous polymers by click chemistry and evaluating the interface properties by Raman spectroscopy and electrochromatography. Surf Interface Anal 46:1009–1013. https://doi.org/10.1002/sia.5493
42. Tennakone K, Tilakaratne CTK, Kottegoda IRM (1995) Photocatalytic degradation of organic contaminants in water with TiO_2 supported on polythene films. J Photochem Photobiol A 87:177–179. https://doi.org/10.1016/1010-6030(94)03980-9
43. Fischer K, Schulz P, Atanasov I, Latif AA, Thomas I, Kühnert M, Prager A, Griebel J, Schulze A (2018) Synthesis of high crystalline TiO_2 nanoparticles on a polymer membrane to degrade pollutants from water. Catalysts 8:376. https://doi.org/10.3390/catal8090376
44. Liu Z, Liu R, Yi Y, Han W, Kong F, Wang S (2019) Photocatalytic degradation of dyes over a xylan/PVA/TiO2 composite under visible light irradiation. Carbohydr Polym 223:115081. https://doi.org/10.1016/j.carbpol.2019.115081
45. Yu J, Pang Z, Zheng C, Zhou T, Zhang J, Zhou H, Wei Q (2019) Cotton fabric finished by $PANI/TiO_2$ with multifunctions of conductivity, anti-ultraviolet and photocatalysis activity. Appl Surf Sci 470:84–90. https://doi.org/10.1016/j.apsusc.2018.11.112
46. Shoueir K, Kandil S, El-hosainy H, El-Kemary M (2019) Tailoring the surface reactivity of plasmonic $Au@TiO_2$ photocatalyst bio-based chitosan fiber towards cleaner of harmful water pollutants under visible-light irradiation. J Clean Prod 230:383–393. https://doi.org/10.1016/j.jclepro.2019.05.103
47. Di Mauro A, Cantarella M, Nicotra G, Pellegrino G, Gulino A, Brundo MV, Privitera V, Impellizzeri G (2017) Novel synthesis of ZnO/PMMA nanocomposites for photocatalytic applications. Sci Rep 7:40895. https://doi.org/10.1038/srep40895
48. Arslan O, Topuz F, Eren H, Biyikli N, Uyar T (2017) Pd nanocube decoration onto flexible nanofibrous mats of core-shell polymer-ZnO nanofibers for visible light photocatalysis. New J Chem 41:4145–4156. https://doi.org/10.1039/C7NJ00187H
49. Benhabiles O, Galiano F, Marino T, Mahmoudi H, Lounici H, Figoli A (2019) Preparation and characterization of TiO2-PVDF/PMMA blend membranes using an alternative non-toxic solvent for UF/MF and photocatalytic application. Molecules 24(4):724. https://doi.org/10.3390/molecules24040724
50. Nakatani H, Hamachi R, Fukui K, Motokucho S (2018) Synthesis and activity characteristics of visible light responsive polymer photocatalyst system with a styrene block copolymer containing TiO_2 gel. J Colloid Interface Sci 532:210–217. https://doi.org/10.1016/j.jcis.2018.07.119
51. Ding Q, Miao YE, Liu T (2013) Morphology and photocatalytic property of hierarchical polyimide/ZnO fibers prepared via a direct ion-exchange process. ACS Appl Mater Interfaces 512:5617–5622. https://doi.org/10.1021/am4009488
52. Lefatshe K, Muiva CM, Kebaabetswe LP (2017) Extraction of nanocellulose and in-situ casting of ZnO/cellulose nanocomposite with enhanced photocatalytic and antibacterial activity. Carbohydr Polym 164:301–308. https://doi.org/10.1016/j.carbpol.2017.02.020
53. Rajeswari A, Christy EJS, Pius A (2018) New insight of hybrid membrane to degrade Congo red and Reactive yellow under sunlight. J Photochem Photobiol B 179:7–17. https://doi.org/10.1016/j.jphotobiol.2017.12.024
54. Fischer V, Lieberwirth I, Jakob G, Landfester K, Muñoz-Espí R (2013) Metal oxide/polymer hybrid nanoparticles with versatile functionality prepared by controlled surface crystallization. Adv Funct Mater 23:451–466. https://doi.org/10.1002/adfm.201201839

55. Morselli D, Campagnolo L, Prato M, Papadopoulou EL, Scarpellini A, Athanassiou A, Fragouli D (2018) Ceria/gold nanoparticles in situ synthesized on polymeric membranes with enhanced photocatalytic and radical scavenging activity. ACS Appl Nano Mater 1:5601–5611. https://doi.org/10.1021/acsanm.8b01227

56. Melinte V, Stroea L, Chibac-Scutaru AL (2019) Polymer nanocomposites for photocatalytic applications. Catalysts 9(12):986. https://doi.org/10.3390/catal9120986

57. Hareesh K, Sunitha DV, Dhamgaye VP, Dhole SD, Bhoraskar VN, Phase DM (2019) Synchrotron X-ray radiation assisted synthesis of Ag/polycarbonate and Au/polycarbonate polymer matrix and its pollutant degradation application. Nucl Instrum Meth B 447:100–106. https://doi.org/10.1016/j.nimb.2019.03.051

58. Chibac AL, Buruiana T, Melinte V, Mangalagiu I, Buruiana EC (2015) Tuning the size and the photocatalytic performance of gold nanoparticles in situ generated in photopolymerizable glycomonomers. RSC Adv 5:90922–90931. https://doi.org/10.1039/C5RA14695J

59. Ghasimi S, Prescher S, Wang ZJ, Landfester K, Yuan J, Zhang KAI (2015) Heterophase photo-catalysts from water-soluble conjugated polyelectrolytes: an example of self-initiation under visible light. Angew Chem Int Ed 54:14549–14553. https://doi.org/10.1002/anie.201505325

60. Kumar R, Travas-Sejdic J, Padhye LP (2020) Conducting polymers-based photocatalysis for treatment of organic contaminants in water. Adv Chem Eng 4:100047. https://doi.org/10.1016/j.ceja.2020.100047

61. Lee SL, Chang C-J (2019) Recent developments about conductive polymer based composite photocatalysts. Polym 11(2):206. https://doi.org/10.3390/polym11020206

62. Yuan X, Floresyona D, Aubert P-H, Bui T-T, Remita S, Ghosh S, Brisset F, Goubard F, Remita H (2019) Photocatalytic degradation of organic pollutant with polypyrrole nanostructures under UV and visible light. Appl Catal B 242:284–292. https://doi.org/10.1016/j.apcatb.2018.10.002

63. Floresyona D, Goubard F, Aubert PH, Lampre I, Mathurin J, Dazzi A, Ghosh S, Beaunier P, Brisset F, Remita S, Ramos L, Remita H (2017) Highly active poly(3-hexylthiophene) nanos-tructures for photocatalysis under solar light. Appl Catal B 209:23–32. https://doi.org/10.1016/j.apcatb.2017.02.069

64. El-Mekkawi DM, Galal HR, Abd EL, Wahab RM, Mohamed WAA (2016a) Photocatalytic activity evaluation of TiO2 nanoparticles based on COD analyses for water treatment applica-tions: a standardization attempt. Int J Environ Sci Technol 13:1077–1088. https://doi.org/10.1007/s13762-016-0944-0

65. Ma BC, Ghasimi S, Landfester K, Vilela F, Zhang KAI (2015) Conjugated microporous polymer nanoparticles with enhanced dispersibility and water compatibility for photocatalytic applications. J Mater Chem A 3(31):16064–16071. https://doi.org/10.1039/C5TA03820K

66. Peng Y, Zhao M, Chen B, Zhang Z, Huang Y, Dai F, Lai Z, Cui X, Tan C, Zhang H (2018) Hybridization of MOFs and COFs: a new strategy for construction of MOF@COF core–shell hybrid materials. Adv Mater 30(3):1705454. https://doi.org/10.1002/adma.201705454

67. He S, Yin B, Niu H, Cai Y (2018) Targeted synthesis of visible-light-driven covalent organic framework photocatalyst via molecular design and precise construction. Appl Catal B 239:147–153. https://doi.org/10.1016/j.apcatb.2018.08.005

68. Li Y, Liu M, Chen L (2017) Polyoxometalate built-in conjugated microporous polymers for visible-light heterogeneous photocatalysis. J Mater Chem A 5(26):13757–13762. https://doi.org/10.1039/C7TA03776G

69. Ma L, Liu Y, Liu Y, Jiang S, Li P, Hao Y, Shao P, Yin A, Feng X, Wang B (2019) Ferrocene-linkage-facilitated charge separation in conjugated microporous polymers. Angew Chem Int Ed 58(13):4221–4226. https://doi.org/10.1002/anie.201813598

70. Wang F, Li FL, Xu MM, Yu H, Zhang JG, Xia HT, Lang JP (2015) Facile synthesis of a Ag(I)-doped coordination polymer with enhanced catalytic performance in the photodegradation of azo dyes in water. J Mater Chem A 3:5908–5916. https://doi.org/10.1039/C5TA00302D

71. Xu XX, Cui ZP, Gao X, Liu XX (2014) Photocatalytic activity of transition-metal-ion-doped coordination polymer (CP): photoresponse region extension and quantum yields enhancement via doping of transition metal ions into the framework of CPs. Dalton Trans 43:8805–8813. https://doi.org/10.1039/C4DT00435C

72. Reddy KR, Karthik KV, Prasad SBB, Soni SK, Jeong HM, Raghu AV (2016) Enhanced photocatalytic activity of nanostructured titanium dioxide/polyaniline hybrid photocatalysts. Polyhedron 120:169–174. https://doi.org/10.1016/j.poly.2016.08.029

73. Zhao J, Biswas MRUD, Oh WC (2019) A novel $BiVO_4$-GO-TiO_2-PANI composite for upgraded photocatalytic performance under visible light and its non-toxicity. Environ Sci Pollut Res 26:11888–11904. https://doi.org/10.1007/s11356-019-04441-6

74. Sangareswari M, Meenakshi Sundaram M (2017) Development of efficiency improved polymer-modified TiO_2 for the photocatalytic degradation of an organic dye from wastewater environment. Appl Water Sci 7:1781–1790. https://doi.org/10.1007/s13201-015-0351-6

75. Che J, Bae N, Noh J, Kim T, Yoo PJ, Shin TJ, Park J (2016) Poly(3-hexylthiophene) nanoparticles prepared via a film shattering process and hybridization with TiO_2 for visible-light active photocatalysis. Macromol Res 27:427–434. https://doi.org/10.1007/s13233-019-7071-y

76. Yang CX, Dong WP, Cui GW, Zhao YQ, Shi XF, Xia XY, Tang B, Wang WL (2017) Highly-efficient photocatalytic degradation of methylene blue by PoPD-modified TiO_2 nanocomposites due to photosensitization-synergetic effect of TiO_2 with PoPD. Sci Rep 7:3973. https://doi.org/10.1039/C7RA02423A

77. Qin R, Hao L, Liu Y, Zhang Y (2018) Polyaniline-ZnO hybrid nanocomposites with enhanced photocatalytic and electrochemical performance. ChemistrySelect 3:6286–6293. https://doi.org/10.1002/slct.201800246

78. Zhang T, Xing G, Chen W, Chen L (2020) Porous organic polymers: a promising platform for efficient photocatalysis. Mater Chem Front 4:332–353. https://doi.org/10.1039/c9qm00633h

79. Wang T-X, Liang H-P, Anito DA, Ding X, Han B-H (2020) Emerging applications of porous organic polymers in visible-light photocatalysis. J Mater Chem A 8:7003–7034. https://doi.org/10.1039/d0ta00364f

80. Li SL, Xu Q (2013) Metal–organic frameworks as platforms for clean energy. Energy Environ Sci 6:1656–1683. https://doi.org/10.1039/C3EE40507A

81. Toyaoa T, Saitoa M, Horiuchi Y, Mochizukib K, Iwatab M, Higashimurab H, Matsuoka M (2013) Efficient hydrogen production and photocatalytic reduction of nitrobenzene over a visible-light-responsive metal–organic framework photocatalyst. Catal Sci Technol 3:2092–2097. https://doi.org/10.1039/C3CY00211J

82. Zhang T, Lin WB (2014) Metal–organic frameworks for artificial photosynthesis and photocatalysis. Chem Soc Rev 43:5982–5993. https://doi.org/10.1039/C4CS00103F

83. Wang CC, Li JR, Lv XL, Zhang YQ, Guo GS (2014) Photocatalytic organic pollutants degradation in metal–organic frameworks. Energ Environ Sci 7:2831. https://doi.org/10.1039/C4EE01299B

84. Wu Y, Luo HJ, Wang H (2014) Synthesis of iron(III)-based metal–organic framework/graphene oxide composites with increased photocatalytic performance for dye degradation. RSC Adv 4:40435–40438. https://doi.org/10.1039/C4RA07566H

85. Xu XX, Yang HY, Li ZY, Liu XX, Wang XL (2015) Loading of a coordination polymer nanobelt on a functional carbon fiber: a feasible strategy for visible-light-active and highly efficient coordination-polymer-based photocatalysts. Chem Eur J 21:3821–3830. https://doi.org/10.1002/chem.201405563

86. Zhang CH, Ai LH, Jiang J (2015) Graphene hybridized photoactive iron terephthalate with enhanced photocatalytic activity for the degradation of rhodamine B under visible light. Ind Eng Chem Res 54:153–163. https://doi.org/10.1021/ie504111y

87. Wang J, Ni X (2008) Photoresponsive polypyrrole-TiO_2 nanoparticles film fabricated by a novel surface initiated polymerization. Solid State Commun 146:239–244. https://doi.org/10.1016/j.ssc.2008.02.022

88. Eskizeybek V, Sari F, Gülce H, Gülce A, Avci A (2012) Preparation of the new polyaniline/ZnO nanocomposite and its photocatalytic activity for degradation of methylene blue and malachite green dyes under UV and natural sun lights irradiations. Appl Catal B 119:197–206. https://doi.org/10.1016/j.apcatb.2012.02.034

89. Chen X, Li H, Wu H, Wu Y, Shang Y, Pan J, Xiong X (2016) Fabrication of TiO_2@PANI nanobelts with the enhanced absorption and photocatalytic performance under visible light. Mater Lett 172:52–55. https://doi.org/10.1016/j.matlet.2016.02.134

90. Li X, Jiang G, He G, Zheng W, Tan Y, Xiao W (2014) Preparation of porous PPyTiO$_2$ composites: improved visible light photoactivity and the mechanism. Chem Eng J 236:480–489. https://doi.org/10.1016/j.cej.2013.10.057

91. Li M, Zhao H, Lu Z-Y (2020) Porphyrin-based porous organic polymer, Py-POP, as a multifunctional platform for efficient selective adsorption and photocatalytic degradation of cationic dyes. Micropor Mesopor Mat 292(109774):1–10. https://doi.org/10.1016/j.micromeso.2019.109774

92. Lin Y, Li D, Hu J, Xiao G, Wang J, Li W, Fu X (2012) Highly efficient photocatalytic degradation of organic pollutants by PANI-modified TiO$_2$ composite. J Phys Chem C 116(9):5764–5772. https://doi.org/10.1021/jp211222w

93. Hasnat MA, Uddin MM, Samed A, Alam SS, Hossain S (2007) Adsorption and photocatalytic decolorization of a synthetic dye erythrosine on anatase TiO$_2$ and ZnO surfaces. J Hazard Mater 147:471–477. https://doi.org/10.1016/j.jhazmat.2007.01.040

94. Yang Y, Wen J, Wei J, Xiong R, Shi J, Pan C (2013) Polypyrrole-decorated Ag-TiO$_2$ nanofibers exhibiting enhanced photocatalytic activity under visible-light illumination. ACS Appl Mater Interfaces 5(13):6201–6207. https://doi.org/10.1021/am401167y

95. Bui D-P, Pham HH, Cao TM, Pham VV (2020) Preparation of conjugated polyvinyl chloride/TiO$_2$ nanotubes for Rhodamine B photocatalytic degradation under visible light. J Chem Technol Biotechnol 95:2707–2714. https://doi.org/10.1002/jctb.6466

96. Krehula LK, Stjepanovic J, Perlog M, Krehula S, Gilja V, Travas-Sejdic J, Hrnjak-Murgic Z (2019) Conducting polymer polypyrrole and titanium dioxide nanocomposites for photocatalysis of RR45 dye under visible light. Polym Bull 76:1697–1715. https://doi.org/10.1007/s00289-018-2463-2

97. Khanam Z, Sadon NA, Adam F (2014) Synthesis and characterization of a novel paramagnetic polyaniline composite with uniformly distributed metallic nanoparticles sandwiched between polymer matrices. Synth Met 192:1–9. https://doi.org/10.1016/j.synthmet.2014.03.001

Application of Hybrid Polymeric Materials as Photocatalyst in Textile Wastewater

Hartini Ahmad Rafaie, Norshahidatul Akmar Mohd Shohaimi, Nurul Infaza Talalah Ramli, Zati Ismah Ishak, Mohamad Saufi Rosmi, Mohamad Azuwa Mohamed, and Zul Adlan Mohd Hir

Abstract Understanding the physicochemical properties of the hybrid polymeric materials is essential to explore its practicality as photocatalyst in the remediation of textile wastewater. The integration of various types of inorganic materials including metal, non-metal, metal oxide, and metal sulfide into the polymer matrix via versatile fabrication and modification processes could endow several enhancements in terms of morphological, optical, electronic, chemical, and physical characteristics. This chapter will discuss the fabrication and modification of hybrid polymer photocatalyst and overview the application of the hybrid polymer incorporated with the inorganic nanomaterials to give a better insight and information based on their enhanced physicochemical properties and photocatalytic performances.

Keywords Hybrid polymer · Inorganic nanomaterials · Photocatalyst · Textile wastewater · Water treatment

1 Introduction

The United Nations Environment Program (UNEP) believed that textile dyeing is the second largest contributor to water pollution globally. The hazardous waste discharge into the water body from the textile industry is very alarming and had contributed to severe water pollution, which ultimately devastating the nature and negative impact to humans and aquatic life. A recent study reported that textile dye was found in

H. A. Rafaie · N. A. M. Shohaimi · N. I. T. Ramli · Z. I. Ishak · Z. A. M. Hir (✉)
Faculty of Applied Sciences, Universiti Teknologi MARA Pahang, 26400 Bandar Tun Abdul Razak Jengka, Pahang, Malaysia
e-mail: zuladlan@uitm.edu.my

M. S. Rosmi
Department of Chemistry, Faculty of Science and Mathematics, Universiti Pendidikan Sultan Idris, 35900 Tanjung Malim, Perak, Malaysia

M. A. Mohamed
Centre for Advanced Materials and Renewable Resources (CAMARR), Faculty of Science and Technology, Universiti Kebangsaan Malaysia, 43600 UKM Bangi, Selangor, Malaysia

© The Author(s), under exclusive license to Springer Nature Singapore Pte Ltd. 2022   101
A. Khadir and S. S. Muthu (eds.), *Polymer Technology in Dye-containing Wastewater*,
Sustainable Textiles: Production, Processing, Manufacturing & Chemistry,
https://doi.org/10.1007/978-981-19-1516-1_5

vegetables and fruits [118]. Definitely, this finding is very alarming and very unsafe to the human as well as animal health. Moreover, there were numerous toxic chemicals released as liquid waste from the textile industry that include dyes, chlorine, heavy metals, formaldehyde, and others.

There are various approaches that have been implemented in the effort to mitigate the issue related to textile wastewater, such as adsorption [57], membrane separation [161], photocatalysis [87], photoelectrocatalysis [147], and many more. Among these mitigation methods, photocatalysis and photoelectrocatalysis are considered the most reliable, clean, and green approaches since they only required photon energy to trigger photocatalytic/photoelectrocatalytic degradation reaction. Besides, photocatalysis/photoelectrocatalysis can also mineralize the organic pollutants in wastewater into carbon dioxide and water molecules as end results. Despite all of the advantages, the removal of textile wastewater using photocatalysis in a particulate suspension system encounters difficulty in recollecting and recovering the photo-catalyst in the water after the photocatalytic treatment. Somehow, it requires an additional step to recollect the photocatalyst suspension to reuse and recycle the photocatalytic materials to treat the organic pollutants from textile wastewater. Thus, the integration of photocatalytically active inorganic nanomaterials within the poly-meric matrix becomes a promising strategy to overcome the drawback mentioned above [41, 89, 90].

Combining photocatalyst and polymeric material as a hybrid photocatalytic polymer significantly improved overall performance in treating textile wastewater. The synergistic interaction between photocatalyst and polymer undeniably enhanced the removal efficiency of hazardous substances via two-step processes, which usually involved: (1) The adsorption of pollutants into/onto polymeric material via surface interaction due to the presence of polymeric porous structure and (2) photodegrada-tion reaction by the photoactive photocatalyst which presence within the surface as well as within the pore of the polymer. These processes are relatively spontaneous, and sometimes the two-step processes occurred concurrently.

In this chapter, the fabrication and modification of photocatalytic hybrid polymer photocatalyst will be introduced, and several parameters that affect the fabrication and modification processes will be discussed in detail. Besides, the examples of various kinds of photocatalytic material (including metal, non-metal, metal oxide, and metal sulfide) incorporated within numerous polymeric materials and their application in treating textile wastewater will be highlighted and deliberated.

2 Fabrication and Modification of Hybrid Polymer Photocatalyst

To date, photoactive materials have grown more popular in a variety of applications including photocatalytic degradation of pollutants [10] water splitting [1], organic synthesis [153] photoreduction of carbon dioxide [59]. Photoactive materials based on heterogeneous semiconductor photocatalysis have been proposed as a low-cost method of removing harmful pollutants and providing long-term sustainable approach. The uses of semiconductor oxide as photocatalyst such as titanium dioxide, zinc oxide, iron oxide, tungsten oxide, sulfides, and ferrites have widely reported and applied in wastewater treatment [48, 78, 103].

The determination of the best photocatalyst for organic molecule degradation using sunlight centers on two factors; oxidation potential and band gap energy [38]. In general, when semiconductor being exposed to light irradiation with a particular wavelength consists of energy higher than its band gap, photocatalysis begins when electrons are transported from valence band to the conduction band, resulting in the generation of positive holes. Migration of these charged species on the photocatalyst surface leads to redox reaction. It is best to take an example on traditional photocatalyst such as titanium dioxide, which has proven to be the most promising as it serves as strong oxidizing ability and long-term photostability. However, this oxide is photocatalytically active under ultraviolet light thus limits the effectiveness under solar spectrum. Another drawback is the high recombination rate of the photoexcited electron–hole pairs.

Currently, researchers are moving forward to look into photocatalyst materials that work within wavelengths of 380–700 nm (visible region). The amounts of energy available from solar radiation can be used more efficiently at this wavelength. Visible light-driven photocatalyst can be engineered to enhance the photoactivity by modifying the most important feature such as band gap energy and specific surface area to encounter the photoexcited electrons and holes recombination issues [12]. Thus, modification and fabrication of semiconductors are ideal strategies for enhancing the photocatalytic performance and addressing all the issues that arose.

A number of pioneering works have published on modification, construction, and design of composite photocatalysts via several routes, for example, metal or non-metal doping [22], compositing with other semiconductors oxide [33], compositing with conductive materials such as graphene [71] or carbon nanotubes, and sensitization conductive polymers or dyes [99]. Assembling nanoparticle semiconductor photocatalysts onto polymeric material and polymer blends as matrix materials have widely been studied and reviewed [46]. The polymeric materials can be employed as an excellent substrate for photocatalyst immobilization thanks to its chemical inertness, low costs, mechanical strength, low density, and durability despite other common support materials include glass, activated carbon, silica etc. Several versatile immobilization techniques will be highlighted in this section with respect to modification and fabrication of polymer photocatalyst. Explanation on factor's influencing photocatalytic activity and its efficiency as well as reusability of the catalyst will be discussed.

2.1 Immobilization of Nanoparticles onto Polymeric Materials

The transformation, deactivation, and, eventually, mineralization of environmentally persistent chemicals are the main goals of photocatalysis in water treatment. To destroy organic molecules, the photocatalyst may use sunlight and air to create a variety of active species, such as strong and non-selective oxidant hydroxyl radicals. The most outstanding photocatalyst that has been frequently reported in literature is titanium dioxide, whether in suspended form or immobilized into or onto the polymeric materials. The main reason for immobilizing titanium dioxide on/into polymeric materials structures is due to instability in suspension mode and the powder catalyst also prone to scatter the incident light thus contributed to the ineffective light absorption. Furthermore, concerns with post-degradation treatment, such as expensive filtration procedures and precipitation have contributed to several immobilization techniques to immobilize the powder photocatalyst, thus post-degradation expenses and time may be reduced or avoided. The photoactivity of the prepared materials is determined based on the amount of available active catalyst particles immobilized on the surface of the substrate. Since the past decades, photocatalytic membrane technology has attracted enormous interest in the industry and academia as it offers a cost-effective alternative as water treatment, which consists of a versatile method for its usage and shows a robust capability of pollutant removal. Photocatalytic membrane also serves as a highly desirable multi-modal functionalities design that can potentially overcome the disadvantages associated with conventional membranes. Conventional membranes possess a few drawbacks such as membrane fouling, limited operational boundary, and ineffectiveness to degrade pollutants, thus limiting its practical application.

Polyvinylidene fluoride, polysulfone, polyether sulfone, polyacrylonitrile, polyvinyl alcohol, polystyrene, polyaniline, polyvinylpyrrolidone, and polyethylene are some of the materials that have been effectively served as good candidates as polymeric support to enhance the photocatalytic activities [60, 69, 95, 141]. Furthermore, the polymeric membrane is also available in a variety of designs to enhance membrane filtering performance. These include flat sheet, spiral, and hollow membrane forms, each of which has an impact on the structure and features of effective membranes. Porous polymer monoliths are a novel category of materials that have been extensively utilized in separation science and catalysis because of its distinctive structural characteristics. They are primarily made up of cross-linked organic materials, which consist of available volume, which allow for the easy diffusion of bulk compounds in the aqueous phase, and subsequently the degradation rate is enhanced. The concept of heterogeneous photocatalysis assisted by porous polymer monoliths is still considered as a new area of research, with just a few papers published [111, 166]. Zhang et al. investigated the incorporation of titanium dioxide into a divinylbenzene-based polymer monolith [166]. The results indicate some improvements due to specific features such as increased adsorption capacity, fast recovery, and high reusability. Recently, Sompalli et al. reported on the preparation of Cr_2O_3-Ag_2O nanocomposite, which uniformly distributed on a polymer monolith by means of solvothermal-assisted polymerization approach [129]. The optimum ratio of Cr_2O_3-Ag_2O (60:40) nanocomposite supported by a polymer monolith exhibits outstanding photoactivity by mineralizing the Reactive Brown-10 dye pollutant using 300 W/cm^2 visible irradiation with 98.2% degradation after 1 h of reaction.

2.2 Grafting Method

Grafting is a popular approach to provide a polymeric material, a range of functional groups. As polymeric material undergoes surface modification, binding sites for the photocatalyst will be generated (adhesion). Essawy et al. prepared a low-density polyethylene-grafted-poly(4-vinylpyridine-co-acrylamide) (LDPE-g-P(4-VP/AAm)) copolymers supported titanium dioxide using γ-radiation graft copolymerization, which requires two concurrent processes [30]. Firstly, irradiation on the polymer substrate's surface will create active sites on or near the surface followed by monomer polymerization on these sites. The radiation grafting method is useful to alter the physicochemical properties of the polymeric materials and titanium dioxide supported onto will lead to easy separation and safely disposing of textile wastewaters into the freshwater bodies. Titanium dioxide-supported LDPE-g-(4-VP/AAm) copolymer membranes could absorb Remazol red RB-133 and Reactive Blue 2 textile azo dyes after 67 and 90 min of immersion into the dye's solution. Different dyes' absorption rate can be explained by various composites of LDPE-g-(4-VP/Aam) copolymers of the membrane. Interestingly, the polymer photocatalysts were also capable to adsorb high concentration of the dyes in comparison to pristine copolymers. The efficient photocatalytic ability that can be observed from photobleaching rate of both dyes showed that impregnated Remazol Red RB-133 has given remarkably higher photobleaching rate than Reactive Blue 2. Despite the nature absorption ability of the two molecules, moderate titanium dioxide content can be best explained on the higher photobleaching rate of impregnated Remazol red RB-133 with respect to the basis of adsorption competitions. High concentrations of titanium dioxide particles will be led to particle's aggregation, thus significantly decrease the amount of available active sites on the membrane's surface.

Besides, surface modification of zinc oxide nanoparticles with polystyrene via grafting polymerization can alleviate the aggregation owing to steric repulsive interactions among particles and increase the compatibility between zinc oxide nanonoparticles with the organic matrix [42]. The changes of hydrophilicity of nanoparticles surface to extreme hydrophobicity led to low methyl orange degradation. Furthermore, the lengthy polymer chains coated on the surface of zinc oxide have obstructed dye molecules absorption. Moreover, the grafted polystyrene layers have significantly interfered the contact of polystyrene-grafted zinc oxide with air. This is because even when electrons or holes are formed at the internal surface of zinc oxide, they seldom approach the modified particles' outer surface, resulting in photocatalytic degradation failure of methyl orange.

Synthesizing of semi-interpenetrating chitosan/polyaniline composites that using chitosan as natural biopolymer provides a broad range of applications in wastewater treatment as chitosan is non-toxic and biodegradable biopolymer. Integrating polyaniline into a chitosan enhancing the adhesion, mechanical strength, absorption ability, and biocompatibility. Moreover, by incorporating cobalt oxide nanocubes into the chitosan-grafted polyaniline nanocomposite via in situ oxidative polymerization showed that 88% methylene blue dye can be degraded upon 180 min of irradiation under ultraviolet light [123]. Cross-linked polymeric network with high porous structures was clearly observed on the surface of chitosan/polyaniline composite, whilst cobalt oxide nanocubes were embedded within the polymer matrix. As a result, methylene blue degradation will improve by generating electrons and holes in polyaniline. The amine and hydroxyl groups in the chitosan chain promote dye adsorption on the catalyst's surface, allowing for strong interactions with dye molecules through the formation of hydrogen bond.

Le et al. synthesized a novel-biobased polymeric microsphere by biomass-derived phenylpropenes trans-anethole and N-phenylmaleimide. They were hydrolyzed to transform anhydride functional groups on the surface to carboxyl functional groups, which allowed rapid grafting amino-modified titanium dioxide nanoparticles [66]. The photocatalyst microspheres showed good performance as photocatalyst for degradation of 50 mg/L Rhodamine B and 100 mg/L tetracycline with almost 95% and 97% degradation rates, respectively. Overall, immobilization of semiconductor nanoparticles onto conducting polymers and other materials through grafting method have offered numerous benefits, for example, energy-efficient, cost-effective, low toxicity, and environmentally friendly, feature its promising pragmatic applications.

2.3 Electrospinning Method

Electrospinning is a technique to fabricate ultrathin fibers having diameters with micrometer and nanometer dimensions. Formed ultrathin fibers have relatively high surface area, sufficient enough to tolerate with a great amount of photocatalysts with minor modifications needed on the fabrication process [43]. Electrospinning has been widely used to fabricate an electrospun mixed matrix membrane containing a mixture of photocatalyst and polymer. This technique requires a jet of polymer solution, which has been electrically charged prior to produce ultrathin fibers on a grounded collector. A high voltage ranging from 10 to 40 kV can be applied to surpass the surface tension on the fluid droplet at the tip of the syringe. Then, the charged jet fluid stretches from the tip of the syringe and deposits onto the ground collector. The incorporation of photocatalyst nanoparticles into the polymeric materials produces polymer nanocomposites with better mechanical strength, resistance to wear, and thermally stable [82]. Moreover, varying the polymer composition, nanoparticle type, photocatalyst loading, and applied voltage could also influence the morphological, structural, and physicochemical properties of photocatalyst/polymer nanocomposites [43, 44].

A titanium dioxide/polyvinylpyrrolidone nanofiber photocatalytic membrane can be prepared by mixing titanium dioxide into the electrospinning dope solution [165]. The formed nanofiber membrane exhibits high specific surface area and showed 72% efficiency of Rhodamine B degradation, higher than pristine titanium dioxide film with only 44% efficiency. Same result was observed by Linh and co-workers who synthesized a titanium dioxide/polyvinyl alcohol nanofiber membrane [70]. Higher surface area was observed for 50 wt.% titanium dioxide content nanofibers, which have smaller fibers size. In this case, it is best to highlight that good photocatalytic performance upon methyl orange degradation was due to the higher concentration of titanium dioxide in nanofiber. Consequently, increasing the surfaces' active sites of the polymeric membranes can be attained.

In 2015, Panthi et al. fabricated polyacrylonitrile electrospun nanofibers incorporated with silver carbonate nanoparticles for visible light response. This method serves a few advantages, for example, straightforward synthesis, great environmental stability, and good processibility meanwhile silver carbonate possess minimal tendency of recombination effect. A series of electric voltage were applied to prepare the nanofiber solution. From the reported result, applied electric voltage of 18 kV demonstrated greater performance towards the methyl orange, methylene blue, and Rhodamine B dyes degradation (within 30, 35, and 50 min, respectively) under visible light irradiation [105]. The applied electric voltage improved the fiber morphology. No presence of nanoparticles was observed on the nanofiber's surface, indicated that the silver carbonate nanoparticles were homogeneously distributed and confined inside the polymeric nanofiber. Because of its hydrophobic characteristic and low density, this nanofiber mat would benefit from easy separation from the solution as well as solving photocatalyst corrosion and secondary pollution issues. As a result, these will encourage its industrial applicability, particularly in the open water surface. Yar et al. have also synthesized titanium dioxide/zinc oxide/polyacrylonitrile electrospun nanofibers for malachite green degradation under ultraviolet irradiation [156]. The highest photoactivity of the nanocomposites was ascribed by the synergetic impact between the hexagonal wurtzite zinc oxide and rutile titanium dioxide phases within the polymer matrix. The composite's stepwise energy level structure with coupled mechanisms such as reduced electron–hole pair recombination [107] makes these hybrid nanofibers membranes a potential photocatalyst for removing organic pollutants in water.

In another study, Lee et al. synthesized bipolymer structure consisting of polyvinylpyrrolidone and polyvinylidene fluoride incorporated with commercialized titanium dioxide [62]. Due to the hydrophobic properties of polyvinylidene fluoride, nonpolar organic pollutants can easily fit on the surface where titanium dioxide photocatalyst is anchored. Interestingly, the pore volume of the fiber mat rose significantly when the polyvinylpyrrolidone concentration was raised from 0.37 to 0.68 cc/g. Porosity promoted by sacrificial removal of polyvinylpyrrolidone showed higher removal rate of methylene blue dye, together with thicker mat providing higher surface area and access of titanium dioxide. In their study, a simultaneous adsorption ("bait-hook") and degradation ("destroy") mechanism was proposed, which can be useful for treating moderately clear water that does not restrict ultraviolet penetration, as well as turbid wastewaters that restrict light penetration. Overall, nanoparticle semiconductor immobilization prepared from electrospinning method provides various potential advantages over its conventional slurry system particularly energy-efficient, which is the crucial aspect influencing the viability of photocatalytic treatment [27]. Furthermore, post-processes for removing the catalyst can be eliminated.

2.4 Entrapping Method

The semiconductor-based photocatalyst is entrapped inside the porous polymer matrix/membrane, which improves substrate adsorption and hence favors interaction with the photoexcited active species. Many studies have been reported on modifying and fabricating polymeric membranes entrapped semiconductors to remove organic wastewater pollutants such as 4-nitrophenol [9, 91], congo red [92], humic acid [158], etc. The organic nature of the membranes enables for organic molecule adsorption, which improves the photodegradation rate, which is generally based on mass transfer processes. The membrane's thickness, morphological, nanoparticle/polymer, and solvent/polymer ratios are crucial for the membrane performance and significantly affect photocatalytic efficiency.

Molinari et al. prepared polysulfone or cellulose triacetate membranes by entrapping polycrystalline titanium dioxide via phase inversion for congo red degradation [92]. Hybrid system that consists of continuous reactor combined with a modified membrane approach was used. However, there was a very low degradation efficiency of the modified polymeric membrane matrix which attributed by the diffusion resistance for the dye to reach the catalyst surface due to the existence of bilayers polymer membrane surrounding the catalyst's particle. Difficult penetration of ultraviolet light on the membrane surface has led to low degradation efficiency. Furthermore, ultraviolet irradiated together with the presence of titanium dioxide has induced photocatalytic degradation of cellulosic membrane. The weakness of anatase-entrapped membrane was mainly due to anatase-catalyzed degradation of the membrane matrix under light [134]. Using styrene, a precursor to polystyrene, Sultanova et al. prepared an organic viologen-polymer intercalated with inorganic materials such as noble metal nanoparticle (Pt, Pd or Au) and titanium dioxide. The modified composites showed a good photoactivity towards methylene blue degradation in aqueous phase with ultraviolet (364 nm) and direct sunlight irradiations operated at 24 °C and –2 °C (autumn sun exposure), respectively [131].

Polyvinylidene fluoride, a highly fluorinated polymer, offers a good resistant towards photocatalytic degradation [136]. Based on this fact, Damodar et al. modified polyvinylidene fluoride membrane by entrapped titanium dioxide via phase inversion method and was tested for photoactivity against Reactive Black 5 dye [23].

Almost 100% of dye removal was attained after 60 min for 1–4% polyvinylidene fluoride/titanium dioxide membrane as compared to 0% titanium dioxide membrane. The higher removal rate of the dye on titanium dioxide entrapped membranes can be related to the ultraviolet/titanium dioxide photocatalytic activity. Similarly, titanium dioxide embedded polyvinylidene fluoride membrane was also reported by Tahiri et al. for the removal of Brilliant Green and Indigo Carmin dyes [136]. The anatase phase of titanium dioxide (granule) was randomly dispersed in the membrane matrix, thus porosity and pore size have both increased significantly. In this work, the polyvinylidene fluoride/titanium dioxide membrane was initially conditioned in ethanol to improve membrane wettability and adsorption capacity. In aqueous solutions, the adsorption capacity of ethanol conditioned-polyvinylidene fluoride/titanium dioxide membranes was better for Brilliant Green (0.26 mg/g) than for Indigo Carmin (0.018 mg/g) dyes.

Recently, membrane separation technology coupled with advanced oxidation process has developed the so-called super-engineered polymers, and they have gained great demand in the environmental remediation area. The super-engineered polymers offer dual functional separation membrane followed by organic dyes degradation offered some benefits including high chemical stability, excellent separation efficacy, easy-care, robust, etc. The fabrication of super-engineering polymeric membranes with combination of titanium dioxide, for example not only alleviates the problematic low adsorption and recovery, however they are also able to improve the operating process for environmental protection. Liu et al. fabricated titanium dioxide hybrid porous membrane through phase inversion using amphiphilic polymer material, namely, sulfonated polyarylene ether nitrile assisted by polyethylene glycol as pore-forming mediator [73]. The wettability of the hybrid membrane was enhanced as the titanium dioxide content was increased, indicating that the nanoparticles were effectively deposited in the porous membranes [163]. Under optimized conditions, almost a complete methylene blue adsorption was attained by the hybrid porous membrane, which can be recycled up to three runs. The entrapment of titanium dioxide within the polymeric matrix membrane was beneficial for water remediation, thanks to its ability for photocatalytic degradation, easy recycling, and straightforward procedure.

2.5   Dip-coating Method

The dip-coating technique involves submerging the support in a gel solution and then pulling out the plate out at a fixed pace using a device with an adjustable motor to control the rate. This technique can be used to coat nanoparticles in a thin or thick film on the support [130]. Generally, polymeric substrates are often hydrophobic, which lead to difficulty in obtaining a well-adhered and crack-free coating [127]. For instance, titanium dioxide nanoparticles are hardly anchored on the hydrophobic surface of polypropylene. Therefore, the use of plasma technology in combination to dip-coating process to pretreat the polymeric surfaces has considerable potential, and it has already been employed to embed titanium dioxide on several support materials.

Plasma pre-treatments are well recognized for increasing the materials' surface tension, which improves the adhesion of several coatings by wet techniques. Daniel prepared a photocatalyst by immobilizing titanium dioxide on polypropylene film that had been priorly treated with corona discharge under standard atmospheric pressure [24]. The results showed that 47% of methylene blue dye degradation was able to obtain using the hybrid film after 180 min of ultraviolet irradiation. As compared to unfunctionalized polypropylene film, no photoactivity was reported under the pre-determined conditions. Surface morphology study showed that there is the presence of both single particles having diameter of 25 nm and particle accumulations with an average diameter ranging from 50 to 100 nm were observed on the hybrid film surface. The introduction of polar functional groups such as hydroxyl, carbonyl, and carboxyl groups on the corona discharge-induced polypropylene surface has significantly improved the chemical composition of the polypropylene surface due to accessible of hydrophilic suspension of titanium dioxide [63]. The photocatalyst was stable enough as it can be recycled up to fifth cycle with a slight methylene blue degradation difference observed as compared to the first cycle.

Recently, Tuna and Simsek developed a photo-active polyester filter immobilized with p-type semiconductor, namely, $LaFeO_3$ perovskites for the use in photocatalytic reactor for antibiotic and dyes degradation (indigo carmine, orange II and tartrazine) under visible light irradiation by dip-coating method followed by hydrothermal treatment [140]. From the surface morphological study, it showed that the perovskite deposition on the polyester filter remained a fiber-like structure after perovskite catalyst was immobilized into it. This would allow the organic pollutant to engage effectively with the surface during the photocatalytic reaction. The perovskite composite can degrade 99% Indigo Carmine through adsorption followed by photocatalytic reaction. Likewise, 41% Orange II dye degradation was reported through only adsorption. The variation in the removal efficacies can be associated with dye adsorption abilities. The hydrogen bond formed between the dyes and the photocatalyst surface helps in the dye's adsorption ability, which is a key step for physisorption. Moreover, the carbonyl bond enhanced the extent of the hydrogen bond, resulting in the robust interaction between the pollutants and the composite photocatalyst. Among the tested dyes, only Indigo Carmine contains carbonyl group in its structure, which explained why it exhibited almost complete decomposition in the presence of perovskite/polymer-based material. The photocatalyst was stable; however, no recyclability study has been reported on the dye's molecules.

2.6 Photosensitizer-Induced Polymerization

The construction of polymeric photosensitizers coupling with hybrid photocatalysts could beneficial in the photocatalytic water treatment process [99]. This type of photocatalyst provides tremendous advantageous especially for the removal of low concentration hazardous chemical compounds in water resources, which are hardly to be removed by currently available technologies. Furthermore, polymeric photosensitizer can easily be eliminated from the aqueous solution in which secondary pollution can be avoided. In this view, the construction of environmentally safe materials and/or biodegradable photosensitizers could become the promising solution to mitigate the polluted water. Generally, photosensitizers can be linked to the monomer through covalent bonding, which later further fused with the polymer through various polymerization methods that include ionic, graft, radical, or can also be constructed using modified synthetic/natural macromolecules that have already been formed.

Varying the polymeric backbone, a wide range of polymers may be created and modified according to the potentiality of the intended photocatalysis process by considering the types of chromophores compounds linked and their ratio [28]. To ensure the polymer's solubility in aqueous environments, the hydrophilic–hydrophobic components ratio should be carefully studied [34]. For instance, the introduction of strong polar N $\rightarrow$ O bond on the hydrophobic organic framework of the hyperbranched polyimide N-oxide photocatalyst which responsive in the visible region has significantly change the hydrophobicity and polarity characteristic, as well as modify the electronic structure of the fabricated polymer photocatalyst. The N-site oxidation of the polyimide has caused strong interaction with polar compounds, degrading almost 90% of methyl orange dye while the pristine polyimide degrade less than 30% of the dye [155]. Moreover, the intense color of the hybrid photocatalyst as compared to that the pristine polyimide was due to introduction of oxygen atoms, which act as auxochromes. An auxochrome compound can be bonded to a chromophore that changes the chromophore's ability to absorb light altogether with enlarged electron density.

Amphiphilic co-polymeric photosensitizers appeared to be very effective at degrading water pollutants. Photosensitizers such as chromophores, e.g., porphyrin or naphthalene may efficiently degrade the particularly persistent organic contaminants including chlorinated aromatic pollutants, polycyclic aromatic hydrocarbons, phenols, organic dyes, cyanides, pesticides, and ketones [76]. For instance, starch-based photoactive water-soluble modified polymers that contain porphyrin [98], dextran [100], and chitosan [97] have been synthesized and well-studied. These polymeric systems demonstrated as potential photosensitizers for exhibiting the reactivity of organic compounds in water without polluting the environment. Recently, Li et al. reported on porphyrin-based porous organic polymer, which can be considered as an effective strategy by integrating adsorption and photocatalysis approaches [64]. The robust π–π interactions between porphyrin molecules can be reduced by incorporating porphyrin structure into the amorphous structure of porous organic polymer, thus avoid porphyrin molecules from interfering the photocatalytic efficacy [35].

Khajone et al. investigated the use of good metallophtalocyanines photosensitizer, namely, Fe-Phthalocyanines, which became entangled with benzimidazolium moiety that had been carboxyl functionalized [55]. In addition, the intense color of phtalocyanines, which are used as pigments, makes them an ideal candidate for oxidation and reduction reaction. With 10, 15, and 12.5 mg catalyst loadings, a complete methylene blue, methyl violet, and congo red dye degradation were obtained in 60 min with atmospheric air. In this study, the small quantity of hydrogen peroxide (3%, 0.1 ml) led to an increase in the amount of superoxide radical anions and hydroxyl radicals resulting in a rapid degradation rate. The availability of these radicals is responsible for the deterioration of dyes [15]. Moreover, a heterogeneous folding-like appearance was observed on the ligand's surface, and metal complex was distributed evenly all over the surface. The performance of the hybrid polymer/photosensitizer showed 95.34, 92.48, and 85.88% of methylene blue, methyl violet, and congo red dyes degradation, indicating good efficiency of fabricated photocatalyst.

2.7   In-situ Polymerization of Conductive Polymer

Conductive polymers offer almost infinite architectural diversity, complex, and multifunctional surface chemistry, and great mechanical robustness, making them ideal for long-term applications. The polymers also help the materials' photocatalytic activity by promoting the pre-adsorption of target molecules onto the surfaces [85]. Therefore, to produce polymer-semiconductor composites that have the potential to serve as photocatalysts for the decomposition of organic contaminants under visible irradiation, it is critical to choose the right polymer. For instance, polypyrrole is known as a polymer with outstanding conductivity, excellent redox ability, environmentally safe, and easy to fabricate under several parameters [36]. All these properties are the significant attributes that can be employed and modified to improve the photocatalytic performance in aqueous solution.

Wang et al. incorporated the polypyrrole with titanium dioxide-coated fly ash cenosphere in degrading methylene blue dye. The low density of the hybrid photocatalyst may help the photocatalyst to stay afloat while maximumly absorb high photon energy from the light source [144]. In another study, the polypyrrole-modified titanium dioxide can also be carried out by oxidizing the nanoparticles solution containing pyrrole via in situ chemical oxidation method [31]. This method is preferable due to easy operation, simple, and high reproducibility. Moreover, this type of photocatalyst was reported to produce higher amount of photoexcited electron–hole pair than pristine titanium dioxide and consequently exhibited enhanced photoactivity. It was also reported that the optimum ratio of polypyrrole: titanium dioxide (1:100) yields energy gap of 3.08 eV and the degradation activity against Rhodamine B reached 97% in a period of 8 h as compared to pristine titanium dioxide. However, increasing the ratio showed a decreased of photocatalytic efficiency. The excessive enwrapping of titanium dioxide by the polypyrrole layer could hamper the composites' light absorption. Besides, the thickening of the polypyrrole layer could also prevent the interaction between photoexcited electron hole pairs, thus less production of reactive radicals [145].

Other than titanium dioxide, graphitic carbon nitride is also known as an excellent material in absorbing photons energies and used them to convert the hazardous organic pollutants into harmless species [109]. This material possesses a unique nature including good thermal and chemical stability, nontoxic and suitable band gap energy position, which has become hotspots in several applications such as energy storage, chemical production, sensors, and pollution mitigation. The appropriate position of valence band and conduction band at 1.3 and -1.4 eV, respectively, is promising in absorbing the photon energies with small energy gap of 2.7 eV. Nonetheless, the pure graphitic carbon nitride also impeded from several drawbacks like most photocatalyst such as rapid recombination behavior of photoexcited electron–hole pairs and ineffective light utilization.

To overcome this limitation, Han et al. analyzed the performance of polypyrrole-modified graphitic carbon nitride to be operated against visible irradiation [37]. The results indicate that the polypyrrole particle size was significantly smaller than the graphitic carbon nitride, which is represented by the smooth bulk layer structure. At optimum ratio of 0.75 polypyrrole/graphitic carbon nitride, the nanocomposites displayed tiny amorphous structure belong to polypyrrole deposited on the graphitic carbon nitride surface with no obvious agglomeration. Meanwhile, the increase of contact area may be due to the heterojunction formed between graphitic carbon nitride and polypyrrole [45], thus, may also help to accelerate the migration of charge carrier and impede the recombination effect. The degradation of methylene blue dye achieved at 90% after 2 h of exposure under visible irradiation. In another study, the anchoring of silver nanoparticles into the polypyrrole-modified graphitic carbon nitride matrix boosted the photocatalytic performance due to silver acted as electron transfer mediator, allowing superior charge carrier separation and migration [71].

Other conducting polymers such as polyaniline could also be used in photocatalysis due to its exceptional electrochemical behavior, environmentally safe, easy operating, high chemical stability, and cost-effective. The combination of polyaniline with titanium dioxide could also endow the materials with narrow band gap acting as photosensitizer with improved performance under visible irradiation [154]. The band gap energy of polyaniline-modified titanium dioxide was reported to be 2.85 eV indicating the extension of light absorption characteristic in the visible range and better photo-response as compared to pristine titanium dioxide or polyaniline alone [114]. The incorporation of polyaniline also prevented the agglomeration of titanium dioxide without affecting its crystalline structure in the photocatalytic reaction. The synergistic effect between titanium dioxide and polyaniline contributed to the enhanced photoactivity up to five cycles and stability up to 15 h of operation.

In another study, polyaniline-doped tin oxide-diatomite hybrid photocatalysts showed higher degradation efficiency (96%) when 1 g/L of dosage was used within 1 h [3]. The combination of tin oxide, polyaniline, and diatomite was found to have a favorable effect on band gap energy. Recently, Mukhtar Mohammed et al. prepared cuprous oxide/zinc oxide/polyaniline using the combination solvothermal and in-situ polymerization approaches [95]. Three types of oxidants were also added in the photocatalytic reaction, namely, ammonium persulfate, ammonium persulfate/potassium dichromate, and ammonium persulfate/potassium permanganate. The presence of oxidants is importance attributed to its redox potential, which affects the structure, morphology, and activity of the synthesized photocatalyst [79]. Among the three composite photocatalysts, potassium permanganate is a better fit for ammonium persulfate as second element of composite oxidant than potassium dichromate, as evidenced by its photocatalytic activity and photoluminescence analysis. The cuprous oxide/zinc oxide/polyaniline–ammonium persulfate/potassium permanganate composite showed to have the small crystallite sizes (18.50 nm), lowest bandgap (2.68 eV), lowest photoluminescence intensity, indicating the lowest rate of carrier charges recombination. Moreover, its photocatalytic activity is substantially higher due to an increased surface area, which offers plenty active sites for adsorption. It is noteworthy that larger pore size and pore volume are two crucial factors that help in photocatalyst ionic diffusion and charge transfer. In return, effective degradation activity was reported with 100% Congo red dye degradation after 30 min.

The construction of core/shell nanostructures has been emerged as an excellent alternative towards metal oxide nanoparticles. Because of their high catalytic activity and ease of separation, magnetic nanoparticles have been widely employed as cores. This increases the catalyst's recyclability. Some appropriate magnetic nanocore materials can also be used such as Fe_3O_4, $BaFe_2O_4$, $NiFe_2O_4$, and $CoFe_2O_4$. Specifically, ferrite-based nanocomposites have been proven as good photocatalyst [77] and when combined with organic or polymer nanostructures, its band gap can be tuned, thus diverse its potential applications in electronics, photonics, information storage, catalysis, etc. For instance, Mosali et al. studied the performance of preformed monodisperse cobalt ferrite (15–20 nm) and silver nanoparticles (~10 nm) coupling with polyaniline [94]. The outstanding photocatalytic efficiency against methylene blue dye under sunlight suggests that the hybrid photocatalyst could be a useful photocatalyst for removing azo dyes from industrially polluted water.

2.8   Metal-Incorporated Polymeric Materials as Photocatalyst

Another example of commonly used hybrid polymeric material in wastewater treatment is an incorporation of metal for modification in the polymeric photocatalyst. The incorporation of metal in the polymeric material known as doping process would improve the photocatalytic process compared by using the polymer and metal individually [56]. Silver zero-valent nanoparticle is found to be extremely toxic to the microorganism such as virus, bacteria, and fungi. Because of its toxic properties, silver nano-zero-valent always uses as an antimicrobial agent in the disinfection of wastewater. The proper catalytic mechanism of silver in wastewater treatment had not been reported yet and still under debate. In the immobilization of nanoparticle photocatalyst into polymeric-based material, the polymer membrane first will allow light penetration before it being absorbed by the photocatalyst. The penetration of light into the polymer membrane is shown in Fig. 1. The physicochemical characteristics of the photocatalyst and membrane surfaces determine the contact between the photocatalyst and polymer-based material [160].

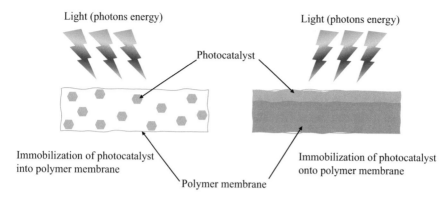

Fig. 1 Incorporation of photocatalyst into polymer-based material with light penetration

Silver nanoparticle has the ability to penetrate the bacterial cell wall and caused death to the bacterial cell [108]. The release of silver ion, Ag^+ by the silver nanoparticle will react with the thiol group and several vital enzymes, which then will inactivate them and then disturb the normal behavior of the cell. The direct application of silver nanoparticle could generate some disadvantages such as the aggregation of the particle in the wastewater could reduce its effectiveness when use for longer period of time [65]. The combination of silver nanoparticle with polymer or other filters such as ceramics/membrane could be promising techniques in wastewater treatment since the efficiency of silver nanoparticle as antimicrobial agent could improve the water quality.

Recently, there are several zero-valent metal nanoparticles had been studied including iron, nickel, zinc, and aluminum. The application of these metal in photocatalytic wastewater treatment had drawn researcher attention nowadays since its offer large surface area which able to drive high catalytic activity and/or sorption capability. Besides, small dimension of zero-valent metal makes its migration in water easier. Table 1 summarized the reactivity series and standard reduction potential of zero-valent nanoparticle that commonly used in wastewater treatment.

Table 1 Reactivity and standard reduction potential of several metals

		Metal	Standard reduction potential E° (V)
Most Reactive		Al	- 1.676
		Zn	- 0.762
		Fe	- 0.440
Less Reactive		Ni	- 0.257

In comparison, aluminum has the highest reactivity unlike iron, nickel, and zinc but lowest standard reduction potential, which reveals aluminum as the stronger reducing agent [74]. Nickel has the lowest reactivity but highest standard reduction potential that makes nickel the weakest reducing agent by comparing to the other three metals in the table. Iron and nickel metals have moderate properties as a zero-valent nanoparticle but both of these metals offer several advantages in application of wastewater treatment such as its offer excellent adsorbent properties, oxidation, and precipitation with the presence of dissolved oxygen and it much cheaper if compared to the other transition metal.

Polymer-based material for wastewater treatment had been extensively studied and reported previously due to its adsorptive capability through blending, crosslinking, and having a large surface area. Incorporation of zero-valent nanoparticle to the polymeric material could offer great performance in the wastewater treatment and desalination process [11]. Besides, the polymeric-based material is highly stable and could improve the weaknesses of individual metal, which easily aggregate in aqueous solution hence reduce the photocatalyst effectiveness. In the field of wastewater treatment process, polymer-based material is used to form coagulation from all suspended material in aqueous water.

The schematic diagram of metal incorporated polymer-based material is shown in Fig. 2. Based on the diagram, incorporation of metal to the polymeric material displays the high efficiency in photocatalytic degradation of impurities/pollutant from wastewater. Doping of metal to the polymer-based material contributes to the larger pore size and improves surface area and stability of the photocatalyst [101].

The mechanism of the reaction involves: (i) the accumulation of the irradiation energy by electron conduction from the metal nanoparticle; (ii) formation of electron with high energy on the metal nanoparticle surface; (iii) activation of electron of target molecules on the particle's surface which then will initiate the chemical reaction [85]. Photodegradation of pollutant from wastewater involves electron transfer combined with redox reaction. Metal incorporates into polymer-based material greatly influenced the stability of the material where it acts as a stabilizer itself. Besides, it provided different functionality than its individual monomer material [120].

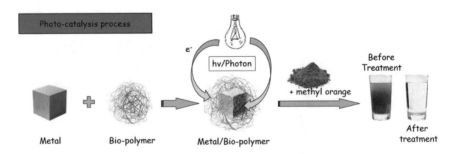

Fig. 2 Schematic diagram for photocatalytic process using metal incorporated polymer-based material

In a recent work, nanoscale zero-valent iron (nZVI) was extensively used as a photocatalyst in wastewater treatment. Nanoscale zero-valent iron is now being explored for the rehabilitation of contaminated wastewaters, waterways, soils, and sediments due to their unique chemical, catalytic, electronic, magnetic, mechanical, and optical capabilities [102]. The application of nanoscale zero-valent iron is more favorable due to its straightforward synthesis via reduction of ferric chloride by sodium borohydride in an aqueous solution with the presence of nitrogen for inert atmosphere. Comparing to the other zero-valent nanoparticle, iron is much cheaper and offers lower energy consumption. Iron atom carries a zero charge, which acts as a bulk reducing agent in converting oxidized material that polluted the water and transformed it into immobilized suspended material. Fe^{2+} ion will be released in the reaction, which will be further oxidized to Fe^{3+} ion [122]. The mechanism of this reaction occurs through chemical reduction process where the contaminant/impurities will be adsorbed to the surface of iron. Because of that, it is very important for the photocatalyst to have large surface area and pore diameter.

Besides iron, zinc also could be considered as an effective metal to be incorporated into polymeric-based material as photocatalyst in the degradation of pollutant from wastewater treatment. Based on Table 1, zinc has lower standard reduction potential like iron if compared to aluminum and nickel, which assist in the degradation of contaminant in wastewater. A study on the comparison of Fe^0 and Zn^0 for environmental application on removal of carbon tetrachloride was reported by Tratnyek et al. in 2014. Based on the study, it was found that zero-valent zinc nanoparticle is more effective in the reduction of carbon tetrachloride level in the wastewater compared to Fe^0. Zero-valent zinc results in more rapid degradation of carbon tetrachloride rather than zero-valent iron where it only involves a stage of oxidation from Zn^0 to Zn^{2+} [139].

Another toxic compound predominantly found in soil, sediment and groundwater called polychlorinated dibenzo-p-dioxins, was previously investigated its degradation using four types of zero-valent metals, namely, zero-valent aluminum, zero-valent iron, zero-valent nickel and zero-valent zinc [13]. Zero-valent metals are crucial in determining the degradation pathway, appearance, and dispersion of contaminants, as well as the overall toxicity. Based on the study, it was found that only zero-valent zinc effectively reduces polychlorinated dibenzo-p-dioxins due to its selectivity in cleavage of the chlorine bond in octachlorodibenzo-p-dioxin.

Modification of polymer surface by adding metals as dopant also reduced the band gap energy thus makes electron easier to excite to higher energy level of conduction band by introducing illumination or solar irradiation. However, not all conductivity materials are ideal for use as dopants; for example, platinum and ruthenium are useless as dopants, whereas chromium, cobalt, magnesium, manganese, indium, and galium can increase the surface area of a photocatalyst [8]. Comparing to the bulk materials, nanomaterials play an important role in wastewater treatment as it will assist in increasing their optical properties, magnetic, electrical, and chemical reactivity. Due to the development of oxidizing species at the surface of the material, which expands its oxidizing capability, pollutants from wastewater can be successfully destroyed utilizing nanomaterial doped polymer photocatalyst [115].

3 Nonmetal-Incorporated Polymeric Materials as Photocatalyst

Conducting polymer composites and their derivatives, including poly(3,4-ethylenedioxythiophene), polyaniline, pyrrole, polyfuran, poly(p-phenylenevinylene), and polythiophene are extensively used in wastewater treatment owing to its biodegradability potential [58]. To date, the studies on applications of polymer nanocomposites seem to be widespread due to its superior characteristics such as ease of synthesis, facile functionalization process, and can be attained at a reasonable price. However, regardless of its advantages, polymer alone has not yet regarded as a star in dye adsorption applications, considering that the potency of these materials can be maximized when it exists in a hybrid form. On the other hand, metals and semiconductor groups possess sets of merits as well, such as high redox potential, good stability, nontoxicity, low-cost as well as environmentally friendly features. For these reasons, photocatalytic degradation is frequently reported to be associated with metals or their complexes, and most semiconductor materials [162]. To date, there are limited numbers of reports on nonmetal-based and metal-free hydrogels that have been published. Hence, in this subchapter, a discussion on potential nonmetal and nonmetal-incorporated polymeric materials applied in photocatalysis will be highlighted.

In a recent review, five different nonmetal materials have been introduced where, the special characteristics of these nonmetal materials are emphasized [61]. The review highlighted on various two-dimensional nonmetal-based materials, which also mentioned the incorporation of metal. However, as an introduction, we will only focus on the characteristics of the nonmetal materials as a single constituent. To begin with, black phosphorus, hexagonal boron nitride, covalent organic frameworks, graphene, and polymeric graphitic carbon nitride are amongst the nonmetal materials that have the photocatalysis potency. The potential features of each abovementioned nonmetal material are summarized in Fig. 3. All these materials are expected to perform better as a photocatalyst when combined with materials that can enable photoinduced charge carriers to participate in surface reactions. The large specific surface area feature is not only associated with the availability of active sites but also affiliated with shorter diffusion length of photogenerated charge carriers [88, 133]. Thereby receding the probability of charge carrier's recombination. Apart from that, another vital characteristic that defines the photocatalyst potency is the band gap energy. A desirable band gap position is often attributed to the expansion of light absorption range. Often, this can be achieved via bandgap engineering. By introducing defect as well as lattice disarrangement, the electronic states will be narrowed or served as mid-gap states to enable photoexcited electrons hence broadened the light absorption to near infrared region [47]. Generally, common methods utilized to tune the photocatalytic efficiency are heterojunction, metal doping, nonmetal doping, and co-doping [132].

With the aim of obtaining materials with superior performance, the study of polymeric photocatalysts has been extended from its individual constituent to its

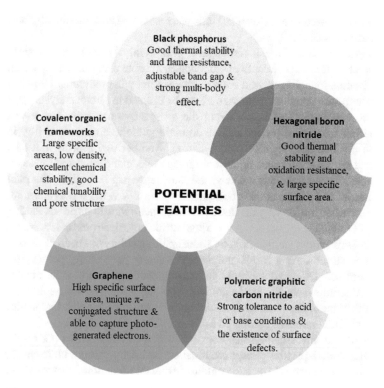

Black phosphorus
Good thermal stability
and flame resistance,
adjustable band gap &
strong multi-body
effect.

**Covalent organic
frameworks**
Large specific
areas, low density,
excellent chemical
stability, good
chemical tunability
and pore structure

**POTENTIAL
FEATURES**

**Hexagonal boron
nitride**
Good thermal
stability and
oxidation resistance,
& large specific
surface area.

Graphene
High specific surface
area, unique π-
conjugated structure &
able to capture photo-
generated electrons.

**Polymeric graphitic
carbon nitride**
Strong tolerance to acid
or base conditions &
the existence of surface
defects.

Fig. 3 The potential features of nonmetal materials

hybridized form. In 2008, a study has caught the attention of many researchers where a polymer/nano-clay hybrid photocatalyst material was successfully synthesized [128]. In most cases, the formation of the polymer/nano-clay composites is predominantly by intercalation or exfoliation. Intercalated nanocomposites are proceeding from the insertion of well-ordered polymer chain in the interlayer gap between the clay sheets. In contrast, exfoliated nanocomposites are associated with the segregation of single clay due to clay dispersion in the polymer matrix [93]. In this work, methoxyethylacrylamide monomer is produced by the reaction between acryloyl chloride and methoxyethylamine. The analogous transformation resulted from the said reaction is referring to the nucleophilic substitution reaction. The final product was then obtained via bulk polymerization using N,N-methylenebisacrylamide crosslinker, potassium persulfate initiator, and tetramethylethylenediamine accelerator. Furthermore, the clay-poly(methoxyethyl)acrylamide samples had a strong photocatalytic capacity in the removal of lead (II) ions, with an adsorptivity of 81.02 mg g^{-1}. Also, it has been found that the adsorption is spontaneously occurred at pH of 4.0–6.0. However, the resulting adsorption is still depending to a few factors including the initial dye concentration, reaction time, and temperature. The pseudo-second-order kinetic model and the Langmuir isotherm model were well fitted to the data obtained.

It was also revealed that the lead (II) ions were adsorbed endothermically within the synthesized materials [128].

The effectiveness of metal-free organic photocatalyst can either be tested manually via laboratory experiments or computational route using the density-functional theory calculations. Recently, a computational survey in predicting the lists of non-metal composite potential material that can be utilized in photocatalysis application has been exploited. The capability of the prospective materials was assessed by considering three vital properties, which highlights the material's ability in splitting the positive and negative charge carrier. Another two aspects that have been evaluated are the appropriate highest occupied molecular orbital and lowest unoccupied molecular orbital energy levels, and the energy gap that allows the photocatalyst material to perform at their optimum level. In this work, a polymeric material, graphitic carbon nitride was selected as the reference owing to its pre-existed reported ability. In reference to the graphitic carbon nitride optoelectronic behavior, a relative comparison centered around the solar water splitting ability has been made on several polyimide strands. As a result, polyimide structure particularly poly(pyromellitic dianhydride-*co*-4,4'-oxydianiline) and benzophenone tetracarboxylic dianhydride–4,4'-oxydianiline have acquired a notable potential in electron–hole splitting. Moreover, both of the aforementioned organic strands perform more effectual in ultraviolet and visible light range as compared to the reference sample [106].

Recently, an investigation has been conducted in chlorpyrifos (O, O-diethylO-3,5,6- trichloro-2-pyridyl phosphorothioate removal by using a chitosan-graphitic carbon nitride catalyst. Chitosan contains invaluable functional groups specifically amino and hydroxyl groups, which evinced an outstanding electrostatic interaction behavior [132]. From the study, it was revealed that parameters such as the initial concentration, pH variation, contact time, and dosage have significant impact on the photocatalytic activities of the chitosan-graphitic carbon nitride catalyst. The degradation efficiency of the sample was quantified by exposing it under a 300 W xenon lamp. The degradation of chlorpyrifos was ascertained to be as high as 85%, suggesting its potentiality in aqueous solution treatment. Without chitosan, the catalyst exhibited only 72% degradation efficiency. This metal-free carbon-based catalyst was found to have a satisfactory resistant towards recombination effects, owing to the presence of chitosan. In term of reusability properties, the chitosan-graphitic carbon nitride catalyst maintained its stability even after five consecutive reusable tests. The study concluded that when compared to other previously reported study that utilized chlorpyrifos as target pollutant, the as-synthesized chitosan-graphitic carbon nitride sample was superior than wheat-straw derived biochar reported back in 2015 [143]. Where, the reported degradation efficiency was only 75% [146].

Other than graphitic carbon nitride, graphene and reduced graphene oxides were also found to be frequently associated to the incorporation of chitosan biopolymer. In a recent study, graphene/chitosan and reduced graphene oxide draped chitosan were prepped by using a facile co-precipitation method. Herein, paraquat, which is a common wastewater toxic chemical was employed as the target pollutant. As is known, pollutants were not completely removed from aqueous solution system via adsorption. The inability of adsorbent in degrading the pollutants entirely may cause byproducts generation. Even worse, the toxicity degree of the produced byproducts is higher [18]. Hence, high surface area reduced graphene oxide was utilized in the study to prevent detrimental impact of electron–hole recombination. The vitality of specific surface area to the photocatalytic activity was revealed when the results show that bare graphene oxide and reduced graphene oxides seem to have smaller active area equated to the samples with chitosan biopolymer. Evidently, the as-synthesized chitosan/reduced graphene oxide sample manifested 281 m^2/g specific area versus 74 m^2/g area for reduced graphene oxide alone. The greater surface area of chitosan/reduced graphene oxide photocatalyst can be imputed to the availability of chitosan moiety, which then prevented the restacking of the carbon layers. Upon degradation test, the results depicted a plausible 85.49% paraquat degradation through 60 min of degradation time. The same sample maintained its degradation efficiency at 82% after five consecutive tests, implying to a reasonable catalyst's stability [142].

In another study, a one-dimensional/zero-dimensional photocatalyst material was innovated to degrade oxytetracycline hydrochloride and chromium pollutant. In this specific work, black phosphorus quantum dots were coalesced with tubular graphitic carbon nitride polymer for further exploration. In this work, the functionality of black phosphorus in the generation of holes was first discovered via computational approach: density functional theory. The aptness of black phosphorus as cocatalyst is also correlated to the broad adjustable band gap characteristic [61]. Based on the attained output, when black phosphorus was combined with the pertinent band gap structure properties of graphitic carbon nitride polymer, the combination seems to work synergistically. The output displays a noteworthy photocatalytic activity in the degradation of target pollutant. The efficiency of oxytetracycline hydrochloride and chromium metal removal was found to be at 0.0276 min^{-1} and 0.0404 min^{-1}, respectively. A detailed study on the photoreduction of both target pollutants was conducted under visible light irradiation. The black phosphorus quantum dots/tubular graphitic carbon nitride polymer photocatalyst manifests 81.05% and 94.71% degradation efficiency for oxytetracycline hydrochloride and chromium, reciprocally. Without the presence of catalyst, the concentration of the target pollutant barely evolved, which demonstrates that the oxytetracycline hydrochloride and chromium samples do not experience photolysis automatically [148].

Polyaniline-based photocatalyst material is gaining much attention in the photocatalysis appliance. For instance, a degradation study of methylene blue dye using polyaniline/mesoporous silica has been conducted more than a decade ago [5]. To date, the search of nonmetal co-catalyst alternatives for semiconductor-based photocatalyst is still evolving [61]. In a typical degradation test, the as-prepared polyaniline/mesoporous silica samples exhibit superior efficiency contrast to polyaniline singly. The conjugated organic/inorganic nanocomposite demonstrates approximately 70% of methylene blue dye deterioration over 120 min of visible light exposure. On the other hand, bare polyaniline shows roughly 26% of target pollutant degradation. The synergistic effects between polyaniline and mesoporous silica were justified by the movement of electrons to the conduction band of mesoporous silica.

Reciprocally, the negative charge carrier jumps from mesoporous silica to polyaniline, leaving behind a hole. Hence, effective positive–negative charge separation on the boundaries of conjugated material contributes to hinder the recombination process [5].

In 2012, a novel graphene/polyaniline nanocomposite has been prepared via polymerization under in situ condition. Graphene was known for its substantial surface area, which endows a propitious catalytic activity. The integration of graphene with polyaniline conducting polymer has undoubtedly improved the electrocatalytic activity of the nanocomposites. In this work, Rose Bengal dye was exploited as the target pollutant and was degraded by 56% under xenon lamp irradiation in a period of 180 min. Compared to pristine polyaniline with only 9% of degradation, the photocatalytic efficiency of the combined nanocomposite seems to be more convincing. Also, the outcome is implying to the role of graphene in conceding electron–hole charge separation. The validity of the degradation test result was substantiated by conducting another catalytic experiment without light irradiation. Thereupon, within 180 min, the deterioration of the Rose Bengal dye appeared to be at only 4%, suggesting that the target pollutant is unable to self-degrade [6].

Besides graphene, carbon nanotube is also of great significance in the photocatalysis application. Both aforementioned carbon blocs are recognized as a material with high active area, which capacitates them to be a good photocatalyst candidate. However, the degradation performance of carbon as a single constituent is perpetually lower compared to its hybrid form. Similarly, polymeric materials perform better with the presence of co-catalyst. The key idea that makes the compositing theory works is the band structure of both materials involved in the conjugation. The structure will determine the excitation path of the electron, thereby affects the photodegradation process. Herein, a composite comprises of polyaniline and single-walled carbon nanotube was well prepped by a group of researchers via in situ polymerization method [16]. With the attempt to find the optimized polymer-to-carbon composition, various weight ratios of polyaniline to single-walled carbon nanotubes were prepared. The photodegradation test demonstrates a significantly high degradation rate for 2% single-walled carbon nanotube composition, compared to pristine polyaniline, 1%, and 4% carbon addition. The efficiencies obtained were 98.6% and 94.35% for Rose Bengal dye and methyl orange target pollutants, respectively [16]. The performance efficiency and experimental conditions for the degradation of heavy metal ions and dyes using polymeric materials are listed in Table 2.

4 Metal Oxide-Incorporated Polymeric Materials as Photocatalyst

Photocatalysts based on polymeric materials have recently piqued the scientific community's interest due to their advantages of flexibility, formability, light in weight, and cost-effective towards textile dye wastewater treatment. Conductive

Table 2 Data on degradation of dyes/organic pollutants by nonmetal-incorporated polymeric materials nanocomposites.

Photocatalyst	Pollutant	Light source	Degradation percentage (%)	Reaction time (min)	References
Chitosan/graphitic carbon nitride	Chlorpyrifos	300 W xenon lamp	85	70	[143]
Chitosan/reduced graphene oxide	Paraquat	Visible light	85.5	60	[142]
Black phosphorus quantum dots/tubular graphitic carbon nitride polymer	Oxytetracycline hydrochloride Chromium	Visible light	81.1 94.7	60	[148]
Polyaniline/mesoporous silica	Methylene blue dye	300 W xenon lamp	70	120	[5]
Polyaniline/graphene	Rose Bengal dye	300 W xenon lamp	56	180	[6]
Polyaniline/single-walled carbon nanotube	Rose Bengal dye Methyl orange dye	Visible light	98.6 94.4	10	[16]

polymer and thermoplastic polymer-based are appropriate for a number of applications due to its transparency to visible light, strong mechanical qualities, chemical durability, and low cost. Besides, due to their versatility, ease of functionalization, and adjustable features, these photocatalysts have been studied for water decontamination to date. From the polymer point of view, due to aforementioned advantages, conductive polymeric-based materials and highly transparent thermoplastic polymer are considered as acceptable adsorbents, photocatalysts, membranes, and supporting beds compared to conventional compositions. Indeed, the ease of processability of these polymeric-based photocatalysts through wet chemical/aqueous method makes it easy to employ for the adsorption, photodegradation, and membrane separation of hazardous heavy metal ions and organic chemicals.

It is challenging to remove pollutants from water by using pristine polymers due to their limited efficiency, despite their numerous advantages. To circumvent this constraint, researchers are combining conductive polymers with organic and inorganic components. Polyaniline, polypyrrole, polythiophene, polyacetylene, polyphenylene, polyphenylene vinylene, poly (3,4-ethylene dioxythiophene), and poly(methyl methacrylate) are some of the most common-based polymeric photocatalysts [32, 50, 81, 135]. The low softening and melting temperatures of polymers, on the other hand, are hurdles to the creation of composite materials with inorganic elements as photocatalysts. Furthermore, because these polymers are hydrophobic and poorly dispersed in aqueous solution, the addition of an organic solvent is required to improve their dispersibility during the photocatalytic reaction [25].

Photocatalysts with metal oxide integrated into polymer display significant efficacy for the photodegradation of azo dyes derived from textile wastewater effluent. This is due to their enhanced surface qualities, specific physical properties, and good mechanical properties [19, 39, 40]. As a heterogeneous photocatalyst, photocatalysts based on metal oxide-incorporated polymer are also attracting immense research for water treatment due to characteristic their strong electron–electron interaction, low cost, adaptability, and long-term viability in the environment of having strong electron–electron interaction in addition to their compositional simplicity. Several metal-oxides that have been used to incorporate with polymer-based photocatalysts including titanium dioxide, zinc oxide, tin oxide, iron (III) oxide, cerium oxide, and tungsten trioxide have been explored to date for photocatalysis applications. These metal oxides have been loaded together with various types of polymeric materials to prepare composite photocatalysts. The photocatalytic performance of common metal oxide incorporated in polymeric material photocatalyst will be addressed in this subchapter.

Titanium dioxide is a substantial metal oxide substance that has been introduced to incorporate in polymeric-based photocatalysts for organic pollutant degradation and removal. Due to its beneficial features, such as inexpensive, large surface area, good chemical stability, low toxicity, and high activity, titanium dioxide is usually regarded as the benchmark photocatalytic material [52, 119]. A few literatures have reported on the synergetic effect of titanium dioxide integrated into polymeric photocatalyst on photocatalytic activity. Because of its high conductivity, simple manufacturing process, and outstanding environmental resistance, polyaniline is one of the most widely studied conducting polymers [29]. The existence of a significant number of amine and imine groups in the polyaniline structure has made this polymer emerged as a promising adsorbent for wastewater treatment alternative to conventional adsorbents [93]. The combination of polyaniline and titanium dioxide has been found to give an alternative way to enhance the photocatalytic activities. Muheb Sboui investigated the effect of titanium dioxide-polyaniline/Cork composite on the photocatalytic activity of methyl orange dye and other contaminants assisted by solar irradiation [121]. The titanium dioxide-polyaniline composites supported on cork and have been created using a simple impregnation technique. From the results, they observed that the methyl orange degradation under sunlight irradiation by the combination of titanium dioxide-polyaniline composites gave a higher degradation rate over titanium dioxide and exhibited an outstanding photostability after four successive runs. This combination of p-type polyaniline and titanium dioxide shows an increase in photocatalytic activities, which could be owing to the synergistic impact between polyaniline and titanium dioxide promoting charge separation efficiency [26]. According to the findings, the combination of titanium dioxide-polyaniline/Cork composite offers up the possibility of developing new hybrid composites with improved photocatalytic activity and stability.

A new biotemplate material containing tin oxide particles can also be intercalated into polyaniline matrix using two different ways in a study [51]. Biotemplate chitosan-tin oxide particles were made by chemical precipitation and subsequently built on a polyaniline matrix using an in-situ chemical oxidative polymerization approach. Introduction of polyaniline has been used as versatile matrix for embedding or dispersing metal nanoparticles in composite nanomaterial synthesis. The samples were prepared at different molar concentrations of tin oxide (0.5 and 0.75 M) added into polyaniline and were labeled as BSP0.50 and BSP 0.75. The photocatalytic performance of methylene blue and RY-15 dye has been measured using polyaniline, BSP0.50 and BSP0.70 photocatalysts exposed under direct sunlight. They discovered that under 210 min, the photocatalytic activity of BSP-0.75 nanocomposites outperformed BSP-0.50 and chitosan-polyaniline, implying that the tin oxide doping boosted the photocatalytic activity of chitosan-polyaniline. They also postulated that the direct sunlight exposure to the BSP-0.75 nanocomposite created electrons and holes on the biotemplate-tin oxide that were then transferred to the conductive polyaniline matrix. The electrons formed an oxygen radical in the nanocomposite when they reacted with oxygen, while the holes produced hydroxyl radicals [7]. The radicals react in the oxidative degradation of photocatalyst with RY-15 and methylene blue dyes.

Another semiconductor metal oxide that is favorable to match or hybrid with polymeric-based photocatalyst is zinc oxide. Zinc oxide with a direct band gap of 3.37 eV significantly played an important role for the degradation of textile dye pollution [53]. Because of its non-toxicity, long-term stability, and high electron mobility, zinc oxide has become one of the most well-known photocatalysts [157]. Incorporation of zinc oxide into polymeric-based photocatalyst is recently received a lot of attention. Many conducting polymers are thought to be good whole conductors. These conductive polymers operate as surface capping agents or stabilizers when combined with metals or semiconductor nanoparticles [29]. Volkan Eskizeybek et al. synthesized new polyaniline/zinc oxide nanocomposite through in situ chemical oxidative polymerization of aniline. Photocatalytic studies of polyaniline and polyaniline/zinc oxide nanocomposite were carried out utilizing methylene blue and malachite green dyes as typical dye pollutant under ultraviolet light and natural sunlight irradiation. They noticed and reported that adding zinc oxide nanoparticles to the polyaniline increased photocatalytic effectiveness when exposed to ultraviolet and natural sunlight irradiation. Under ultraviolet or natural sunlight irradiation, Fig. 4 displays the time-dependent absorption spectra of methylene blue and malachite green dyes by polyaniline/zinc oxide nanocomposites catalyst. The band gap of polyaniline is 2.81 eV, which is narrower than the 3.3 eV of zinc oxide, implying that only ultraviolet light can excite the zinc oxide nanoparticles to form electron–hole pairs, demonstrating a high charge recombination rate of the photo-induced electron/hole pairs [14]. Hence, the synergistic effect of polyaniline/zinc oxide nanocomposites with capability of higher absorption at visible region can be generated by producing higher amount of electron–hole pairs under sunlight, which could result in efficient photocatalytic activity. Thus, under the exposure of sunlight, the synergistic effect of polyaniline/zinc oxide nanocomposites with higher absorbance in visible regions can be activated and formed more electron–hole pairs, resulting in efficient photoactivity [29].

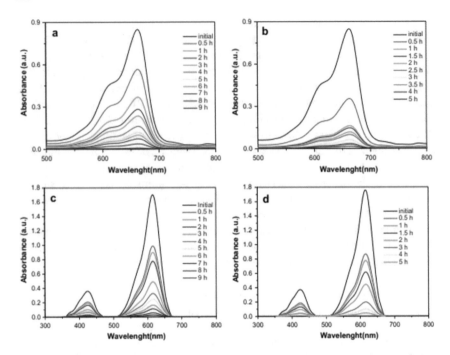

Fig. 4 UV–vis absorption spectra of polyaniline/zinc oxide nanocomposites containing methylene blue and malachite green dyes for various irradiation durations under **a** ultraviolet light for methylene blue, **b** sunlight for methylene blue, **c** ultraviolet light for malachite green, and **d** sunlight for malachite green [29]

A similar photocatalytic enhancement of polyaniline/zinc oxide also discovered by Zor and Budak [168]. Under ultraviolet visible light irradiation, the photocatalytic activities of polyaniline and polyaniline/zinc oxide on the degradation of Congo red dye were investigated. Polyaniline/zinc oxide retained higher photoactivity than that of pure polyaniline in visible-light photodegradation by degrading 100% congo red dye with photocatalytic degradation rate constants of 0.0431 and 0.051 min^{-1}. These findings can be contributed to the absorption capability of polyaniline/zinc oxide that significantly increased due to the modification of zinc oxide nanoparticle with polyaniline. Moreover, the polyaniline/zinc oxide is stated to have a less band gap compared to pure zinc oxide and this indicates that the interaction between polyaniline and zinc oxide is within the visible region of the solar spectrum by providing electron transfer to zinc oxide. Besides, the composite of polyaniline/zinc oxide exhibited an increasing in the active surface area result in an enhancement in the photocatalytic activities.

Due to its effectiveness in visible light, the mechanical and chemical stability of poly(methyl methacrylate) is widely used in a number of applications. It is also considered an inexpensive, hydrophobic polymer that works well with food and beverages, as well as a photocatalyst for textile effluent treatment. The integration of active metal-oxide materials with polymeric is a promising method that has gotten a lot of attention in recent years due to its low weight and facile formability. Alessandro Di Mauro et al. reported zinc oxide/poly(methyl methacrylate) nanocomposites synthesized via sonication and solution casting method [84]. The methylene blue degradation in water under ultraviolet light irradiation was used to assess the photocatalytic activity of the samples. Figure 5 shows the SEM images of zinc oxide/poly(methyl methacrylate) nanocomposites, which showed that the poly(methyl methacrylate) films were transparent, while the zinc oxide/poly(methyl methacrylate) composites were white due to the presence of zinc oxide. According to photocatalytic experiments, zinc oxide/ poly(methyl methacrylate) powders caused 60% of the methylene blue to degrade after 4 h of light irradiation as compared to poly(methyl methacrylate) film. They also tested the XRD pattern of zinc oxide/poly(methyl methacrylate) nanocomposites after photocatalytic measurement to verify the stability of the photocatalyst and found that the photocatalytic experiments did not induce any structural modification. From the findings, the incorporation of zinc oxide nanomaterials in a poly(methyl methacrylate) polymeric matrix is one of the alternative and significant tools for treating textile wastewater.

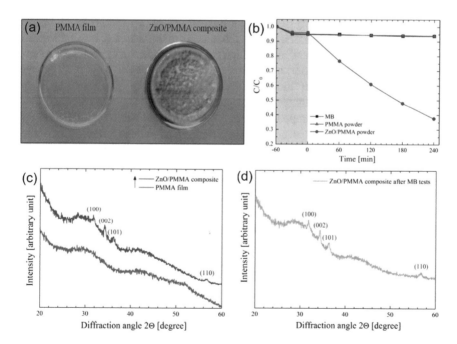

Fig. 5 a Photographs of poly(methyl methacrylate) film (left side) and zinc oxide/poly(methyl methacrylate) composite (right side) in Petri dishes, **b** photodegradation of pure methylene blue, poly(methyl methacrylate) powders, and zinc oxide/poly(methyl methacrylate) powders under ultraviolet light irradiation, **c** and **d** XRD spectra of zinc oxide/poly(methyl methacrylate) composite before and after photodegradation measurement [84]

Incorporation of metal oxide to polymeric materials has emerged as a promising candidate in the textile wastewater treatment. The synergistic effect between this composite demonstrated that it could function as an active photocatalyst. Polymeric-based photocatalysts offer processing convenience, structural flexibility, photoconductivity, and tunable electronic properties, while metal oxide coupling adds additional benefits such as high carrier mobility, band gap tuning, thermal and mechanical stability, and good magnetic properties. Combination within these two materials also enhanced interface interaction and increased the surface area compared to individual constituent.

5   Metal Sulfide-Incorporated Polymeric Materials as Photocatalyst

The environmental stability, chemical inertness, ultraviolet radiation resistance, mechanical stability, low pricing, and ease of access are all aspects that influence the choice of polymer supports for immobilization of photoactive materials. Besides, exerting polymer nanocomposites as photocatalysts enables for easier material separation and reuse, eliminating the need for post-treatment separation operations and reducing the procedure's costs. This has been the solution to a difficulty with the use of nanoparticle photocatalyst, in which the nanoparticles' tiny size causes some challenges in their separation and reuse, as well as probable dangers to ecosystems and human well-being from the nanoparticles' potential release into the water bodies [167]. This disadvantage of nanoparticles can be avoided by immobilizing nanoparticles onto a larger support material, resulting in the construction of modified nanocomposite. Moreover, several studies have been done on polymer-based photocatalytic materials, which include inorganic nanoparticles with high catalytic activity under ultraviolet or visible light, such as titanium dioxide, zinc oxide, cerium (IV) oxide, or plasmonic (silver, gold, platinum and palladium) [21, 49, 110, 112, 138, 150]. However, there are limited numbers of reports on metal sulfide-incorporated polymer as a photocatalytic material. The review emphasizes on the preparation, photocatalytic activity, approaches for increasing photocatalytic efficiency under visible light, and hybrid polymer catalyst reuse. This review will describe the metal sulfide incorporated polymer including conducting polymer as well as biodegradable polymer. Six types of metal sulfide that will be explored are cadmium sulfide, zinc sulfide, tungsten sulfide, molybdenum sulfide, copper sulfide, and bismuth sulfide.

Cadmium sulfide has a large surface area and tiny bandgap of 2.4 eV, which make it function well in the visible region [67]. Hence, cadmium sulfide is regarded as one of the most appealing visible light-driven photocatalysts. Moreover, the morphological properties of the photocatalyst are well recognized to have a significant impact on photocatalytic performance. The high recombination rate of photogenerated electron–hole pairs, on the other hand, reduces the photocatalytic effectiveness of this material [4]. Several methods, such as combining cadmium sulfide with polymer, might be promising in overcoming this challenge due to large surface area, which can help to limit cadmium sulfide nanoparticle aggregation on the surface and manage their size and shape, hence enhancing photocatalytic activity. The porous polymelamine–formaldehyde contains high amount of triazine and amine functional groups, as well as a large surface area and exceptional photoactivity [149]. Polymelamine–formaldehyde works well to capture carbon dioxide, adsorbent for organic dye and hazardous metal removal when compared to alternative polymer supports [137]. As a result, polymelamine–formaldehyde is predicted to be an excellent choice for photocatalyst support. Li et al. precipitated cadmium sulfide nanoparticles onto a porous polymelamine–formaldehyde for the first time. The large specific surface area of the as-prepared cadmium sulfide/polymelamine–formaldehyde composite will then lead to a high dye adsorption capability. The cadmium sulfide/polymelamine–formaldehyde composite outperforms pure cadmium sulfide particles in terms of adsorption ability and photocatalytic effectiveness. Polymelamine–formaldehyde has high surface area of 943 m^2 g^{-1}, which makes the dye to be easily adsorbed by cadmium sulfide/polymelamine–formaldehyde, increasing the degradation efficiency. By using modified cadmium sulfide/polymelamine–formaldehyde sample, 99% of Rhodamine B was degraded after 3 h of visible light irradiation. In addition, the cadmium sulfide/polymelamine–formaldehyde sample is stable during the recycle runs, indicating that it has a high potential for photocatalysis to remove organic pollutants [67].

Polylactic acid fibers have also been incorporated with cadmium sulfide. Polylactic acid fibers were synthesized using an electrospinning method and used as a photocatalyst support. Chit Seng Ho et al. proposed the use of cadmium sulfide/polylactic acid fibers to degrade the methylene blue [20]. As the quantity of cadmium sulfide was raised from 1 wt% to 5 wt%, the diameter of the fibers shrank, which might be due to an increased in polymer solution electrical conductivity. Cadmium sulfide (3 wt%)/polylactic acid fibers have the best degradation performance of methylene blue, with a 90% removal rate. Polylactic acid fibers provided photocorrosion stability for cadmium sulfide, resulting in improved degradation performance for composite cadmium sulfide/polylactic acid fibers. However, it is found that the removal efficiency of cadmium sulfide/polylactic acid fibers reduced to 60% after five cycles. The kinetics of methylene blue removal follows pseudo-first-order with the highest kinetics constant was found from Cadmium sulfide (3 wt%).

Meanwhile, S. Sharma et al. used a conductive polymer called polyaniline to integrate cadmium sulfide [124]. Polyaniline is a semiconductor, which has electron delocalization that improves its conducting behavior. Using simple chemical technique, cadmium sulfide nanoflower and nanorods have been incorporated into polyaniline, and it has been shown that cadmium sulfide nanoflower/polyaniline has better degradation efficiency compared to cadmium sulfide nanorods/polyaniline. This proves that the morphologies of cadmium sulfide nanorods have a synergistic effect with polyaniline in which more interfacial sites between cadmium sulfide nanoflowers and polyaniline were created that leads to better charge carrier separation efficiency. The reaction followed pseudo-first-order with rate constant of 0.304

and $0.379 h^{-1}$ for polyaniline/cadmium sulfide nanoflower and polyaniline/cadmium sulfide nanorods, respectively.

Cadmium sulfide has also been combined with a biodegradable polymer called starch. Carboxymethyl starch is a biodegradable polymer that is environmentally safe and cheap. Xue et al. used in situ formation method to synthesize cadmium sulfide/carboxymethyl starch nanocomposite with different ratios of cadmium sulfide and carboxymethyl starch [152]. As carboxymethyl starch has grafted carboxyl group, cadmium sulfide, a frequent photocatalyst for visible light excitation, was chosen as the supporting material as it was able to efficiently interact with ion cadmium (II) and bond cadmium sulfide particles well in the starch matrix [17, 72]. This combination of cadmium sulfide/carboxymethyl starch film did not only illustrate efficient removal of various organic dyes, such as methylene blue and crystal violet, but this photocatalyst material also has good selectivity in degradation of methylene blue, crystal violet, and rhodamine B. This difficult film's preparation procedure was improved. Cadmium sulfide/carboxymethyl starch with a sufficient cadmium sulfide (38 wt%) showed 86.72% removal of methylene blue and 81.03% removal of crystal violet within 2 h, indicating good efficacy for both dyes. After five cycles of degradation, cadmium sulfide/carboxymethyl starch exhibits very little decrease in removal rates and maintains selectivity for these two dyes [152]. The cationic groups on methylene blue and crystal violet may attach efficient electrostatic interactions with the negatively carboxyl groups of carboxymethyl starch, affecting their high affinities. However, the anionic groups on rhodamine B had an electrostatic repulsion to the starch matrix. The significantly varied affinities of several dyes for carboxymethyl starch resulted in strong adsorption preferences and high selectivity in degradation. pH and several coexisting inorganic anions have little effect on the selectivity of cadmium sulfide/carboxymethyl starch. Furthermore, this complex film did not need to be regenerated and could be reused with no significant decrement in removal capacity.

The following researcher paired cadmium sulfide with two types of polymers: chitosan and polyaniline. Chitosan is easily available in nature, biodegradable and possesses an amino and hydroxyl group on its which enables them to have a chelating activity with metals ion. Thus, it has a potential to be a candidate to produce nanocomposite with cadmium sulfide. Pure cadmium sulfide's pH sensitivity, on the other hand, is a clear disadvantage that has limited its industrial use [164]. To curb this limitation, chitosan-based composites with better adsorption properties and acid tolerance have been created. Polyaniline, a conducting polymer, is the potential candidate for chitosan-based composite due to its better environmental sustainability and cost-effectiveness as well as its chelating characteristics produced by the electron-donating group and delocalized conjugated structure. Rasoulifard et al. prepared the nanocomposites of chitosan/polyaniline, chitosan/cadmium sulfide and chitosan/polyaniline/cadmium sulfide to degrade the reactive blue 19 dye [113]. Without the presence of visible light, removal efficiency of chitosan/polyaniline and chitosan/polyaniline/cadmium sulfide composites on reactive blue 19 was 60 and 79%, respectively. The degradation efficiency reached 99% in 60 min under visible light irradiation by chitosan/polyaniline/cadmium sulfide, suggesting strong photoactivity and high degradation rate. Degradation efficiency for chitosan/cadmium sulfide was 66% under the same working conditions. Here, the degradation of reactive blue 19 is induced by the formation of unstable cationic dye radicals as well as the attack by hydroxyl radical. The adsorption and degradation activity of chitosan/polyaniline/cadmium sulfide were taken into account in the development of a unique kinetic model for dye removal efficiency prediction. After five rounds of visible light irradiation, chitosan/polyaniline/cadmium sulfide maintained strong photocatalytic activity, indicating that the nanocomposite is stable.

Another semiconductor metal sulfide that is favorable to embed with polymeric-based photocatalyst is copper (II) sulfide. Copper (II) sulfide is a p-type semincon-ductor that has a narrow band gap between 1.55 and 2.16 eV. There are different structural kinds of copper sulfide, which can be produced by varying the concentration ratio of copper and sulfur, such as hexagonal plates of (Cu_7S_4-CuS), digenite (Cu_9S_5), covellite (CuS), and chalcocite (Cu_2S) [86, 104]. Here, Nie et al. embedded covellite on electropun polyacrylonitrile nanofibers. Polyacrylonitrile captured interest as it is easy to prepare, has enormous surface area and also chemically stable. The electro-spinning process is a flexible method for fabricating continuously long fibers with various shapes. Covellite incorporated into electrospun polyacrylonitrile nanofibers can improve catalyst reusability while also preventing covellite agglomeration during the catalytic process. Covellite nanoparticles were dispersed homogeneously on the surface of electrospun polyacrylonitrile nanofibers using a simple hydrothermal approach. Covellite/polyacrylonitrile exhibits good photocatalytic activity in methy-lene blue degradation with the presence of hydrogen peroxide. The performance of the photocatalyst increases significantly as polyacrylonitrile itself can degrade only 35% of methylene blue while covellite/polyacrylonitrile can degrade up to 95.4% in 40 min. The photocatalyst also maintains the degradation performance not below than 80% even after five times, which prove that covellite/polyacrylonitrile is a good photocatalyst in Fenton reaction [96].

Next, covellite also has been incorporated into cellulose material. Cellulose-based aerogel has high tendency to form a bonding with a functional group like amino and carboxyl group as it has many hydroxyl group on its structure [54, 151]. Saeed et al. used wet chemical deposition method to induce the growth of covellite in the outer and inner surface of cellulose-based aerogel [116]. Covellite/cellulose-based aerogel composite has a porous structure. When compared to pure covellite, the photocat-alytic performance of covellite/cellulose-based aerogel was significantly improved. The methylene blue dye was degraded in just 6 min with the help of hydrogen peroxide (97% degradation). In the presence of hydrogen peroxide, electron–hole pair recombination is hindered, which lead to an improvement in methylene blue degradation. Covellite/cellulose-based aerogel also has been discovered as a good visible light photocatalyst without the presence of hydrogen peroxide. Under visible light irradiation, covellite/cellulose-based aerogel has successfully degraded 94.15% of methylene blue, compared to only 67.4% for pure covellite.

Also, covellite also was mixed with polyvinyl alcohol and gellatine. Abdullah A. Al-Kahtani utilized gamma irradiation to enhance copolymerization for the synthesis of gelatin/covellite/polyvinyl alcohol [2]. The nanocomposite has been used for the degradation of Rhodamine B under visible irradiation. The result showed that the efficiency of gelatin/covellite/polyvinyl alcohol to degrade rhodamine B reduced as the concentration of rhodamine B increased but increased when the pH is increased in basic condition. Also, the degradation rate increased when the amount of catalyst used increase, but only up to the optimized condition of 0.25 g. Increase the catalyst dosage higher than that lead to the reduction in photocatalytic performance due to the agglomeration of the catalyst itself. This photocatalyst follows pseudo-first-order reaction with k = 2.614×10^{-2}. The materials also show good performance as the efficiency remains above 80% even after five times of usage.

Molybdenum disulfide is also among the metal sulfide that incorporated into polymer. Two-dimensional molybdenum disulfide has many advantages in term of its physicochemical properties as it possesses good crystallinity, high surface area as well as better separation of charge carrier. However, molybdenum disulfide has a major drawback due to their poor electronic conductivity, which limits their photocatalytic properties. To make it worse, molybdenum disulfide has large interlayer stacking and week interlayer bonding [159]. Thus, few efforts have been done to improve this weakness including the synthesis of nanocomposite molybdenum disulfide with good conductivity materials such as graphene and carbon nanotubes [75]. Nevertheless, the decrease in interconnected conductive pathway resulted in a moderate performance in terms of conductivity. Thus, the researcher explored the possibility of incorporating the molybdenum disulfide into polymeric material such as polyaniline and polydimethylsiloxane. Saha et al. reported a simple approach in the synthesis of molybdenum disulfide/polyaniline nanocomposites as a photocatalytic material in degradation of methylene blue [117]. Polyaniline is considered to have similar properties as p-type inorganic semiconductor thus the combination of molybdenum disulfide and polyaniline will increase the electrical conductivity of nanocomposite as well as reduce the electron hole recombination. This also will prevent the agglomeration of molybdenum disulfide and resulting in increment in the surface area. The results show that the photocatalytic reaction and adsorption follow first-order adsorption model with $k_1 = 0.0163$ g mg^{-1} min^{-1}. The synergistic effect between molybdenum disulfide–polyaniline resulted in the increase of photocatalytic efficiency where 30% of methylene blue degraded in less than 3 h compared to only 15% degradation for pure molybdenum disulfide.

In the meantime, tungsten disulfide has been combined with a polymer to be employed as a photocatalytic material. Masimukku et al. utilized ultrasonic vibration method in a dark environment to synthesize single-layer tungsten disulfide nanoflower/polydimethylsiloxane for the degradation of Rhodamin B dye [83]. They achieved Rhodamine B degradation in a dark environment for the first time using polydimethylsiloxane/tungsten disulfide nanoflower photocatalyst with maximum degradation of 99% in 90 min. The kinetic study shows that polydimethylsiloxane/tungsten disulfide nanoflower has the fastest degradation rate with a rate constant of 0.13 (ppm s^{-1}) with maximum degradation rate of 6624 ppm L mole$^{-1 \, s-1}$.

Also, the recyclability test showed that the performance of the photocatalyst remains good even after ten cycling tests with 90% degradation. This proves that the utilization of piezo catalytic technology without the presence of light can be an effective and alternative way of changing the mechanical energy into chemical energy to degrade the dye wastewater.

Bismuth sulfide is also one of the metal sulfides, which was embedded with polymeric material for the degradation of textile dyes. Bismuth sulfide has received a lot of interest as a typical photocatalyst because of it has narrow band gap, nontoxic, and cheap [80]. Various types of bismuth sulfide with different morphologies have been synthesized to be used in wastewater treatment. The utilization of solvothermal process to synthesize bismuth sulfide with the shape of nanorods to degrade the organic dye can also be employed. However, the degradation performance of bismuth sulfide itself is poor under visible light due to fast recombination rate. Researcher took an initiative to combine bismuth sulfide with gold, bismuth oxychloride, and bismuth subcarbonate to improve the photoactivity [68, 126]. However, their complicated synthesis process can't match the photocatalytic performance. Thus, researcher searches for another alternative by embedding the bismuth sulfide with polymer such as polyvinyl pyrrolidone. Shen et al (2020) showed that bismuth sulfide/polyvinylpyrrolidone possesses high degradation rate of rhodamine B with 93.9% in 30 min (k = 0.07675 min^{-1}) which is far better than any bismuth-based semiconductors previously described. It is revealed that embedding polyvinylpyrrolidone on the surface of tungsten disulfide led to better absorption of visible light and accelerates electron–hole separation, which improve the photocatalytic performance. Shen et al. revealed the performance of bismuth sulfide/titanium dioxide heterojunction/polymer fiber for the methylene blue degradation under visible irradiation [125]. Here, polysulphone/styrene-maleic anhydride copolymer fibers have been produced using electrospinning method before being embedded with bismuth sulfide/titanium dioxide. The photocatalytic performance of bismuth sulfide/titanium dioxide/polysulphone/styrene-maleic anhydride is far better than the bismuth sulfide/titanium dioxide and polysulphone/styrene-maleic anhydride itself; 95.3%, 69.9%, and 27.75%, respectively. This is believed due to narrower bandgap of bismuth sulfide/titanium dioxide as well as the phenomenon of adsorption together with migration and photodegradation under the visible light. The results also showed that the crystallinity of the photocatalytic material is maintained even after five cycling tests.

6 Conclusion

There are several fabrication methods that can be employed to prepare different types of hybrid polymeric materials as photocatalyst. As the photoactivity of pristine polymeric materials is not effective for industrial applicability, the use of a variety of nanoparticles or inorganic fillers introduced into the polymer matrix has recently become the focus of study. Moreover, the photocatalyst in suspension mode alone could impede its commercialization due to several technical aspects such as ineffective absorption of light due to agglomeration, inhomogeneous distribution, and difficulty in recovering the photocatalyst after the reaction has been completed. Besides, the different types of preparation methods may influence the physicochemical characteristics and photocatalytic performance of the resultant polymer-based photocatalyst in treating various types of textile wastewater. The data obtained from the previous experimental studies might differ from others, which depend on the parameters used during the process or photocatalyst preparation. This chapter discussed several issues to use different types of inorganic nanoparticles such as metal, nonmetal, metal-oxide, and metal-sulfide as the most versatile additives for polymer matrix insertion. The applications of modified polymer photocatalysts have also been concluded in terms of morphological characteristics, structural and optical properties, photocatalytic activity, dye removal, and textile wastewater treatment.

References

1. Miyoshi A, SN, KM (2018) Water splitting on rutile TiO2-based photocatalysts. Chem A Eur J 24:18204–18219
2. Al-Kahtani AA (2017) No Title. J Biomater Nanobiotechnol 8:66–82
3. Akti F (2018) Photocatalytic degradation of remazol yellow using polyaniline–doped tin oxide hybrid photocatalysts with diatomite support. Appl Surf Sci 455:931–939. https://doi.org/10.1016/j.apsusc.2018.06.019
4. Alipour A, Lakouarj MM (2019) Photocatalytic degradation of RB dye by CdS-decorated nanocomposites based on polyaniline and hydrolyzed pectin: isotherm and kinetic. J Environ Chem Eng 7(102837).https://doi.org/10.1016/J.JECE.2018.102837
5. Ameen S, Im Y, Jo CG, Kim YS, Shin H (2011) Synthesis and characterization of polyaniline / MCM-41 nanocomposites and their photocatalytic activity. J Nanosci Nanotechnol 11:541–545. https://doi.org/10.1166/jnn.2011.3161
6. Ameen S, Seo H, Akhtar MS, Shik H (2012) Novel graphene / polyaniline nanocomposites and its photocatalytic activity toward the degradation of rose Bengal dye. Chem Eng J 210:220–228. https://doi.org/10.1016/j.cej.2012.08.035
7. Amir MNI, Julkapli NM, Hir ZAM, Hamid SBA (2016) Effects of layers and ratio Cs-TiO2/Glass photocatalyst towards removal of methylene orange via adsorption-photodegradation process. Malaysian J Chem 18:46–59
8. Anjum M, Miandad R, Waqas M, Gehany F, Barakat MA (2019) Remediation of wastewater using various nano-materials. Arab J Chem 12:4897–4919. https://doi.org/10.1016/j.arabjc.2016.10.004
9. Artale MA, Augugliaro V, Drioli E, Golemme G, Grande C, Loddo V, Molinari R, Palmisano L, Schiavello M (2001) Preparation and characterisation of membranes with entrapped TIO$_2$ and preliminary photocatalytic tests. Ann Chim 91:127–136
10. Chen B, Meng YH, Sha JW, Zhong C, WH, NZ, (2018) Preparation of MoS2/TiO2 based nanocomposites for photocatalysis and rechargeable batteries: progress, challenges, and perspective. Nanoscale 10:34–68
11. Berber MR (2020) Current advances of polymer composites for water treatment and desalination. J Chem. https://doi.org/10.1155/2020/7608423
12. Binas V, Venieri D, Kotzias D, Kiriakidis G (2017) Modified TiO2 based photocatalysts for improved air and health quality. J Mater 3:3–16. https://doi.org/10.1016/j.jmat.2016.11.002
13. Bokare V, Jung JL, Chang YY, Chang YS (2013) Reductive dechlorination of octachlorodibenzo-p-dioxin by nanosized zero-valent zinc: modeling of rate kinetics and

congener profile. J Hazard Mater 250–251:397–402.https://doi.org/10.1016/j.jhazmat.2013.02.020

14. Chai B, Wang X, Cheng S, Zhou H, Zhang F (2014) One-pot triethanolamine-assisted hydrothermal synthesis of Ag/ZnO heterostructure microspheres with enhanced photocatalytic activity. Ceram Int 40:429–435. https://doi.org/10.1016/j.ceramint.2013.06.019
15. Chang N, Zhang H, Shi M-S, Li J, Yin C-J, Wang H-T, Wang L (2018) Regulation of the adsorption affinity of metal-organic framework MIL-101 via a TiO2 coating strategy for high capacity adsorption and efficient photocatalysis. Microporous Mesoporous Mater 266:47–55. https://doi.org/10.1016/j.micromeso.2018.02.051
16. Chatterjee MJ, Ghosh A, Mondal A, Banerjee D (2017) Polyaniline–single walled carbon nanotube composite–a photocatalyst to degrade rose bengal and methyl orange dyes under visible-light illumination. RSC Adv 7:36403–36415. https://doi.org/10.1039/C7RA03855K
17. Chen R, Zhang Y, Shen L, Wang X, Chen J, Ma A, Jiang W (2015) Lead(II) and methylene blue removal using a fully biodegradable hydrogel based on starch immobilized humic acid. Chem Eng J 268:348–355. https://doi.org/10.1016/J.CEJ.2015.01.081
18. Chen Z, Zhang S, Liu Y, Alharbi NS, Rabah SO, Wang S, Wang X (2020) Synthesis and fabrication of g-C3N4-based materials and their application in elimination of pollutants. Sci Total Environ 731(139054). https://doi.org/10.1016/j.scitotenv.2020.139054
19. Chijioke-Okere MO, Mohd Hir ZA, Ogukwe CE, Njoku PC, Abdullah AH, Oguzie EE (2021) TiO2/Polyethersulphone films for photocatalytic degradation of acetaminophen in aqueous solution. J Mol Liq 338116692). https://doi.org/10.1016/j.molliq.2021.116692
20. Ho CS, Abidin NHZ, Nugraha MW, N.S.S., Fathilah Ali, MOHD Dzul Hakim Wirzal, Laksmi Dewi Kasmiarno, and S.A.A., (2020) Fibers and Polymers. Fibers Polym. 21:1212–1221
21. Choe HR, Han SS, Kim Y-I, Hong C, Cho EJ, Nam KM (2021) Understanding and improving photocatalytic activity of Pd-loaded BiVO4 microspheres: application to visible light-induced suzuki-miyaura coupling reaction. ACS Appl Mater Interfaces 13:1714–1722. https://doi.org/10.1021/acsami.0c15488
22. Daghrir R, Drogui P, Robert D (2013) Modified TiO2 for environmental photocatalytic applications: a review. Ind Eng Chem Res 52:3581–3599. https://doi.org/10.1021/ie303468t
23. Damodar RA, You SJ, Chou HH (2009) Study the self cleaning, antibacterial and photo-catalytic properties of TiO2 entrapped PVDF membranes. J Hazard Mater 172:1321–1328. https://doi.org/10.1016/j.jhazmat.2009.07.139
24. Daniel C (2021) Composite polymeric films with photocatalytic properties 86:571–576. https://doi.org/10.3303/CET2186096
25. Di Mauro A, Farrugia C, Abela S, Ref Alo P, Grech M, Falqui L, Nicotra G, Sfuncia G, Mio A, Buccheri MA, Rappazzo G, Brundo MV, Scalisi EM, Pecoraro R, Iaria C, Privitera V, Impellizzeri G (2020) Ag/ZnO/PMMA nanocomposites for efficient water reuse. ACS Appl Bio Mater 3:4417–4426. https://doi.org/10.1021/acsabm.0c00409
26. Dinoop lal S, Sunil Jose T, Rajesh C, Anju Rose Puthukkara P, Savitha Unnikrishnan K, Arun KJ (2021) Accelerated photodegradation of polystyrene by TiO2-polyaniline photocatalyst under UV radiation. Eur Polym J 153, 110493.https://doi.org/10.1016/j.eurpolymj.2021.110493
27. Dong H, Zeng G, Tang L, Fan C, Zhang C, He X, He Y (2015) An overview on limitations of TiO2-based particles for photocatalytic on the degradation of organic pollutants and the corresponding counter-measures. Water Resour 79:128–146
28. Drozd D, Szczubiałka K, Kumorek M, Kepczynski M, Nowakowska M (2014) Photoactive polymer–nanoclay hybrid photosensitizer for oxidation of phenol in aqueous media with the visible light. J Photochem Photobiol A Chem 288:39–45. https://doi.org/10.1016/j.jphotochem.2014.04.025
29. Eskizeybek V, Sarı F, Gülce H, Gülce A, Avcı A (2012) Preparation of the new polyaniline/ZnO nanocomposite and its photocatalytic activity for degradation of methylene blue and malachite green dyes under UV and natural sun lights irradiations. Appl Catal B Environ 119–120:197–206. https://doi.org/10.1016/j.apcatb.2012.02.034

30. Essawy AA, Ali AEH, Abdel-Mottaleb MSA (2008) Application of novel copolymer-TiO2 membranes for some textile dyes adsorptive removal from aqueous solution and photocatalytic decolorization. J Hazard Mater 157:547–552. https://doi.org/10.1016/j.jhazmat.2008.01.072

31. Gao F, Hou X, Wang A, Chu G, Wu W, Chen J, Zou H (2016) Preparation of polypyrrole/TiO2 nanocomposites with enhanced photocatalytic performance. Particuology 26:73–78. https://doi.org/10.1016/j.partic.2015.07.003

32. Gilja V, Vrban I, Mandić V, Žic M, Hrnjak-Murgić Z (2018) Preparation of a PANI/ZnO composite for efficient photocatalytic degradation of acid blue. Polymers (Basel) 10:1–17. https://doi.org/10.3390/polym10090940

33. Gómez-Cerezo MN, Mu͂noz-Batista MJ, Tudela D, Fernández-García M, Kubacka A (2014) Composite Bi2O3–TiO2 catalysts for toluene photo-degradation: ultraviolet and visible light performances. Appl Catal B Environ 156:307–313

34. Guillet JE, Burke NAD, Nowakowska M, Paone S (1998) Polymer catalysts for important photoelectron transfer reactions. Macromol Symp 134:41–49. https://doi.org/10.1002/masy.19981340106

35. Guo P, Chen P, Liu M (2012) Porphyrin assemblies via a surfactant-assisted method: from nanospheres to nanofibers with tunable length. Langmuir 28:15482–15490. https://doi.org/10.1021/la3033594

36. Xiao HM, SYF (2014) Synthesis and physical properties of electromagnetic polypyrrole composites via addition of magnetic crystals. Cryst Eng Comm 16:2097–2112

37. Han H, Fu M, Li Y, Guan W, Lu P, Hu X (2018) In-situ polymerization for PPy/g-C3N4 composites with enhanced visible light photocatalytic performance. Cuihua Xuebao/Chinese J Catal 39:831–840. https://doi.org/10.1016/S1872-2067(17)62997-8

38. Hathway TL (2009) Titanium dioxide photocatalysis : studies of the degradation of organic molecules and characterization of photocatalysts using mechanistic organic chemistry. Iowa State University

39. Hir ZAM, Abdullah AH, Zainal Z, Lim HN (2018) Visible light-active hybrid film photocatalyst of polyethersulfone–reduced TiO2: photocatalytic response and radical trapping investigation. J Mater Sci 1–16.https://doi.org/10.1007/s10853-018-2570-3

40. Hir ZAM, Abdullah AH, Zainal Z, Lim HN (2017) Photoactive hybrid film photocatalyst of Polyethersulfone-ZnO for the degradation of methyl orange dye: kinetic study and operational parameters. Catalysts 7:1–16. https://doi.org/10.3390/catal7110313

41. Hir ZAM, Moradihamedani P, Abdullah AH, Mohamed MA (2017) Immobilization of TiO2 into polyethersulfone matrix as hybrid film photocatalyst for effective degradation of methyl orange dye. Mater Sci Semicond Process 57:157–165. https://doi.org/10.1016/j.mssp.2016.10.009

42. Hong RY, Li JH, Chen LL, Liu DQ, Li HZ, Zheng Y, Ding J (2009) Synthesis, surface modification and photocatalytic property of ZnO nanoparticles. Powder Technol 189:426–432. https://doi.org/10.1016/j.powtec.2008.07.004

43. Hoogesteijn von Reitzenstein N, Bi X, Yang Y, Hristovski K, Westerhoff P (2016) Morphology, structure, and properties of metal oxide/polymer nanocomposite electrospun mats. J Appl Polym Sci 133:43811–43819

44. Hui KC, Suhaimi H, Sambudi NS (2021) Electrospun-based TiO2 nanofibers for organic pollutant photodegradation: a comprehensive review. Rev Chem Eng https://doi.org/10.1515/revce-2020-0022

45. Jin J, Liang Q, Ding CY, Li ZY, SX (2017) Simultaneous synthesis-immobilization of Ag nanoparticles functionalized 2D g-C3N4 nanosheets with improved photocatalytic activity. J Alloy Compd 691:763–771

46. Javed S, Noreen R, Kamal S, Rehman S, Yaqoob N, Abrar S (2020) Chapter 2 - Polymer blends as matrix materials for the preparation of the nanocomposites. In Mahmood Zia K, Jabeen F, Anjum MN, Ikram SBT-B (eds) Micro and nano technologies. Elsevier, pp 21–54. https://doi.org/10.1016/B978-0-12-816751-9.00002-7

47. Jiang L, Yang J, Zhou S, Yu H, Liang J, Chu W, Li H, Wang H, Wu Z, Yuan X (2021) Strategies to extend near-infrared light harvest of polymer carbon nitride photocatalysts. Coord Chem Rev 439(213947).https://doi.org/10.1016/j.ccr.2021.213947

48. Byrappa K, Subramani AK, Ananda S, Rai KML, Dinesh R, MY (2006) Photocatalytic degradation of rhodamine B dye using hydrothermally synthesized ZnO. Bull Mater Sci 29:433–438
49. Kalathil S, Khan MM, Ansari SA, Lee J, Cho MH (2013) Band gap narrowing of titanium dioxide (TiO2) nanocrystals by electrochemically active biofilms and their visible light activity. Nanoscale 5:6323–6326. https://doi.org/10.1039/C3NR01280H
50. Kanth N, Xu W, Prasad U, Ravichandran D, Kannan AM, Song K (2020) Pmma-tio2 fibers for the photocatalytic degradation of water pollutants. Nanomaterials 10:1–8. https://doi.org/10.3390/nano10071279
51. Karpuraranjith M, Thambidurai S (2016) Biotemplate-SnO2 particles intercalated PANI matrix: enhanced photo catalytic activity for degradation of MB and RY-15 dye. Polym Degrad Stab 133:108–118. https://doi.org/10.1016/j.polymdegradstab.2016.08.006
52. Katančić Z, Chen WT, Waterhouse GIN, Kušić H, Lončarić Božić A, Hrnjak-Murgić Z, Travas-Sejdic J (2020) Solar-active photocatalysts based on TiO2 and conductive polymer PEDOT for the removal of bisphenol A. J Photochem Photobiol A Chem 396(112546).https://doi.org/10.1016/j.jphotochem.2020.112546
53. Kaur J, Bansal S, Singhal S (2013) Photocatalytic degradation of methyl orange using ZnO nanopowders synthesized via thermal decomposition of oxalate precursor method. Phys B Condens Matter 416:33–38. https://doi.org/10.1016/j.physb.2013.02.005
54. Keshipour S, Khezerloo M (2018) Au-dimercaprol functionalized cellulose aerogel: synthesis, characterization and catalytic application. Appl Organomet Chem 32.https://doi.org/10.1002/aoc.4255
55. 55. Khajone VB, Balinge KR, Patle DS, Bhagat PR (2019) Synthesis and characterization of polymer supported Fe-phthalocyanine entangled with carboxyl functionalized benzimidazolium moiety: a heterogeneous catalyst for efficient visible-light-driven degradation of organic dyes from aqueous solutions. J Mol Liq 288(111032). https://doi.org/10.1016/j.molliq.2019.111032
56. Khan F, Zahid M, Hanif MA, Tabasum A, Mushtaq F, Noreen S, Mansha A (2021) Photocatalytic polymeric composites for wastewater treatment, Aquananotechnology 467–490,https://doi.org/10.1016/b978-0-12-821141-0.00005-7
57. Krishnamoorthy M, Ahmad NH, Amran HN, Mohamed MA, Kaus NHM, Yusoff SFM (2021) BiFeO3 immobilized within liquid natural rubber-based hydrogel with enhanced adsorption-photocatalytic performance. Int J Biol Macromol 182:1495–1506. https://doi.org/10.1016/j.ijbiomac.2021.05.104
58. Kumar R, Travas-Sejdic J, Padhye LP (2020) Conducting polymers-based photocatalysis for treatment of organic contaminants in water. Chem Eng J Adv 4(100047).https://doi.org/10.1016/j.ceja.2020.100047
59. Wei LF, Yu CL, Zhang QH, HL, YW (2018) TiO2-based heterojunction photocatalysts for photocatalytic reduction of CO2 into solar fuels. J Mater Chem 6:22411–22436
60. Ge L, Han CC, JL (2012) In situ synthesis and enhanced visible light photocatalytic activities of novel PANI–g-C3N4 composite photocatalysts. J Mater Chem 22:11843–11850
61. Lai C, An N, Li B, Zhang M, Yi H, Liu S, Qin L, Liu X, Li L, Fu Y, Xu F, Wang Z, Shi X, An Z, Zhou X (2021) Future roadmap on nonmetal-based 2D ultrathin nanomaterials for photocatalysis. Chem Eng J 406(126780).https://doi.org/10.1016/j.cej.2020.126780
62. Lee CG, Javed H, Zhang D, Kim JH, Westerhoff P, Li Q, Alvarez PJJ (2018) Porous Electrospun fibers embedding TiO2 for adsorption and photocatalytic degradation of water pollutants. Environ Sci Technol 52:4285–4293. https://doi.org/10.1021/acs.est.7b06508
63. Leroux F, Campagne C, Perwuelz A, Gengembre L (2008) Polypropylene film chemical and physical modifications by dielectric barrier discharge plasma treatment at atmospheric pressure. J Colloid Interface Sci 328:412–420
64. Li M, Zhao H, Lu ZY (2020) Porphyrin-based porous organic polymer, Py-POP, as a multifunctional platform for efficient selective adsorption and photocatalytic degradation of cationic dyes. Microporous Mesoporous Mater 292(109774).https://doi.org/10.1016/j.micromeso.2019.109774

65. Li X, Lenhart JJ, Walker HW (2012) Aggregation kinetics and dissolution of coated silver nanoparticles. Langmuir 28:1095–1104. https://doi.org/10.1021/la202328n

66. Li X, Raza S, Liu C (2021) Preparation of titanium dioxide modified biomass polymer microspheres for photocatalytic degradation of rhodamine-B dye and tetracycline. J Taiwan Inst Chem Eng 122:157–167. https://doi.org/10.1016/j.jtice.2021.04.040

67. Li X, Wang Y, Xie Y, Yin S, Lau R, Xu R (2017) CdS nanoparticles loaded on porous polymelamine–formaldehyde polymer for photocatalytic dye degradation. Res Chem Intermed 43:5083–5090. https://doi.org/10.1007/s11164-017-3048-7

68. Liang N, Zai J, Xu M, Zhu Q, Wei X, Qian X (2014) Novel Bi2S3/Bi2O2CO3 heterojunction photocatalysts with enhanced visible light responsive activity and wastewater treatment. J Mater Chem A 2:4208–4216. https://doi.org/10.1039/C3TA13931J

69. Sun L, Shi Y, Li B, Xiaochen Li YW (2013) Preparation and characterization of Polypyrrole/TiO2 nanocomposites by reverse microemulsion polymerization and its photocatalytic activity for the degradation of methyl orange under natural light. Polym Compos 34:1076–1080. https://onlinelibrary.wiley.com/doi/10.1002/pc.22515

70. Linh NTB, Lee KH, Lee BT (2011) Fabrication of photocatalytic PVA-TiO2 nano-fibrous hybrid membrane using the electro-spinning method. J Mater Sci 46:5615–5620. https://doi.org/10.1007/s10853-011-5511-y

71. Liu J, Wang H, Antonietti M (2016) Graphitic carbon nitride "reloaded": emerging applications beyond photocatalysis. Chem Soc Rev 45:2308–2326

72. Liu Q, Li F, Lu H, Li M, Liu J, Zhang S, Sun Q, Xiong L (2018) Enhanced dispersion stability and heavy metal ion adsorption capability of oxidized starch nanoparticles. Food Chem 242:256–263. https://doi.org/10.1016/J.FOODCHEM.2017.09.071

73. Liu X, Wang L, Zhou X, He X, Zhou M, Jia K, Liu X (2021) Design of polymer composite-based porous membrane for in-situ photocatalytic degradation of adsorbed organic dyes. J Phys Chem Solids 154(110094).https://doi.org/10.1016/j.jpcs.2021.110094

74. Lu Q, Yu Y, Ma Q, Chen B, Zhang H (2016) 2D transition-metal-dichalcogenide-nanosheet-based composites for photocatalytic and electrocatalytic hydrogen evolution reactions. Adv Mater 28:1917–1933. https://doi.org/10.1002/adma.201503270

75. Lv H, Liu Y, Tang H, Zhang P, Wang J (2017) Synergetic effect of MoS2 and graphene as cocatalysts for enhanced photocatalytic activity of BiPO4 nanoparticles. Appl Surf Sci 425:100–106. https://doi.org/10.1016/J.APSUSC.2017.06.303

76. Nowakowska M, Szczubiałka K, SZ (1996) Photosensitized dechlori- nation of polychlorinated phenols 2. Photoinduced by poly(sodium styrenesulphonate-Co-N-vinylcarbazole) dechlorination of pentachloro- phenol in watere. J Photochem Photobiol A Chem 97:1–2

77. Qasim M, Asghar K, Singh BR, Prathapani S, Khan W, Naqvi AH, Das D (2015) No Title. Spectrochim. Acta Part A Mol Biomol Spectrosc 137(1348)

78. Soltaninezhad M, Aminifar A (2011) Study nanostructures of semiconductor zinc oxide (ZnO) as a photocatalyst for the degradation of organic pollutants. Int J Nano Dimens 2:137–145

79. Saeb MR, Zarrintaj P, Khandelwal P, NPSC (2019) Synthetic route of polyaniline (I): Conventional oxidative polymerization, Elsevier Inc

80. Ma L, Zhao Q, Zhang Q, Ding M, Huang J, Liu X, Liu Y, Wu X, Xu X (2014) Controlled assembly of Bi2S3 architectures as Schottky diode, supercapacitor electrodes and highly efficient photocatalysts. RSC Adv 4:41636–41641. https://doi.org/10.1039/C4RA07169G

81. Ma Y, Xu Y, Ji X, Xie M, Jiang D, Yan J, Song Z, Xu H, Li H (2020) Construction of polythiophene/Bi4O5I2 nanocomposites to promote photocatalytic degradation of bisphenol a, J Alloys Compd 823(153773).https://doi.org/10.1016/j.jallcom.2020.153773

82. Mangal R, Srivastava S, Archer LA (2015) Phase stability and dynamics of entangled polymer-nanoparticles composites. Nat Commun 6:7198

83. Masimukku S, Hu YC, Lin ZH, Chan SW, Chou TM, Wu JM (2018) High efficient degradation of dye molecules by PDMS embedded abundant single-layer tungsten disulfide and their antibacterial performance. Nano Energy 46:338–346. https://doi.org/10.1016/J.NANOEN.2018.02.008

84. Mauro AD, Cantarella M, Nicotra G, Pellegrino G, Gulino A, Brundo MV, Privitera V, Impellizzeri G (2017) Novel synthesis of ZnO / PMMA nanocomposites for photocatalytic applications. Nat Publ Gr 1–12. https://doi.org/10.1038/srep40895
85. Melinte V, Stroea L, Chibac-Scutaru AL (2019) Polymer nanocomposites for photocatalytic applications. Catalysts 9(986).https://doi.org/10.3390/catal9120986
86. Meng X, Tian G, Chen Y, Zhai R, Zhou J, Shi Y, Cao X, Zhou W, Fu H (2013) Hierarchical CuS hollow nanospheres and their structure-enhanced visible light photocatalytic properties. Cryst Eng Comm 15:5144–5149. https://doi.org/10.1039/C3CE40195B
87. Mohamed MA, Jaafar J, M, Zain MF, Minggu LJ, Kassim MB, Rosmi MS, Alias NH, Mohamad Nor NA, W Salleh WN, Othman MHD (2018a) In-depth understanding of core-shell nanoarchitecture evolution of g-C 3 N 4 @C, N co-doped anatase/rutile: efficient charge separation and enhanced visible-light photocatalytic performance. Appl Surf Sci 436:302–318.https://doi.org/10.1016/j.apsusc.2017.11.229
88. Mohamed MA, Minggu LJ, Kassim MB, Amin NAS, Salleh WN, Salehmin MNI, Nasir MFM, Hir ZAM (2018) Constructing bio-templated 3D porous microtubular C-doped g-C3N4 with tunable band structure and enhanced charge carrier separation. Appl Catal B Environ 236:265–279. https://doi.org/10.1016/j.apcatb.2018.05.037
89. Mohamed MA, Salleh WNW, Jaafar J, Ismail AF, Abd Mutalib M, Jamil SM (2015) Incorporation of N-doped TiO2 nanorods in regenerated cellulose thin films fabricated from recycled newspaper as a green portable photocatalyst. Carbohydr Polym 133:429–437. https://doi.org/10.1016/j.carbpol.2015.07.057
90. Mohamed MA, Salleh WNW, Jaafar J, Ismail AF, Mutalib MA, Sani NAA, Asri SEAM, Ong C.S (2016) Physicochemical characteristic of regenerated cellulose/N-doped TiO2 nanocomposite membrane fabricated from recycled newspaper with photocatalytic activity under UV and visible light irradiation. Chem Eng J 284:202–215.https://doi.org/10.1016/j.cej.2015.08.128
91. Molinari R, Grande C, Drioli E, Palmisano L, Schiavello M (2001) Photocatalytic membrane reactors for degradation of organic pollutants in water. Catal Today 67:273–279
92. Molinari R, Scicchitano G, Pirillo F, Loddo V, Palmisano L (2005) Preparation, characterisation and testing of photocatalytic polymeric membranes with entrapped or suspended TiO2. Int J Environ Pollut 23:140–152
93. Momina KA (2021) Study of different polymer nanocomposites and their pollutant removal efficiency: review. Polymer (Guildf) 217(123453).https://doi.org/10.1016/j.polymer.2021.123453
94. Mosali VSS, Qasim M, Mullamuri B, Chandu B, Das D (2017) Synthesis and characterization of Ag/CoFe2O4/polyaniline nanocomposite for photocatalytic application. J Nanosci Nanotechnol 17:8918–8924. https://doi.org/10.1166/jnn.2017.13903
95. Mukhtar Mohammed A, Mohtar SS, Aziz F, Aziz M, Usman Nasir M (2021) Effects of oxidants on the in-situ polymerization of aniline to form Cu2O/ZnO/PANI composite photocatalyst. Mater Today Proc 2020https://doi.org/10.1016/j.matpr.2021.02.757
96. Nie G, Li Z, Lu X, Lei J, Zhang C, Wang C (2013) Fabrication of polyacrylonitrile/CuS composite nanofibers and their recycled application in catalysis for dye degradation. Appl Surf Sci 284:595–600. https://doi.org/10.1016/J.APSUSC.2013.07.139
97. Nowakowska M, Moczek L, Szczubiałka K (2008) Photoactive modified chitosan. Biomacromol 9:1631–1636. https://doi.org/10.1021/bm800141v
98. Nowakowska M, Sterzel M, Zapotoczny S (2005) Novel water-soluble photosensitizer based on starch and containing porphyrin. Photochem Photobiol 81:1227. https://doi.org/10.1562/2005-03-07-ra-455
99. Nowakowska M, Szczubiałka K (2017) Photoactive polymeric and hybrid systems for photocatalytic degradation of water pollutants. Polym Degrad Stab 145:120–141. https://doi.org/10.1016/j.polymdegradstab.2017.05.021
100. Nowakowska M, Zapotoczny S, Sterzel M, Kot E (2004) Novel water-soluble photosensitizers from dextrans. Biomacromol 5:1009–1014. https://doi.org/10.1021/bm034506w

101. Opoku F, Govender KK, Sittert van CGCE, Govender PP (2017) Recent progress in the development of semiconductor-based photocatalyst materials for applications in photocatalytic water splitting and degradation of pollutants. Adv Sustain Syst 1.https://doi.org/10.1002/adsu.201700006

102. Oprčkal P, Mladenovič A, Vidmar J, Mauko Pranjić A, Milačič R, Ščančar J (2017) Critical evaluation of the use of different nanoscale zero-valent iron particles for the treatment of effluent water from a small biological wastewater treatment plant. Chem Eng J 321:20–30. https://doi.org/10.1016/j.cej.2017.03.104

103. Sathishkumar P, Pugazhenthiran N, Mangalaraja RV, Asiri AM, S.A., (2013) ZnO supported CoFe2O4 nanophoto- catalysts for the mineralization of Direct Blue 71 in aqueous environments. J Hazardous Mater 252–253:171–179

104. Pal M, Mathews NR, Sanchez-Mora E, Pal U, Paraguay-Delgado F, Mathew X (2015) Synthesis of CuS nanoparticles by a wet chemical route and their photocatalytic activity. J Nanoparticle Res 17:301. https://doi.org/10.1007/s11051-015-3103-5

105. Panthi G, Park M, Park SJ, Kim HY (2015) PAN electrospun nanofibers reinforced with Ag2CO3 nanoparticles: highly efficient visible light photocatalyst for photodegradation of organic contaminants in waste water. Macromol Res 23:149–155. https://doi.org/10.1007/s13233-015-3032-2

106. Park J, Soon A, Lee J (2020) A computational survey of metal-free polyimide-based photocatalysts within the single-stranded polymer model. Mol Catal 497(111184).https://doi.org/10.1016/j.mcat.2020.111184

107. Pei CC, Leung WW-F (2013) Enhanced photocatalytic activity of electrospun TiO2/ZnO nanofibers with optimal anatase/rutile ratio. Catal Commun 37P:100–104. https://doi.org/10.1016/j.catcom.2013.03.029

108. Prabhu S, Poulose EK (2012) Silver nanoparticles: mechanism of antimicrobial. Int Nano Lett 2:32–41

109. Qi K, Liu S, yuan, Zada A (2020) Graphitic carbon nitride, a polymer photocatalyst. J Taiwan Inst Chem Eng 109:111–123. https://doi.org/10.1016/j.jtice.2020.02.012

110. Qin R, Meng F, Khan MW, Yu B, Li H, Fan Z, Gong J (2019) Fabrication and enhanced photocatalytic property of TiO 2 -ZnO composite photocatalysts. Mater Lett 240:84–87. https://doi.org/10.1016/j.matlet.2018.12.139

111. Arrua RD, Strumia MC, CIAI (2009) Macroporous monolithic polymers: preparation and applications. Materials (Basel) 2:2429–2466

112. Rashad MM, Ismail AA, Osama I, Ibrahim IA, Kandil AHT (2014) Photocatalytic decomposition of dyes using ZnO doped SnO2 nanoparticles prepared by solvothermal method. Arab J Chem 7:71–77. https://doi.org/10.1016/J.ARABJC.2013.08.016

113. Rasoulifard MH, Seyed Dorraji MS, Amani-Ghadim AR, Keshavarz-Babaeinezhad N (2016) Visible-light photocatalytic activity of chitosan/polyaniline/CdS nanocomposite: kinetic studies and artificial neural network modeling. Appl Catal A Gen 514:60–70. https://doi.org/10.1016/J.APCATA.2016.01.002

114. Razak S, Nawi MA, Haitham K (2014) Fabrication, characterization and application of a reusable immobilized TiO2–PANI photocatalyst plate for the removal of reactive red 4 dye. Appl Surf Sci 319:90–98. https://doi.org/10.1016/j.apsusc.2014.07.049

115. Review WA, Yaqoob AA, Parveen T, Umar K (2020) Role of Nanomaterials in the treatment of. Water 2020(12):495. https://doi.org/10.3390/w12020495

116. Saeed RMY, Bano Z, Sun J, Wang F, Ullah N, Wang Q (2019) CuS-functionalized cellulose based aerogel as biocatalyst for removal of organic dye. J Appl Polym Sci 136, 47404. https://doi.org/10.1002/app.47404

117. Saha S, Chaudhary N, Mittal H, Gupta G, Khanuja M (2019) Inorganic–organic nanohybrid of MoS2-PANI for advanced photocatalytic application. Int Nano Lett 9:127–139. https://doi.org/10.1007/s40089-019-0267-5

118. Sakamoto M, Ahmed T, Begum S, Huq H (2019) Water pollution and the textile industry in bangladesh: flawed corporate practices or restrictive opportunities? Sustainability 11:1951. https://doi.org/10.3390/su11071951

119. Sambaza SS, Maity A, Pillay K (2020) Polyaniline-coated TiO2 nanorods for photocatalytic degradation of bisphenol A in water. ACS Omega 5:29642–29656. https://doi.org/10.1021/acsomega.0c00628

120. Sarkar S, Ponce NT, Banerjee A, Bandopadhyay R, Rajendran S, Lichtfouse E (2020) Green polymeric nanomaterials for the photocatalytic degradation of dyes: a review. Environ Chem Lett 18:1569–1580. https://doi.org/10.1007/s10311-020-01021-w

121. Sboui M, Nsib MF, Rayes A, Swaminathan M, Houas A (2017) TiO2–PANI/Cork composite: a new floating photocatalyst for the treatment of organic pollutants under sunlight irradiation. J Environ Sci (China) 60:3–13. https://doi.org/10.1016/j.jes.2016.11.024

122. Sen S, Shah P (2014) Application of nanoscale zero-valent iron for wastewater treatment. Int Conf Multidiscip Res Pract I:0–2

123. Shahabuddin S, Sarih NM, Ismail FH, Shahid MM, Huang NM (2015) Synthesis of chitosan grafted-polyaniline/Co3O4 nanocube nanocomposites and their photocatalytic activity toward methylene blue dye degradation. RSC Adv 5:83857–83867. https://doi.org/10.1039/c5ra11237k

124. Sharma S, Singh S, Khare N (2016) Synthesis of polyaniline/CdS (nanoflowers and nanorods) nanocomposites: a comparative study towards enhanced photocatalytic activity for degradation of organic dye. Colloid Polym Sci 294:917–926. https://doi.org/10.1007/s00396-016-3844-4

125. Shen H, Shao Z, Zhao Q, Jin M, Shen C, Deng M, Zhong G, Huang F, Zhu H, Chen F, Luo Z (2020) Facile synthesis of novel three-dimensional Bi2S3 nanocrystals capped by polyvinyl pyrrolidone to enhance photocatalytic properties under visible light. J Colloid Interface Sci 573:115–122. https://doi.org/10.1016/J.JCIS.2020.03.111

126. Shi Y, Xiong X, Ding S, Liu X, Jiang Q, Hu J (2018) In-situ topotactic synthesis and photocatalytic activity of plate-like BiOCl/2D networks Bi2S3 heterostructures. Appl Catal B Environ 220:570–580. https://doi.org/10.1016/J.APCATB.2017.08.074

127. Shimizu K, Imai H, Hirashima H, Tsukuma K (1999) Low-temperature synthesis of anatase thin films on glass and organic substrates by direct deposition from aqueous solutions. thin solid films, 351. Thin Solid Films 351:220–224

128. Şölener M, Tunali S, Özcan AS, Özcan A, Gedikbey T (2008) Adsorption characteristics of lead(II) ions onto the clay/poly(methoxyethyl)acrylamide (PMEA) composite from aqueous solutions. Desalination 223:308–322. https://doi.org/10.1016/j.desal.2007.01.221

129. Sompalli NK, Mohanty A, Mohan Am, Deivasigamani P (2021) Heterojunction Cr2O3-Ag2O nanocomposite decorated porous polymer monoliths a new class of visible light fast responsive heterogeneous photocatalysts for pollutant clean-up. J Environ Chem Eng 9(104846).https://doi.org/10.1016/j.jece.2020.104846

130. Sonawane RS, Hegde SG, Dongare MK (2003) Preparation of titanium(IV) oxide thin film photocatalyst by sol–gel dip coating. Mater Chem Phys 77:744–750. https://doi.org/10.1016/S0254-0584(02)00138-4

131. Sultanova ED, Nizameev IR, Kholin KV, Kadirov MK, Ovsyannikov AS, Burilov VA, Ziganshina AY, Antipin IS (2020) Photocatalytic properties of hybrid materials based on a multicharged polymer matrix with encored TiO2and noble metal (Pt, Pd or Au) nanoparticles. New J Chem 44:7169–7174. https://doi.org/10.1039/c9nj06413c

132. Suresh R, Rajendran A, Hoang DKA, Vo D-VN, Siddiqui MN, Cornejo-Ponce L (2021) Recent progress in green and biopolymer based photocatalysts for the abatement of aquatic pollutants. Environ Res 199(111324).https://doi.org/10.1016/j.envres.2021.111324

133. Syazwani ON, Mohd Hir ZA, Mukhair H, Mastuli MS, Abdullah AH (2018) Designing visible-light-driven photocatalyst of Ag3PO4/CeO2 for enhanced photocatalytic activity under low light irradiation. J Mater Sci Mater Electron 30:415–423. https://doi.org/10.1007/s10854-018-0306-4

134. Kemp TJ, RAM (2001) Mechanism of action of Titanium dioxide pigment in the photodegradation of poly(vinyl chloride) and other polymers. Prog React Kinet Mech 26:337–374

135. Taghizadeh A, Taghizadeh M, Jouyandeh M, Yazdi MK, Zarrintaj P, Saeb MR, Lima EC, Gupta VK (2020) Conductive polymers in water treatment: a review. J Mol Liq 312.https://doi.org/10.1016/j.molliq.2020.113447

136. Tahiri Alaoui O, Nguyen QT, Mbareck C, Rhlalou T (2009) Elaboration and study of poly(vinylidene fluoride)–anatase TiO2 composite membranes in photocatalytic degradation of dyes. Appl Catal A Gen 358:13–20. https://doi.org/10.1016/j.apcata.2009.01.032

137. Tan MX, Sum YN, Ying JY, Zhang Y (2013) A mesoporous poly-melamine-formaldehyde polymer as a solid sorbent for toxic metal removal. Energy Environ Sci 6:3254–3259. https://doi.org/10.1039/C3EE42216J

138. Tanaka A, Hashimoto K, Kominami H (2017) A very simple method for the preparation of Au/TiO2 plasmonic photocatalysts working under irradiation of visible light in the range of 600–700 nm. Chem Commun 53:4759–4762. https://doi.org/10.1039/C7CC01444A

139. Tratnyek PG, Salter AJ, Nurmi JT, Sarathy V (2010) Environmental applications of zerovalent metals: iron vs Zinc. ACS Symp Ser 1045:165–178. https://doi.org/10.1021/bk-2010-1045.ch009

140. Tuna Ö, Simsek EB (2021) Anchoring LaFeO3 perovskites on the polyester filters for flowthrough photocatalytic degradation of organic pollutants. J Photochem Photobiol A Chem 418.https://doi.org/10.1016/j.jphotochem.2021.113405

141. Ullah H, Tahir AA, Mallick TK (2017) Polypyrrole/TiO2composites for the application of photocatalysis. Sens Actuat B Chem 241:1161–1169. https://doi.org/10.1016/j.snb.2016.10.019

142. Vigneshwaran S, Preethi J, Meenakshi S (2020) Interface engineering of ultrathin multi-functional 2D draped chitosan for efficient charge separation on degradation of paraquat-a mechanistic study. J Environ Chem Eng 8(104446).https://doi.org/10.1016/j.jece.2020.104446

143. Vigneshwaran S, Preethi J, Meenakshi S (2019) Removal of chlorpyrifos, an insecticide using metal free heterogeneous graphitic carbon nitride (g-C3N4) incorporated chitosan as catalyst: photocatalytic and adsorption studies. Int J Biol Macromol 132:289–299. https://doi.org/10.1016/j.ijbiomac.2019.03.071

144. Wang B, Li C, Pang J, Qing X, Zhai J, Li Q (2012) Novel polypyrrole-sensitized hollow TiO 2 /fly ash cenospheres: synthesis, characterization, and photocatalytic ability under visible light. Appl Surf Sci 258:9989–9996. https://doi.org/10.1016/j.apsusc.2012.06.061

145. Wang D, Wang Y, Li X, Luo Q, An J, Yue J (2008) Sunlight photocatalytic activity of polypyr-role–TiO2 nanocomposites prepared by 'in situ' method. Catal Commun 9:1162–1166. https://doi.org/10.1016/j.catcom.2007.10.027

146. Wang P, Yin Y, Guo Y, Wang C (2015) Removal of chlorpyrifos from waste water by wheat straw-derived biochar synthesized through oxygen-limited method. RSC Adv 5:72572–72578. https://doi.org/10.1039/C5RA10487D

147. Wang Q, Cai J, Biesold-McGee GV, Huang J, Ng YH, Sun H, Wang J, Lai Y, Lin Z (2020) Silk fibroin-derived nitrogen-doped carbon quantum dots anchored on TiO2 nanotube arrays for heterogeneous photocatalytic degradation and water splitting. Nano Energy 78(105313).https://doi.org/10.1016/j.nanoen.2020.105313

148. Wang W, Niu Q, Zeng G, Zhang C, Huang D (2020) 1D porous tubular g-C3N4 capture black phosphorus quantum dots as 1D / 0D metal-free photocatalysts for oxytetracycline hydrochlo-ride degradation and hexavalent chromium reduction. Appl Catal B Environ 273(119051 https://doi.org/10.1016/j.apcatb.2020.119051

149. Wang Y, Xie Y, Zhang Y, Tang S, Guo C, Wu J, Lau R (2016) Anionic and cationic dyes adsorption on porous poly-melamine-formaldehyde polymer. Chem Eng Res Des 114:258–267. https://doi.org/10.1016/J.CHERD.2016.08.027

150. Weon S, Kim J, Choi W (2018) Dual-components modified TiO2 with Pt and fluoride as deactivation-resistant photocatalyst for the degradation of volatile organic compound. Appl Catal B Environ 220:1–8. https://doi.org/10.1016/J.APCATB.2017.08.036

151. XiuPing S, Qian L, Lin L, RanJu M, ZhouQi Q, HuiYing G, JuMing Y (2017) Cu2O nanoparticle-functionalized cellulose-based aerogel as high-performance visible-light photo-catalyst. Cellulose 24:1017–1029. https://doi.org/10.1007/s10570-016-1154-0

152. Xue Y, Chang Q, Hu X, Cai J Yang H (2020) A simple strategy for selective photocatalysis degradation of organic dyes through selective adsorption enrichment by using a complex film

of CdS and carboxylmethyl starch. J Environ Manage 274(111184).https://doi.org/10.1016/J.JENVMAN.2020.111184

153. Bai Y, Mora-Sero I, Angelis FD, JB, PW (2014) Titanium dioxide nanomaterials for photovoltaic applications. Chem Rev 114:10095–10130

154. Yang C, Dong W, Cui G, Zhao Y, Shi X, Xia X, Tang B, Wang W (2017) Enhanced photocatalytic activity of PANI/TiO2 due to their photosensitization-synergetic effect. Electrochim Acta 247:486–495. https://doi.org/10.1016/j.electacta.2017.07.037

155. Yang J, Chu S, Guo Y, Luo L, Kong F, Wang Y, Zou Z (2012) Hyperbranched polymeric N-oxide: a novel kind of metal-free photocatalyst. Chem Commun 48:3533–3535. https://doi.org/10.1039/c2cc30308f

156. Yar A, Haspulat B, Üstün T, Eskizeybek V, Avci A, Kamiş H, Achour S (2017) Electrospun TiO2/ZnO/PAN hybrid nanofiber membranes with efficient photocatalytic activity. RSC Adv 7:29806–29814. https://doi.org/10.1039/c7ra03699j

157. Yin L, Zhang D, Wang J, Huang J, Kong X, Fang J, Zhang F (2017) Materials characterization improving sunlight-driven photocatalytic activity of ZnO nanostructures upon decoration with Fe (III) cocatalyst. Mater Charact 127:179–184. https://doi.org/10.1016/j.matchar.2017.03.004

158. Younas H, Shao J, He Y, Fatima G, Jaffar STA, Afridi ZUR (2018) Fouling-free ultrafiltration for humic acid removal. RSC Adv 8:24961–24969. https://doi.org/10.1039/c8ra03810d

159. Yuan Y, Guo Rt, Hong Lf, Ji Xy, Li Zs, Lin Zd, Pan Wg (2021) Recent advances and perspectives of MoS2-based materials for photocatalytic dyes degradation: a review. Colloids Surfaces A Physicochem Eng Asp 611(125836).https://doi.org/10.1016/J.COLSURFA.2020.125836

160. Zakria HS, Othman MHD, Kamaludin R, Sheikh Abdul Kadir SH, Kurniawan TA, Jilani A (2021) Immobilization techniques of a photocatalyst into and onto a polymer membrane for photocatalytic activity. RSC Adv 11:6985–7014. https://doi.org/10.1039/d0ra10964a

161. Zeng H, Yu Z, Shao L, Li X, Zhu M, Liu Y, Feng X, Zhu X (2020) Ag2CO3@UiO-66-NH2 embedding graphene oxide sheets photocatalytic membrane for enhancing the removal performance of Cr(VI) and dyes based on filtration. Desalination 491(114558).https://doi.org/10.1016/j.desal.2020.114558

162. Zhang JY, Mei JY, Yi SS, Guan XX (2019) Constructing of Z-scheme 3D g-C3N4-ZnO@graphene aerogel heterojunctions for high-efficient adsorption and photodegradation of organic pollutants. Appl Surf Sci 492:808–817. https://doi.org/10.1016/j.apsusc.2019.06.261

163. Zhang L, Guan H, Zhang N, Jiang B, Sun Y, Yang N (2019) A loose NF membrane by grafting TiO2-HMDI nanoparticles on PES/β-CD substrate for dye/salt separation. Sep Purif Technol 218:8–19. https://doi.org/10.1016/j.seppur.2019.02.018

164. Zhang L, Sun F, Zuo Y, Fan C, Xu S, Yang S, Gu F (2014) Immobilisation of CdS nanoparticles on chitosan microspheres via a photochemical method with enhanced photocatalytic activity in the decolourisation of methyl orange. Appl Catal B Environ 156–157:293–300. https://doi.org/10.1016/J.APCATB.2014.03.015

165. Zhang X, Xu S, Han G (2009) Fabrication and photocatalytic activity of TiO2 nanofiber membrane. Mater Lett 63:1761–1763. https://doi.org/10.1016/j.matlet.2009.05.038

166. Zhang Y, Wei S, Zhang H, Liu S, Nawaz F, Xiao F-S (2009) Nanoporous polymer monoliths as adsorptive supports for robust photocatalyst of Degussa P25. J Colloid Interface Sci 339:434–438. https://doi.org/10.1016/j.jcis.2009.07.050

167. Zhao X, Lv L, Pan B, Zhang W, Zhang S, Zhang Q (2011) Polymer-supported nanocomposites for environmental application: a review. Chem Eng J 170:381–394. https://doi.org/10.1016/J.CEJ.2011.02.071

168. Zor S, Budak B (2020) Investigation of the effect of PAn and PAn/ZnO photocatalysts on 100% degradation of Congo red under UV visible light irradiation and lightless environment. Turkish J Chem 44:486–501. https://doi.org/10.3906/KIM-1907-30

Synthesis of Pillared Clay Adsorbents and Their Applications in Treatment of Dye Containing Wastewater

Desai Hari and A. Kannan

Abstract Dyes are refectory organic compounds present in aqueous streams and are not easily degradable. Adsorption is one of the efficient and economical processes for the removal of dyes. Many adsorbents have been used for treating dye wastewater but pillared clays are gaining attention in recent times. In this chapter, we have discussed the fundamental features of clays and the desirable outcomes gained after their pillaring with suitable metal polymeric agents. The different compounds, inorganic and organic, incorporated in pillared clays affect the morphological as well as chemical properties of the final product which in turn influences the adsorption performance. The applications of pillared clays for dye removal and the mechanisms involved are discussed. Typical in-house studies are provided in which the improved properties and performance of Al pillared clays are compared to those of raw, i.e., unmodified clays. Finally, application of metal pillared clays for enhanced removal of dyes using different techniques such as photocatalysis and catalytic oxidation is briefly discussed. Areas are suggested where research involving pillared clays may be focused on to aid both fundamental understanding and optimized operation.

Keywords Metal polymers · Pillaring · Organoclays · Pillared clays · Adsorption · Cationic and anionic dyes · Basal spacing · BET surface area

1 Introduction

The textile industry is one of the oldest industries catering to the necessities, comforts, and luxuries of the human race. In 2018, the textile dyes market in the world was $9.4 billion and is expected to reach to value is about $15.5 billion within the next decade. According to Market Report and Forecast [53] the production of dyes in India is 133.52 million tons and is expected to grow by 11% by 2026. Dyeing of clothes became an important component of textile cloth manufacturing as it adds

D. Hari · A. Kannan (✉)
Department of Chemical Engineering, Indian Institute of Technology Madras, Chennai, Tamil Nadu, India
e-mail: kannan@iitm.ac.in

© The Author(s), under exclusive license to Springer Nature Singapore Pte Ltd. 2022   145
A. Khadir and S. S. Muthu (eds.), *Polymer Technology in Dye-containing Wastewater*,
Sustainable Textiles: Production, Processing, Manufacturing & Chemistry,
https://doi.org/10.1007/978-981-19-1516-1_6

esthetic appeal and variety to the clothing material. The quantities of water used in dyeing units are huge. The spent aqueous waste is loaded with heavy metals, dyes, and salts. These may not be directly disposed in water bodies as it may lead to bioaccumulation of toxic substances and these inevitably become unfit for marine life and human consumption. Hence, the effluents from dyeing units need to be treated so that the damage to the environment may be mitigated.

Dye wastewater comprises of refractory organic compounds which being non-biodegradable [49] contribute to the COD value of wastewater. The typical value of COD of dye wastewater is around 12,000 mg/L and BOD value is around 6000 mg/L. If the BOD to COD ratio is less than 0.5, then the wastewater is considered to be non-biodegradable [86]. These dyes being non-biodegradable may not be handled in the secondary stage of the wastewater treatment. The dyes are hence removed in the tertiary treatment stage. Adsorption is an attractive tertiary treatment option owing to its capability in removing organics as well as inorganic pollutants that are present in a very low concentration range [61]. It is also economical, efficient, easy to operate and maintain, and energy-efficient [84]. Further, it does not lead to sludge formation. Based on these advantages, it is a suitable tertiary treatment process for the removal of dyes from wastewater.

The various adsorbents that are commonly used in dye removal include activated carbons, biopolymers (cellulose, chitosan, alginate, etc.), nanomaterial of organic origin, clay minerals, zeolites, metal organic frameworks, industrial waste products (fly ash, waste slurry from processing plants, etc.), and nanomaterial of inorganic origin [84]. Activated carbons are amorphous materials with very high surface area. Depending upon the type of precursor used and treatment method used for activation there may be different functional groups such as hydroxyl, carboxyl, carboxylic anhydride, lactone, lactol, and pyrones on the carbon surface [65]. Biopolymers like chitosan and alginate are useful in the removal of inorganic pollutants such as metal ions due to hydroxyl and nitro groups present in the biopolymer matrix of chitosan and exchangeable ions present in the Ca-alginate complex [89]. Metal organic frameworks are synthetic porous materials with organic ligands coordinatively attached to central metal atoms forming a porous structure. Zeolites are synthetic materials generally made from Al and Si atoms in form of their oxides creating a cage-like porous structure [80]. Among these adsorbents, the choice for the specific treatment requirement is made on the basis of cost, surface area, and suitability for operation in tall packed bed industrial columns, eco-friendliness, ease of processing, and reusability over many adsorption-regeneration cycles. Adsorbent materials involving clays have some distinct features and they meet many of the above requirements. These features of clays and their synthesis will form the scope for further discussions.

2 Fundamentals of Clay and Pillared Clay

2.1 Clay and Clay Minerals

According to Clay Minerals Society, clay is a natural material with (a) very small particle size, (b) which is plastic under hydration, and (c) hardens when calcined at higher temperatures. In general, clay is a finely grained natural rock or soil material which is composed of one or more minerals, metal oxides, and organic matter.

Clay minerals are hydrated phyllosilicates which contain continuous tetrahedral sheets linked to octahedral sheets [14]. When a tetrahedral sheet is linked to an octahedral sheet, such clays are known as 1:1 phyllosilicates (e.g., Kaolinite, Amesite, Odinite, etc.) as can be seen in Fig. 1A. When one octahedral sheet is present in between two tetrahedral sheets, they are known as 2:1 phyllosilicates (e.g. Talc, Saponite, Vermiculite, Biotite, etc.) as can be seen in Fig. 1B. The metals forming the octahedral layer could be typically Al, Mg, Fe, Ni, Zn, and Li and the metals forming tetrahedral layers could be typically Si and Fe. Montmorillonite is a 2:1 type clay which belongs to the Smectite group and is readily available.

Isomorphic substitution involves the replacement of similar atoms with either identical or smaller valence electrons in the clay matrix [14]. When the same valence is replaced there is no change in the net charge on the clay matrix, e.g., replacement of Al^{+3} by Fe^{+3} or Ca^{+2} by Mg^{+2}. However, when there is replacement of atoms with smaller valence states then there arises a net negative charge on the clay layer, e.g., replacement of Al^{+3} by Mg^{+2} or Si^{+4} by Al^{+3}. To balance this negative charge, cations are present in the interlayer space along with water molecules. These cations may

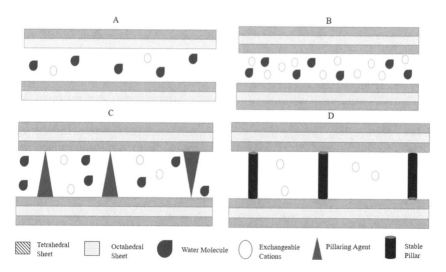

Fig. 1 Schematic representation of clay and pillared clays (**A** 1:1 Phyllosilicates, **B** 2:1 Phyllosilicates, **C** pillaring agent intercalated clay, **D** Clay with stable pillars)

be substituted by other ions from the aqueous solution owing to the unique natural cation exchange property of clay minerals [67].

2.2   Problems Associated with Raw Clay and Solution Through Pillaring

At the molecular level clay may be potentially a good adsorbent owing to its characteristic property of exchanging cations from aqueous solutions. However, the use of clay for industrial applications is beset by many problems which are listed below.

1. Clay layers contain exchangeable cations that are hydrated such as calcium and sodium in the interlayer space, which upon exchange with heavy metal ions in polluted water streams leads to hardness in the treated water.
2. Clay minerals lack permanent porosity, i.e., clay swells upon hydration and collapses upon dehydration. Upon dehydration the distance between the layers of clay reduces drastically. Hence, surface area is not accessible for any transport or reaction processes, leading to poor performance of clay either as an adsorbent or as a catalyst [10, 33, 62].

The swelling characteristics of clay may be arrested by calcination treatment. However, the entire structure may collapse and the surface area will be reduced to a minimum thereby degrading the performance of clays in applications where this property plays an important role. The tendency of clay to form a hard structure upon calcination and leading to permanent porosity may be used without the disadvantage of reduced surface area through the process of pillaring. In this process, the exchangeable cations of the clay layers are replaced with bulky cations in the interlayer space. The metal oxides aggregates of molecular dimensions formed after calcination prevent collapse of the clay layer by forming stable bonds with clay layers [59]. The layer charge gets neutralized by the release of protons thereby imparting Bronsted acidity to the clay material [21].

The metal oxide pillars formed are joined through oxygen bridges to the aluminum and magnesium atoms in the octahedral layer leading to the formation of a rigid cross-linked structure [60]. These measures viz. intercalation of bulky cations through ion exchange, pillaring, and calcination overcome the problems of reduced surface area and enable the clay to acquire permanent porosity. The pore openings thus created are larger than those seen in the usual zeolite cages [13].

Figure 1C represents the schematic diagram of clay with intercalated pillaring agent formed by ion exchange of the pillaring agent with the exchangeable cation naturally present within the clay matrix. Figure 1 D represents the schematic diagram of calcined clay with stable metal pillars which hold the clay layers intact thereby preventing swelling of clays. The following conditions need to be fulfilled for a successful transformation of the raw clay mineral into a pillared one [71].

1. Upon pillaring, the interlayer spacing should increase considerably, which may be characterized in terms of d_{001} spacing obtained through XRD analysis
2. When the pillared material is calcined due to dehydroxylation and dehydration, the interlayer spacing tends to reduce. However, for pillared materials there should not be a significant reduction in interlayer spacing upon calcination
3. The pillared clay must sustain its porosity after ion exchange, calcination, and any other modifications in order to adsorb pollutants.

A pillaring agent is any chemical species which imparts the above-listed characteristics to the parent clay matrix. Depending on the application of the clay material, either as a catalyst or as an adsorbent, the porosity may be tuned by a suitable choice of the pillaring agent. Various pillaring agents which have been used in past include organic cations (polymers and surfactants), organometallic cations, metal clusters, metal oxide sols, and metal polyoxycations [62].

2.3 Process of Pillaring

The process of preparing the pillared clays is facile and comprises of three basic steps as depicted in Fig. 2, namely suspension of clay particles in aqueous solution, preparation of pillaring agent solution, intercalation of pillaring agent in the clay, and finally calcination.

1. Preparation of pillaring agents: Pillaring agents are prepared from their precursors to forms which are suitable for intercalation. In order to achieve specific characteristics like d_{001} spacing, BET surface area, surface charge, etc. of the pillared clay, blends of two or even three metal ions have been used to form a pillaring agent. For example, removal of cationic dyes would require the pillared clay to have a lower point of zero charge and for removal of anionic dyes, a

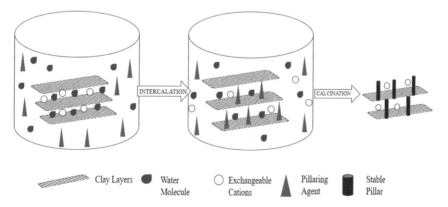

Fig. 2 Steps involved in the pillaring process—intercalation of the metal polyoxy cations followed by the formation of rigid metal oxide pillars through calcination

higher point of zero charge would be preferred. The pillared clays may be further quantified through measurements on surface acidity, crystal planes, surface area, and pore volume. The following factors affect the formation of pillaring species via the hydrolysis technique [11, 33]:

    a.   Nature of precursor salt being used (nitrates, chlorides, phosphates, etc.)

    b.   Degree of hydrolysis, defined as the ratio of the concentration of precursor salts used to the concentration of base, plays a major role. If the ratio is too high, there is a probability of precipitation in the form of metal hydroxides. On the other hand, if the ratio is less, polycations formation may not occur. Hence, an optimum value of hydrolysis ratio is extremely important.

    c.   Operational conditions (temperature, mixing speed, and rate of addition of reactants)

    d.   Aging of the mixed solution.

2.   Intercalation: In this step, there is an exchange of ions naturally present in clay with those of the pillaring agents, e.g., Na^+ ions in clay are exchanged with bulky pillaring agents like Al_{13}^{+7} ions [67]. This may be accomplished through direct mixing of the clay suspension and the solution bearing the pillaring agent or may be enabled through a separating medium such as a dialysis bag. The latter reduces the cost and time of separation for clay and solution but also imparts an extra diffusional resistance for ion exchange to occur. Intercalation is a slow process and the following factors play important roles.

    a.   Time of aging is a crucial parameter in order to achieve the desired degree of intercalation

    b.   Temperature at which intercalation is being carried out

    c.   Concentrations of clay suspension and pillaring species solution

Steps 1 and 2 may be either done in an ex situ mode as described above, i.e., the previously prepared pillaring agent may be added to clay suspension and the intercalation is allowed to take place. The alternative mode of operation involves the in situ mode wherein the metal salt precursor and the hydrolyzing agent are added to the clay suspension directly without forming the polycations first.

The processes of preparation of pillaring agents and their intercalation are however time-consuming and may be accelerated using process intensification techniques such as microwave irradiation [25, 28, 34, 63, 64] and sonication at high frequencies exceeding 20 kHz [24, 46, 75, 79, 82]. These process intensified approaches have proven successful in both the preparations of desired polycations and intercalated clays with analogous properties as those products made from conventional methods. The time required for conventional methods may exceed 24 h at room temperature for both steps 1 and 2. However, the use of microwave and ultrasound reduced the time requirements to around 20 and 30 min, respectively. Typically 100–600 W of microwave power and around 50 kHz ultrasound frequency have been used to intensify the preparation of the polycation and the process of their intercalation.

3. Calcination: Dehydroxylation and dehydration of the intercalated clay occur during this step leading to bond formation between the clay layers and pillaring agents. This ensures permanent porosity. The following factors determine the process effectiveness:

   a. Temperature at which calcination is carried out
   b. Time for calcination
   c. Calcination environment (nitrogen/air/oxygen atmosphere).

After these treatments, the pillared clay has enhanced surface area, higher pore volume and more surface acidity making them suitable for both adsorption and catalysis.

2.4 Types of Pillaring Agents and Their Applications

Broadly pillaring agents could be organic which upon intercalation produce organoclays and inorganic which upon intercalation produce metal oxide pillared clays.

A. Improving organophilic characteristics of clays

Organoclays are prepared through two mechanisms, i.e., one with the use of solvent and the other without solvent [40, 66]. In the former, intercalation occurs through the cation exchange mechanism. When the solvent is absent, the synthesis of organoclay occurs through a solid-state reaction between the clay layers and the organic species to be intercalated. Table 1 summarizes the different organic pillaring agents used in the preparation of organoclay.

Organic pillaring agents such as ionic surfactants increase the d_{001} spacing upon intercalation. The existing cations in clay are exchanged for the surfactant cations in the interlayer region transforming the inherently hydrophilic clay matrix into an organophillic one, thereby increasing its affinity toward organic compounds. Further, the presence of surfactant in the pillaring solution seemed to enable a more uniform pillar distribution [13]. These authors added an alkylated polyethylene oxide surfactant to the pillaring solution and prepared surfactant modified aluminum pillared clay. The modified adsorbent was found to have an improved uptake rate as well as capacity for the nonionic Supranol Yellow 4-GL dye.

Al-pillaring considerably enhanced the basal spacing. It was found to increase from 11.88 nm for Na^+ saturated bentonite to 18.43 nm for surfactant added Al pillared clay. Correspondingly, the BET surface area increased from 110 to 348 m^2/g. The addition of the surfactant did not adversely influence the pillar density and basal spacing. It further contributed to increase in surface area by improving the mesoporosity. The improvement in mesoporosity enabled the dye molecules to diffuse more rapidly into the interlamellar space of the clay to aggregate with themselves and also complex with the Al-pillars.

Table 1 Organic pillaring agents used in the preparation of organoclays

References	Parent clay	Organic pillaring agent
[39]	Sodium montmorillonite Kunipia F	Octadecyl ammonium chloride
[47]	Sodium montmorillonite Kunipia F	Dodecyl trimethylammonium bromide
[87]	Bentonite	Tetradecyl trimethylammonium bromide
[87]	Bentonite	Hexadecyl trimethylammonium bromide
[90]	Montmorillonite	Cetyl trimethyl ammonium bromide
[19]	Sodium montmorillonite Cloisite	Quinolinium and pyridinium salts
[41]	Laponite (synthetic hectorite)	γ-metacryloxypropyl dimethyl methoxy silane
[41]	Laponite (synthetic hectorite)	γ-methacryloxypropyl trimethoxy silane
[68]	Bentonite	Quaternary phosphonium salts like: Tetrabutyl phosphonium bromide Tributyl hexadecyl phosphonium bromide Tributyl tetradecyl phosphonium chloride Tetraphenyl phosphonium bromide Methyl triphenyl phosphonium bromide Ethyl triphenyl phosphonium bromide Propyl triphenyl phosphonium bromide
[77]	Bentonite	Brij 78 Polyethylene glycol ether Dodecyl pyridinium chloride

Dvininov et al. [26] studied the photodegradation of Congo red dye using different clay materials. They used sodium-treated montmorillonite and pillared it with titania in the absence and presence of surfactant. The surfactant CTAB was first added to the clay suspension with the aim of increasing the final interlayer spacing of the pillared clay. The suitably treated clay suspension after surfactant addition was subsequently intercalated with the titanium polyoxy cations. In the absence of the surfactant, the TiO_2 pillars were densely distributed and short. When the surfactant was added first followed by intercalation with metal cations, these species replaced the surfactant cations leading to taller and more isolated pillars. Increasing the surfactant amount increased the basal spacing from 1.23 nm of the raw clay to 1.90 nm for the Ti-pillared clay and this was ascribed to the perpendicular orientation of the organic molecule in the interlayer space.

Zermane et al. [88] modified iron-pillared clay with the 0.5% (w/w) cetyl trimethylammonium bromide (CTAB) surfactant at room temperature. It was used to study the coadsorption of a binary mixture of 4-nitrophenol and a basic yellow 28 dye system. This system is an interesting one as the dye is cationic and nitrophenol dissociates into anions. An unusual synergy rather than competitive adsorption was observed in this system wherein increasing presence of the dye molecules in the solution actually increased the adsorption of the nitrophenol by nearly fourfold. However, the reverse synergy was not that pronounced as increasing presence of nitrophenol had only a slight increase in the adsorption of the dye. A cooperative mechanism was suggested by the authors wherein the neutral/anionic organic combined with the cationic dye molecules and this combined entities actually adsorbed on the modified pillared clay material.

However, when the surfactant-laden clays are heated to higher temperatures during calcination or applied in catalysis at higher temperature above 300 to 400 °C their structures have been found to decompose thereby collapsing the d_{001} or interlayer spacing [10]. Further, the quaternary ammonium salts tend to leach out in presence of dissolved organic pollutants from the clay matrix leading to secondary pollution of the treated water [40]. Qin et al. [70] also observed that surfactant cations weakly bound to the clay tend to leach out from the interlayer space into the solution and cause additional pollution when there are changes in pH. They recommended that grafting of trimethylchlorosilane using solvents of different polarities into Al_{13} pillared montmorillonite led to controlled improvement in the adsorption properties. The previous electrostatic interactions between the sulfonic group of the anionic dye and the metal cations of clay could now be augmented with hydrophobic interactions between the embedded silanes and the aromatic rings of Orange II dye molecules.

B.   Metal polycations as pillaring agents

Metal polycations, when compared to organic intercalating agents are more stable and provide swelling resistance to the clay. Also due to electrostatic interactions between the cations and the anionic portion of organic pollutants such as dyes, their adsorption is enabled (will be discussed in detail in Sect. 3.3). Metal polycations have been the most commonly explored and studied type of pillaring agents. These metal polycations when calcined form strong metal oxide pillars which hold the clay layers together by preventing swelling and collapsing of the structure, and increasing the clays surface area through increase in their basal spacing [33, 67]. The advantages proffered by polycation pillaring agents over other alternatives are briefly summarized below.

The most stable and sustainable pillaring agents are the metal polycations. These polycations modified pillared clays are thermally stable. The metal polycations are prepared from partial hydrolysis of their respective precursors which are metal salts (like Al) using bases such as NaOH and Na_2CO_3 [3]. There are also various other methods for producing metal polycations. The second method is to dissolve Al metal in acid and form Al polycation [4] while the third one involves electrolysis [5]. These may be formed from various metals. Al was the first to be successfully formed as polycations and also the first to be used as a pillaring agent with clays.

There are also other metals like Cr, Fe, Ti, and Zr which have been commonly used as pillaring agents. The elements highlighted in red in the periodic table shown

1 IA	2 IIA	3 IIIB	4 IVB	5 VB	6 VIB	7 VIIB	8	9 VIIIB	10	11 IB	12 IIB	13 IIIA	14 IVA	15 VA	16 VIA	17 VIIA	18 VIIIA
1 H																1 H	2 He
3 Li	4 Be											5 B	6 C	7 N	8 O	9 F	10 Ne
11 Na	12 Mg											13 Al	14 Si	15 P	16 S	17 Cl	18 Ar
19 K	20 Ca	21 Sc	22 Ti	23 V	24 Cr	25 Mn	26 Fe	27 Co	28 Ni	29 Cu	30 Zn	31 Ga	32 Ge	33 As	34 Se	35 Br	36 Kr
37 Rb	38 Sr	39 Y	40 Zr	41 Nb	42 Mo	43 Tc	44 Ru	45 Rh	46 Pd	47 Ag	48 Cd	49 In	50 Sn	51 Sb	52 Te	53 I	54 Xe
55 Cs	56 Ba	57-71 La	72 Hf	73 Ta	74 W	75 Re	76 Os	77 Ir	78 Pt	79 Au	80 Hg	81 Tl	82 Pb	83 Bi	84 Po	85 At	86 Rn
87 Fr	88 Ra	89-103 Ac	104 Rf	105 Db	106 Sg	107 Bh	108 Hs	109 Mt	110 Ds	111 Rg	112 Cn	113 Nh	114 Fl	115 Mc	116 Lv	117 Ts	118 Og

57 La	58 Ce	59 Pr	60 Nd	61 Pm	62 Sm	63 Eu	64 Gd	65 Tb	66 Dy	67 Ho	68 Er	69 Tm	70 Yb	71 Lu
89 Ac	90 Th	91 Pa	92 U	93 Np	94 Pu	95 Am	96 Cm	97 Bk	98 Cf	99 Es	100 Fm	101 Md	102 No	103 Lr

Fig. 3 Periodic table highlighting in red shows the elements that have been used so far as pillaring agents [10]

in Fig. 3 represent the metals that have been already used as pillaring agents to form pillared clays in literature.

Table 2 summarizes different pillaring agents apart from Al used to form pillared clays. Various pillaring agents impart different properties to clay matrix. For a given clay, the effect of type of pillaring agent is manifested in terms of morphological variations in basal spacing, BET surface area, and chemical change in form of point of zero charge as may be seen in Table 2 [36, 56]. These properties may also be influenced by the use of mixed pillaring agents as may be seen in the combination of Fe and Cr metals [7]. A typical procedure often used for preparing the commonly used aluminum polycations is described in the next section.

C. Preparation of Al_{13} polycations: process and reactions

While there are many methods for the preparation of pillaring agents, the most popular one involves the hydrolysis of metal salts. The most commonly used metal polycation is Al in the form of Al_{13} [10]. The following mechanism is postulated for the formation of Al_{13} polycations [29].

The gradual addition of a base, generally NaOH, to a metal salt solution, results in the expected increase in pH of the reactant mixture. Below pH~3, the aluminum cation exists as an aqueo-cation $[Al(OH_2)_6]^{3+}$ [29]. Increase in pH leads to formation of $[Al(OH)(OH_2)_5]^{2+}$ species. Further increase of pH causes hydrolysis to $[Al(OH)_2(OH_2)_4]^+$ cationic monomeric species. Further hydrolysis of simple monomeric hydrolysis products results in uncharged metal hydroxide, $Al(OH)_3$. It may be seen that the oxo-aqua cation's oxidation state reduces with the continued addition of the base. The soluble tetrahedral aluminate $[Al(OH)_4]^-$ is finally formed.

Table 2 Physicochemical properties of pillared clays synthesized from different pillaring agents

References	Host clay	Pillaring agent	Surface area (m^2/g)		pHZC		Basal spacing (d_{001})(nm)	
			UN	PILC	UN	PILC	UN	PILC
[17]	Montmorillonite	Ti polycation	65	216	7.4	6.8	1.05	2.1
[16]	Montmorillonite	Zr polycation	65	226	7.4	5.2	1.05	1.78
[30]	Calcium Bentonite	Cr Polycation	45	184.3	–	–	1.34	1.88
[56]	Bentonite	Fe Trimeric Polycation Mn Trimeric Polycation	52	275 261	7.4	6 6	0.96	1.78 1.71
[6]	Bentonite	Zr Polycation	31.9	42.6	4.2	3.4	–	1.35
[52]	Na Montmorillonite	Fe(III) Modified	1.4	20.7			1.27	1.43
[7]	Bentonite	Fe Polycation	–	–	7.3	3.8	1.53	1.7
[81]	Bentonite	Fe Polycation Fe:Cr = 0.9 Fe:Cr = 0.5 Fe:Cr = 0.1 Cr Polycation	–	157 139 176 202 138		–	–	1.34 1.41 1.46 1.86 1.79
[36]	Smectite	Al Polycation Ti Polycation Cr Polycation Zr Polycation	44.3	176.72 185.40 189.99 198.56	7.32	3.81 3.87 3.88 3.90	1.554	1.909 1.957 1.972 2.004
[72]	Montmorillonite	Al–Fe polycation	41	172	5.2	7.4	–	–
[12]	Kaolinite	Zr Polycation	3.8	13.4	–	–	0.4452	–
[37]	Montmorillonite	Zr Polycations	19.2	35.8	–	–	1.009	1.05

(continued)

Table 2 (continued)

References	Host clay	Pillaring agent	Surface area (m²/g)		pHZC		Basal spacing (d_{001})(nm)	
			UN	PILC	UN	PILC	UN	PILC
[50]	Montmorillonite	Al Polycation Fe Polycation Ti Polycation	54	229 165 249	–	–	1.4	2.0 2.2 2.6
[83]	Montmorillonite	Zr Polycation	128.5	169.2	6.4	3.2	1.26	1.47

The formation of Al dimers and trimers occurs through hydrolysis of monomeric aluminum. The Al dimeric species are formed by OH bridge between two molecules of $[Al(OH)(H_2O)_5]^{2+}$ as seen in Eq. (1) [29].

$$2[Al(OH)(OH_2)_5]^{+2} \rightleftharpoons [Al_2(OH)_2(H_2O)_8]^{+4} + 2H_2O \tag{1}$$

The dimerization of $[Al(OH)(OH_2)_5]^{2+}$ precursor results in edge-sharing dimers. Further hydrolysis by adding one hydrolyzed monomer to the dimer results in a trimeric species, as seen in Eq. (2) [29].

$$[Al_2(OH)_2(H_2O)_8]^{+4} + [Al(OH)_2(OH_2)_4]^{+} \rightleftharpoons [Al_3(OH)_4(H_2O)_9]^{+5} + 3H_2O \tag{2}$$

These trimeric species undergo deprotonation to form $[Al_3O(OH)_3(O_2H_3)_3(H_2O)_3]^{+}$ and the deprotonated product reacts with monomeric species. Four trimer molecules react with monomeric Al species as seen in Eq. (3) [29].

$$[Al(OH_2)_6]^{3+} + 4[Al_3O(OH)_3(O_2H_3)_3(H_2O)_3]^{+}$$
$$\rightleftharpoons \left\{Al\left[[Al_3O(OH)_3(O_2H_3)_3(H_2O)_3]^{+}\right]_4\right\}^{+7} + 6H_2O \tag{3}$$

The final Al_{13} polycations are formed by intramolecular condensation by olation as seen in Eq. (4) [29].

$$\left\{Al\left[[Al_3O(OH)_3(O_2H_3)_3(H_2O)_3]^{+}\right]_4\right\}^{+7} \rightleftharpoons [Al_{13}O_4(OH)_{24}(H_2O)_{12}]^{7+} + 12H_2O \tag{4}$$

The four trimers are coordinated to monomeric aluminum atom to form tetrahedral structure. These steps represent the process of formation of Al_{13} polycations in

Fig. 4 Keggin structure of Al_{13} polycation [Red: O atoms, Pink: Al atoms, White: H atoms, created using Gaussian software]

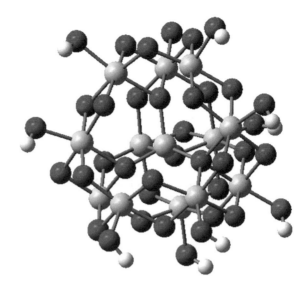

aqueous solutions [29]. Further, other polymeric aluminum polycations such as Al_{30} may be formed during the synthesis process as well [15].

For Al polycations, the hydrolysis ratio (OH^-/Al^{+3}) is optimal at 2.4. The formation of Al polycations may be detected and confirmed from NMR studies which determine the state of the aluminum present in the polycation solution as well as in the clay post-intercalation [31].

Figure 4 shows the atomic arrangement of Al_{13} polycation. This structure is referred to as the Keggin structure. Here the central Al is surrounded by 12 Al atoms forming the tetrahedral structure.

D. Polymerization with other metal precursors for pillaring of clay

Similar to the synthesis of Al polymers, other metal polymers may also be prepared. Ga_{13} polymers are formed through hydrolysis of Ga salts and are stable with structures similar structure to Al_{13} polycations [20].

For Zr polycations, precursor salts such as zirconium nitrates and oxychlorides are used. Naturally zirconium exists as tetramers $[Zr_4(OH)_8(H_2O)_{16}]^{+8}$ in aqueous solutions and hence hydrolysis is not required to be performed [16, 54]. If further polymerization of Zr is required then hydrolysis may be carried out similar to Al.

Fe polycations are also prepared through hydrolysis of Fe nitrate and chloride salts. Fe_{13} polycations $[Fe_{13}O_4(OH)_{24}(H_2O)_{12}]^{7+}$ have been prepared and are structurally similar to Al_{13} polycations, but which are unstable and decompose [62]. Hence, organic complexes such as Fe acetate [55] have been proposed as intercalating agent to form pillared clays. Fe and Sn polycations could be analyzed through Mössbauer Spectroscopy [32].

Ti polycations are prepared through hydrolysis of Ti chlorides and nitrates. Polymer of Ti containing eight atoms of Ti $[TiO_8(OH)_{12}]^{4+}$ is also been prepared and used as intercalating agent to form Ti-pillared clays [62].

Polymers of Cr are prepared through hydrolysis of Cr chlorides and nitrates [69]. For Cr polycations, the hydrolysis ratio (OH^{-1}/Cr^{+3}) is optimal at 2. Polymerization of Cr^{+3} cation leads to the formation of the dimer $[Cr_2(OH)_2(H_2O)_8]^{4+}$, the trimer $[Cr_3(OH)_4(H_2O)_9]^{5+}$, and the tetramers $[Cr_4(OH)_6(H_2O)_{11}]^{6+}$ and $[Cr_4(OH)_5O(H_2O)_{10}]^{5+}$ [31]. Formation of Cr polycations could be characterized through UV analysis [32].

The abovementioned are the few commonly used polymeric metal pillaring agents. Further, organometal complexes such as Mn acetate [57] and Cr acetate [58] have also been employed as pillaring agents to form metal pillared clays. Mixed metal pillaring agents co-hydrolysis is carried out from respective precursor metal salts. For example, to form Fe-Cr-pillared clays, salts of Fe and Cr are mixed and they undergo hydrolysis to form a complex (Fe-Cr) containing the two atoms [81].

E.   Comparison of organo and metal pillared clay structures

A schematic representation of the atomic structures of organo and metal pillared clays has been depicted in Fig. 5. For organoclay, the surfactant hexadecyltrimethylammonium bromide (HDTMA) has been used as the intercalating agent. In metal pillared clays, Al_{13} polycations have been used. As may be seen from Fig. 5, the exchangeable Na ions are substituted by the respective intercalating agent and hence the Cation Exchange Capacity (CEC) of the original clay is reduced post pillaring. It may also be seen that there are no water molecules present in the clay post pillaring, as dehydration occurs during the calcination step. Further, the basal spacing is increased post pillaring in both organoclays and metal pillared clays.

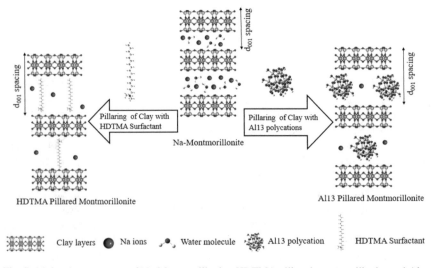

Fig. 5 Molecular structures of Na Montmorillonite, HDTMA pillared montmorillonite and Al_{13} pillared montmorillonite (Generated using Gaussian Software and Crystallographic data from American Mineralogist Crystal Database)

3 Applications of Metal Pillared Clays for Treating Dye-Containing Wastewater

After pillaring the parent clay it was observed that its adsorption capacity had increased. This may be attributed to the increase in basal spacing by intercalation of metal polycations which in turn enhances the surface area. There is a strong correlation found between d_{001} spacing and BET surface area upon pillaring with different metal polycation species. This is illustrated in Fig. 6 where intercalation with Fe [55], Cr [58], and Mn [57] species is involved.

3.1 Application of Al Pillared Clays for Adsorption of Dyes from Textile Wastewater

The beneficial effects of pillaring the clay matrix with Al polycations viz. enhancement in basal spacing, surface area, and pore volume was observed by [1, 13, 15, 31, 38, 44]. The details are summarized in Table 3. Owing to these enhancements, the adsorption capacity increased. Further, pillaring retarded the swelling of the clay as

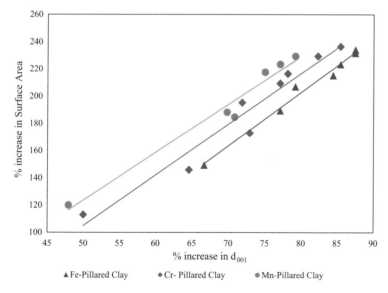

Fig. 6 Correlation between d_{001} spacing and surface area for different pillaring agents applied on the same raw clay

Table 3 Morphological properties of Al pillared clays

References	Host clay	Surface area (m²/g)		Total pore volume (cm³/gm)		Micro pore volume (cm³/gm)		Basal spacing (d_{001})(nm)	
		UN	PILC	UN	PILC	UN	PILC	UN	PILC
[44]	Tunisian Bentonite	61.74	164.64	0.12	–	0.068	–	–	–
[15]	Montmorillonite	94	108	0.21	0.166	0.015	0.029	1.26	1.63
*[15]	Montmorillonite	94	125	0.21	0.173	0.015	0.035	1.26	1.69
[1]	Montmorillonite	43	304	0.08	0.17	0.010	0.100	1.26	1.86
[38]	Bentonite	10.15	200.1	0.03	0.152	–	–	1.54	1.897
[32]	Montmorillonite	9	273	0.047	0.136	0.002	0.097	1.05	1.64
[13]	Bentonite	110.07	220.67	–	–	–	–	1.188	1.8592

* The pillaring agent was Al_{30} polycation and for remaining entries it was Al_{13} polycation
[UN: unmodified clay, PILC: Pillared Interlayer Clay]

well. As the pore size was enlarged by the presence of large polycations, the adsorption of dye molecules was facilitated. Kacha et al. [73] observed that aluminum polymers incorporated clays could remove 72–88% of the industrial dyes. The increase in adsorption capacity post pillaring is summarized in Table 4.

Butman et al. [15] observed that pillaring greatly enhanced the individual adsorption of both acid red and methylene blue dyes on the clay when compared to the un-pillared clay as given in Table 4. However, the best fitting isotherm was different for the two dyes. The adsorption capacity after pillaring nearly doubled for methylene blue dye and by about 3.5 times for acid red dye [1].

Adsorbents are typically evaluated in terms of their equilibrium and kinetic characteristics. The former is a measure of the adsorbent capacity while the latter characterizes its uptake rate for the specific adsorptive present in the aqueous solution. In this section, these characteristics pertaining to the adsorption of different dyes on Al pillared clays are discussed. Equation (5) represents the Langmuir isotherm, Eq. (6) represents the Freundlich isotherm, Eq. (7) represents Henry isotherm, and Eq. (8) represents Redlich Peterson equation.

$$q_e = \frac{q_{max} K_L C_e}{(1 + (K_L\ C_e)}\tag{5}$$

$$q_e = K_F C_e^n\tag{6}$$

$$q_e = K C_e\tag{7}$$

$$q_e = \frac{K_{RP}\ C_e}{(1 + (\alpha_{RP}\beta)}\tag{8}$$

Table 4 Al pillared clay adsorbents for dye removal [*Pillaring agent is Al_{30} and for the rest it is Al_{13}]

References	Pollutant	Operating pH for maximum adsorption	q_m (mg/g) UN	q_m (mg/g) PILC	Isotherm	Kinetics
[44]	Methylene blue	6	62.50	166.67	Langmuir $K_L = 0.66$ L/mg	PFO $k_1 = 0.027$/min
[15]	Acid red Methylene blue	~3 ~13	–	4.4 5.4	Langmuir $K_L = 0.19$ L/mg Freundlich $k_F = 5.26$, n = 5.88	PSO for both dyes $k_2 = 0.210$ mg/g min $k_2 = 0.034$ mg/g min
*[15]	Acid red Methylene blue	~3 ~13	–	4.4 6.8	Langmuir $K_L = 0.12$ L/mg Freundlich $K_F = 5.98$, n = 4.54	PSO for both dyes $k_2 = 0.034$ mg/g min $k_2 = 0.320$ mg/g min
[1]	Reactive black 5 Methylene blue	2 12	12 25.4	40.8 49.4	Langmuir $K_L = 0.0207$ L/mg Langmuir $K_L = 0.0476$ L/mg	– –
[38]	Acid Turquoise blue A Basic fuchsin Basic green	~0.8 ~6.8 ~10	–	2.639 45.54 24.63	Henry's law $K = 0.4271$ L/g Freundlich $K_F = 2.797$, n = 0.9158 Henry's law $K = 10.99$ L/g	PSO for all 3 dyes $k_2 = 0.1367$ mg/g min $k_2 = 0.6448$ mg/g min $k_2 = 1.133$ mg/g min

(continued)

Table 4 (continued)

References	Pollutant	Operating pH for maximum adsorption	q_m(mg/g) UN	q_m(mg/g) PILC	Isotherm	Kinetics
[31]	Acid orange 7 Methylene blue	7 7	–	8.9 21	Langmuir K_L = 0.0047 L/mg Freundlich K_F = 2.2, n = 2.6	PSO for both dyes k_2 = 0.040 mg/g min k_2 = 0.0017 mg/g min
[13]	Supranol yellow 4-GL	~3.1	–	63.93	Langmuir K_L = 0.2 L/mg	PFO 0.079/min

From Table 4 it is evident that Langmuir [1, 13, 15, 31, 44] and Freundlich models [15, 32, 38] are the most applicable to characterize the equilibrium uptakes of the clays and the corresponding isotherm parameters are mentioned in the table. The Langmuir constant K_L describes the strength of the interaction between the adsorbate and adsorbent. The parameter q_m is the maximum capacity of the adsorbent for a particular solute. In case of the Freundlich isotherm, K_F is the constant which is a measure of the adsorbent's affinity for the targeted species. Basic (cationic) dyes seem to have more affinity toward the pillared clay than acid dye as indicated by values of the respective (K_F).

Pseudo-first and pseudo-second-order models (PFO and PSO) are frequently used to describe the kinetics of the adsorption process. Equation (9) represents pseudo-first order kinetic equation and Eq. (10) represents pseudo-second order kinetic equation. PSO is favorable for adsorption of dyes as noted from the works of [15, 32, 38].

$$\frac{dq_t}{dt} = k_1(q_e - q_t) \tag{9}$$

$$\frac{dq_t}{dt} = k_2(q_e - q_t)^2 \tag{10}$$

PFO is favorable for adsorption of dyes as noted from the works of [13, 44]. Constants k_1 and k_2 for PFO and PSO, respectively, determine the rate of the process. Basic dyes have faster kinetics when compared to acid dyes as summarized in Table 4. There is no specific trend between the type of dyes and the PFO or PSO kinetic model appropriate for them. These models are rather simplistic and the q_e values estimated by them are often not in good comparison with the actual equilibrium capacities of the adsorbent when starting from a given initial dye concentration in the aqueous solution. In this connection, models derived from first principles involving external convective mass transfer in the aqueous solution and intra-particle diffusion in the adsorbent would shed more insight into the kinetics as well as interfacial equilibrium loading of the solutes in the adsorbent. However, these models come at the cost of additional computational effort.

3.2 Application of Pillared Clays from Metals Other Than Aluminum for Adsorption of Dyes

Pillaring of clay minerals with a mixture of two different metal polycations has also been carried out, where the first is Al while the second may be different inorganic cations that are added in various molar fractions to improve porosity, thermal stability, adsorptive capacity, and catalytic properties [10]. Further, the porosity of the PC may also be easily be tuned with two metals depending upon its applications. In [15] Al_{13} pillared clays were modified with Ce atoms. Ce was introduced in Al_{13} structure by hydrolyzing Ce and Al salts together. The Ce modified clays were observed to have higher surface area and basal spacing when compared to Al_{13} alone and Al_{30} pillared clays as may be seen from Tables 3 and 5. Also the clay properties could be modified by doping the atoms onto the clays before calcination.

González et al. [35] prepared Ti polycation and used it as a pillaring agent. Further, Ti polycations were doped by incorporating other metals such as Ag, Nd, In, Fe, Cu, and Cr into the clay matrix to form pillared clays with different morphological properties. This was accomplished by adding these dopant metal salts in the Ti salt solution before intercalating the ions into the clay matrix. The major reason for doping was to increase the surface acidity of the Ti-pillared clays. Enhanced acidities were confirmed by evaluating the Lewis and Bronsted acid sites on the clay through pyridine adsorption studies followed by FTIR analysis. The pillared clay obtained with doping of Cu had the highest surface area and micropore volume among the different dopants. Clays doped with Ag, Nd, and In had a relatively lower number of acid sites when compared to those doped with Cr, Cu, and Fe. Pillaring by incorporating different polycations in the clay matrix leads to an increase in basal spacing and corresponding increase in surface area and pore volume. This has been summarized in Table 5, where metals such as Ce, Cr, Ti, Ag, Nd, etc. have been used apart from Al.

Both equilibrium and kinetic aspects for adsorption of various dyes on different metal pillared clays are summarized in Table 6. The affinity coefficient was not distinctly different for acid and basic dyes for pillared clays as it was for Al pillared clays as discussed in Sect. 3.1.

3.3 Major Factors Determining the Performance of Pillared Clays

Factors that influence the intrinsic kinetics and equilibrium capacity of the adsorbent (pH, temperature, physicochemical characteristics of the clay, and mass of adsorbent per unit volume of the treated solution, i.e., the adsorbent dosage and effect of salt concentrations) play important roles in the performance of pillared clay adsorbents. pH especially has a distinct role to play when dealing with aqueous systems.

Results from kinetics and equilibrium studies have been summarized in the previous sections. The effects of pH and salt concentrations are discussed next.

A. Effect of pH on adsorption of ionic dye species on pillared clays

Materials have surface charge owing to the presence of edge functional groups (generally hydroxyl groups in both organic as well as inorganic adsorbents) [85]. As these functional groups are present on the edge of adsorbent, there arises charge on the surface which may change under different pH conditions. The pH when the net charge on the surface is zero is referred to as point of zero charge (pHZC). When the pH of the solution is below pHZC there arises a net positive charge on the surface [85]. When the pH of the solution exceeds the pHZC there is a net negative charge on the surface. The point of zero charge may be different for different adsorbents depending

Table 5 Morphological properties of pillared clays with different intercalating agents

References	Host clay	Pillaring agent	Surface area (m²/g)		Total pore volume (cm³/gm)		Micro pore volume (cm³/gm)		Basal spacing (d$_{001}$)(nm)	
			UN	PILC	UN	PILC	UN	PILC	UN	PILC
[45]	Natural clay of Darbandikhan	Ce-Al$_{13}$	22.39	53.2	0.0652	0.0877	0.00503	0.0121	1.43	2 to 3
[22]	Natural clay	Cr Polycation	41.626	130.555	0.067	0.128	–	–	1.508	1.6298
[15]	Montmorillonite	Al-Ce	94	154	0.21	0.161	0.015	0.033	1.26	2.13
[35]	Montmorillonite	Ti polycation	80	164	–	–	0.009	0.047	0.957	1.860
		(a) Ti Doped with Ag (b) Ti Doped with Nd (c) Ti Doped with In (d) Ti Doped with Fe (e) Ti Doped with Cu (f) Ti Doped with Cr		(a) 219 (b) 211 (c) 235 (d) 300 (e) 329 (f) 272				(a) 0.064 (b) 0.059 (c) 0.072 (d) 0.103 (e) 0.108 (f) 0.082		(a) 1.794 (b) 1.938 (c) 2.046 (d) 1.699 (e) 1.694 (f) 1.603
[43]	Bentonite	Zr Polycation			–	–	–	–	1.0063	1.5684
[48]@	Montmorillonite (a) Tagansk (b) Polyana (c) Latnensk	Al-Co Polycation	(a) 72.9 (b) 70.7 (c) 31.5	(a) 155.5 (b) 72.5 (c) 35.4	(a) 0.063 (b) 0.081 (c) 0.084	(a) 0.084 (b) 0.077 (c) 0.082	(a) 0.016 (b) 0.008 (c) 0.002	(a) 0.065 (b) 0.020 (c) 0.003	–	–
[32]	Montmorillonite	Zr Polycation	9	356	0.047	0.186	0.002	0.128	1.05	1.91
[42]	Bentonite	Fe polycation	7.5	36.8	0.028	0.042			1.47	1.51

@ Tagansk, Polyana, and Latnesk refer to the different sources of clay being used

Table 6 Pillared clays prepared from different combinations of metal cations and used as adsorbents for dye removal

References	Type of pillared clay	Pollutant	pH	q_m (mg/g)		Isotherm	Kinetics
				UN	PILC		
[45]	Ce-Al$_{13}$ pillared clay	Methyl orange	~3.2	–	248.7	Redlich Peterson $K_{RP} = 1.66$ L/mg $\alpha_{RP} = 5.2 \times 10^{-12}$ L/mg $\beta = 5.8$	PSO $k_2 = 0.00045$
[15]	Al-Ce pillared clay	Acid red Methylene blue	~3 ~13	4.4 6	9.1 30.4	Langmuir $K_L = 0.11$ L/mg $K_L = 0.14$ L/mg	PSO for both dyes $k_2 = 0.06$ g/mg min $k_2 = 0.120$ g/mg min
[35]	Ti-pillared clay Ti-pillared clay doped with Ag Ti-pillared clay Doped with Nd Ti-pillared clay Doped with In Ti-pillared clay Doped with Fe Ti-pillared Clay Doped with Cu Ti-pillared clay Doped with Cr	Methylene blue	–	30.9	31.8 39.3 45.2 32.8 32.8 58.2 23.6	– – – – $K_L = 0.022$ mL/mg – $K_L = 0.088$ mL/mg	– – – – PSO $k_2 = 0.24$ g/g min – PSO $k_2 = 0.32$ g/mg min
[43]	Zr-pillared clay	Methyl orange	~3.5	-	44.13	Langmuir $K_L = 0.0228$ L/mg	–
[48]	Al-Co pillared clay Tagansk Al-Co pillared clay Polyana	Azorubine	–	0.05 0.16	0.07 0.26	–	–
[48]	Al-Co pillared clay Tagansk Al-Co pillared clay Polyana	Methylene blue	–	1.17 1.05	1.18 1.18	–	–
[31]	Zr-pillared clays	Acid orange 7 Methylene blue	7 7	–	38 27	Langmuir $K_L = 0.0043$ L/mg Freundlich $K_F = 7$ n = 3.6	PFO for both dyes $k_1 = 0.103$/min $k_1 = 0.019$/min

(continued)

Table 6 (continued)

References	Type of pillared clay	Pollutant	pH	q_m (mg/g)		Isotherm	Kinetics
				UN	PILC		
[42]	Fe-pillared clay	Rohdamine B	5	–	98.62	Langmuir $K_L = 3.63$ L/mg	PSO $k_2 =$ 1.08 mg/mg min

on the functional groups present in them. The addition of positively charged metal polycations that form the metal oxide pillar results in lower point of zero charge. The metal polycations would be highly acidic which in turn increases the overall acidity of the pillared material and hence the pHZC is reduced drastically. This effect is seen in Table 2 [7, 16, 17, 36, 56, 83]. The extent to which the pHZC is reduced depends upon the nature of pillaring species used and the amount of pillaring species present within the clay matrix as may be seen in Table 2. Hence, depending upon the type of application, the appropriate intercalating agent may be chosen.

The pK_A for the adsorptive also plays an important role. The acidic functional groups get substantially deprotonated when the solution pH exceeds the pK_A of the adsorptive. On the other hand, the basic solutes in the solution get substantially protonated below the pK_A. The electrostatic interactions between the adsorptive and adsorbent may influence the extent of adsorption depending on the charges possessed by them at different pH conditions.

For a cationic species the adsorption will be favored at pH values above pHZC as the surface will be negative. Adsorption will be not favored if the system is operated below pHZC due to electrostatic repulsion. This effect was observed for adsorption of methylene blue dye [1, 15, 32], basic Fuchsin and basic green [38], and Supranol Yellow 4-GL dye [13] on Al pillared clays. This effect was also observed for adsorption of methylene blue dye on Ce [15], modified Al pillared clays and Zr-pillared clays [31].

For anionic species, adsorption will be favored at pH values below the pHZC as the surface will be positive and adsorption will not be favored if the system is operated above pHZC. Further, there could be competition between the anionic species and abundant hydroxyl ions in the solution at higher pH leading to reduced adsorption. This effect was observed for adsorption of acid red dye [15], reactive black 5 [1], and Acid Turquoise Blue A [38] over Al pillared clays. This was also observed in adsorption of methyl orange dye on Ce-Al pillared clays [45] and Zr-pillared clays [43], acid red dye [15] on Ce modified Al pillared clays and acid orange 7 dye on Zr-pillared clays [32]. As may be seen from Tables 4 and 6 we observe that the operating pH of adsorption was determined according to the point of zero charge of adsorbent and pKa values in Table 7 of the adsorptive in order to enhance the adsorption performance.

When no charged species are present, adsorption may still occur due to van der Waals and π-π interaction between the clay surface and dye molecules. This was

Table 7 pKa values of dyes

References	Dye	pKa
[18]	Methylene blue	3.8
[18]	Rhodamine B	3.7
[76]	Methyl orange	3.47
[2]	Acid orange 7	11.4 & 1
[76]	Acid red	11
[78]	Basic green	6.5

observed for adsorption of Acid Turquoise Blue A [38], Acid Orange 7 [31] on Al pillared clays.

Figure 7 is a schematic representation of the surface charge at pH when it is above the pHZC and below the pHZC. Also it shows the electrostatic interaction with the charged pollutants under different pH conditions.

Metal ions tend to precipitate out at higher pH. When the clay layers and metal polycations are not held together by oxide bridges then there are chances for leaching out of metals from pillared clay matrix. Leaching of Fe ions was observed from pillared clay when used under higher pH conditions [42]. The dye adsorbed was Rhodamine B. At pH lower than 3, the dye existed at +1 and +2 ionic states. These faced competition with hydrogen ions and the adsorption capacity was lowered. At higher pH the dye exists in the basic form. The lactone group present caused decrease adsorption due to electrostatic repulsion between the dyes and the adsorbent and also resulted in the leaching out of the Fe ions. At around pH 5, the adsorbent was stable as there was no leaching observed and the dye adsorption was also reasonable and hence the operating pH was at 5 as seen in Table 6.

Zermane et al. [88] studied both single and binary adsorption in surfactant modified iron-pillared clay. The adsorption of both nitrophenol and basic yellow 28 dye

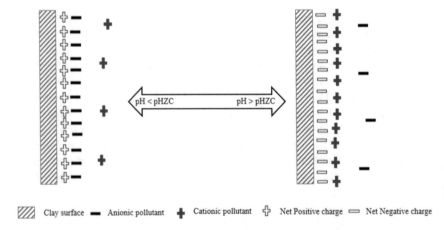

Fig. 7 Adsorbent-adsorptive interaction at different pH conditions relative to pHZC

increased when the pH was increased from 5 to 9. Zeta potential measurements at pH 6.5 indicated a positive charge on the adsorbent indicating that at least until this pH, the phZC of the adsorbent was not exceeded. Non-polar interactions led to adsorption of the undissociated nitro phenol (pKa = 7.15) at pH 5. At pH higher than 7.5, the nitrophenolate ions may have been adsorbed through electrostatic interactions with the positively charged adsorbent. However, the exact point of zero charge was not mentioned in this study. The dye adsorption was also attributed to donor–acceptor interactions between the free electron pairs of the nitrogen and the positively charged adsorbent surface. Also there may have been donor–acceptor interactions between the aromatic groups of the dye and the modified clay as well as hydrophobic interactions involving nonpolar forces between the alkyl groups of the surfactant modified pillared clay with the molecules of the dye.

Even though the adsorption is strongly dependent on phZC, many literature studies have not measured and discussed this important chemical property for pillared clays. More information on this parameter will be very helpful to get accurate insights on the governing mechanisms and help identify suitable pH for maximum adsorption. The closer the operation is to natural pH, the better as adding acid or base to maintain pH will come at additional cost and usage of chemicals.

B. **Effect of salt concentration on adsorption of ionic dye species on pillared clays**

In dyeing, salts are also used in baths. About 80% of the total salts that were used get discharged in wastewater from textile industries with high TDS values of about 5000–7000 mg/L [1]. NaCl and Na_2SO_4 are the most common salts used in textile industries and are present in wastewater. The presence of dissociated salt ions greatly affects the adsorption of ionic dye species. Hence it becomes vital to study the role played by salts in adsorption process.

For cationic dyes such as methylene blue [1], the adsorption capacity of Al pillared clay was increased when NaCl salt concentration was increased from 0.0 to 1.0 mol/L. In the absence of salt content, the maximum adsorption capacity was 49.4 mg/g. When the salt content increased to 1 mol/L, the adsorption capacity increased nearly tenfold to 474.9 mg/g. The authors suggested that the tremendous increase in adsorption capacity was due to formation of dye dimers and trimer at high salt concentrations. Fernández-Pérez and Marbán [27] observed that even tetramers may get formed at higher salt concentrations. They have reported that the adsorption of the dye occurs only on the surface and not inside the pores as the pore size of Al pillared clay is about 1.3 nm and the dye size is about 1.4 nm. As the adsorption occurs on external surface, larger aggregates could be accommodated on the same area and hence the adsorption was higher. At low salt concentrations, there are no aggregates of dye in the solution and hence the amount of dye adsorbed is less.

For anionic dyes such as reactive black 5, the adsorption capacity on Al pillared clay generally decreased upon increasing salt concentration from 0.0 to 1.0 mol/L [1].

The adsorption capacity was 40.8 mg/g when salt concentration was zero increased slightly to 43.0 mg/g at 0.1 mol/L salt concentration but reduced by about 7 times to 6.15 mg/g at 1.0 mol/L salt concentration. At low salt concentration range the negative charge in the clay is neutralized by salt cations and hence we observe the increase in adsorption capacity of anionic dyes up to 0.2 mol/L salt concentration. Similar trend was observed for adsorption of acid orange 7 dye on Al and Zr-pillared clays [32]. When the salt concentration was increased from 0.0 to 0.1 M, adsorption was increased from 4.3 mg/g to 38 mg/g for Zr-pillared clay and from 4.1 to 4.3 mg/g for Al pillared clay. More adsorption capacity was observed in Zr-pillared clays owing to higher surface area when compared to Al pillared clays as may be seen in Tables 5 and 3, respectively. But at very high salt concentration of 1.0 mol/L, the ions block the charged sites on the clay surface and also neutralize the dye molecular charge which leads to reduced electrostatic interaction and reduction in adsorption capacity [1].

Hence, the adsorption capacity may be greatly enhanced or suppressed depending upon the nature of dyes, pH of solution, and the concentration of salt. This dependence may be used to enhance the performance of dye wastewater treatment plants by optimizing the significant process parameters.

4 Typical Results from the In-House Characterization of Pillared Clays

Pillared clay was prepared from bentonite procured from Loba Chemie, India using Al polycations in ultrasound (at 35 kHz) assisted process for 30 min and then calcined at 500 °C for 2 h. Al polycations were prepared through partial hydrolysis of $AlCl_3.6H_2O$ salt with NaOH at $OH^-/Al^{+3} = 2.4$. Pillared clays are usually characterized through XRD, TGA, and EDX analyses. Typical in-house results obtained from Al_{13} pillared clays are presented in this section. From X-ray diffraction (XRD) analysis given in Fig. 8, it is evident that the basal spacing (d_{001}) is increased after pillaring the clay with Al_{13} polycations from 1.12 nm to 1.50 nm.

From Fig. 9 Thermogravimetric Analysis (TGA), it was evident that the prepared pillared clay is thermally stable and the percentage loss of mass is only about 7% whereas for raw clay it is around 17%, which confirms the thermal stability of the Al pillared clay.

From Energy Dispersive X-Ray Analysis (EDX) in Fig. 10 it was seen that the percentage Al was increased post pillaring and the percentage of exchangeable cations such as Na and Mg decreased as these were replaced by Al polycations in the clay matrix.

Figure 11 shows the adsorption capacity for acid orange dye, with pillared clay exhibiting 9 times increase in adsorption when compared to raw clay. Adsorption was carried out at natural pH of dye solution, at 298 K, and dosage of 1 g/L with initial dye concentration of 100 mg/L.

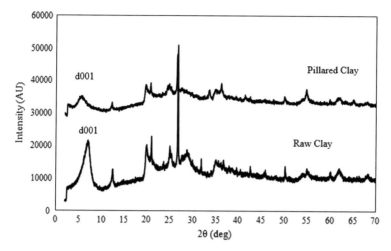

Fig. 8 XRD analysis of raw and pillared clays

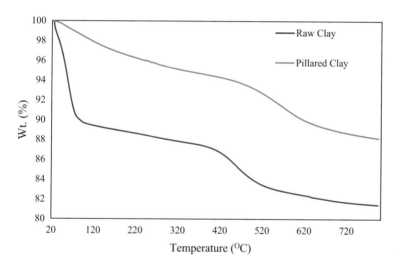

Fig. 9 (TGA) of raw and pillared clays

5 Enhanced Performance of Pillared Adsorbents for Treating Dye Using Pillared Clays

For treating dye wastewater, pillared clays have been augmented with other tech-niques such as photocatalysis [51, 74] and photo Fenton [9, 23]. Catalytic wet air oxidation and catalytic wet peroxide oxidation using pillared clays prepared from Al–Fe, Al-Cu, and Cu metals have been used for the treatment of dyes from textile industries [8]. The adsorption capacity of methyl orange dye was 68.1 mg/g and

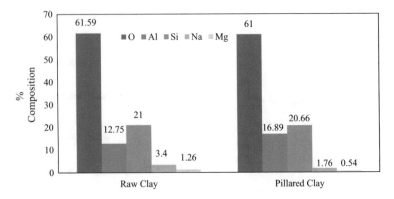

Fig. 10 Comparison of elemental compositions from EDX Analysis of raw and pillared clays

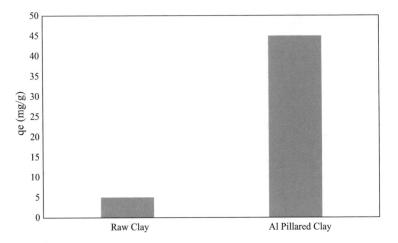

Fig. 11 Comparison of adsorption capacity of raw and pillared clays

248.7 mg/g for acid-activated clay and Ce-Al pillared clay [45]. Tremendous increase in the removal of the dye was enabled when adsorption was accompanied by catalytic degradation of dye molecules due to presence of active metals such as Ce in the pillared clay matrix [45].

Dvininov et al. [26] showed that surfactant-modified Ti-pillared clays exhibited high photodegradation efficiency for the Congo red dye. Rather than the amount of Ti present in the photocatalyst, it was the contact between the dye and the catalyst as well as the adsorptive capability of the photocatalyst which determined the photodegradation.

Ti-pillared clay along with several doped metals such as Cu, Cr, Ag, In, Nd, and Fe was used as a photocatalyst for degradation of methylene blue dye with UV source [35]. With UV light source and the doped and pillared clay, it was observed that almost 99% removal of dye had occurred. Without UV light, the removal was only around 30 to 60% due to physical adsorption as mentioned in Table 6. The doped species were not only contributing as active photocatalysts but also promoting adsorption of dye molecules due to the specific acidic sites on the clay surface. Also, it was observed that there were no secondary peaks of the treated solution in the UV Visible spectrum which may confirm that there was complete degradation of the dye and no intermediates were formed.

6 Conclusions

After introducing the fundamentals of clays, the efficacy of pillaring them with different metals to improve their adsorption properties were discussed. It was found that organic modification on clay was not adequate to form stable pillared material due to lack of thermal stability and possibility of secondary pollution to environment. Metal polycations on the other hand formed stable metal oxide pillars thereby preventing the swelling of clays. There was increase in basal spacing which lead to increase in surface area and pore volume post pillaring and calcination. This increased in surface area and pore volume enhanced the adsorption capacity many folds to the clay mineral. Pillaring not only modified the morphology of clay but also the surface chemistry by reducing the point of zero charge due to addition of more acidic sites from metal pillars. This reduced point of zero charge allowed a wide range of working pH for different nature of dyes.

Pillared clays have been successfully used for adsorption of single dye from simulated solutions. But for applying them in real-life applications multicomponent dyes adsorption studies are important as the wastewater from dyeing or textile industries are expected to contain mixture of dyes. Textile wastewater is often polluted with heavy metal ions. Hence, adsorption studies involving the simultaneous uptake of metal and dye may also be more extensively carried out.

Suitably designed experiments may provide insights into the interactions between the process variables as well as with the dye molecules. More studies are required on synergetic or antagonistic effects during simultaneous adsorption of more than one dye. Response surface methodologies may be applied to optimize pillared clay adsorption processes used to treat both synthetic and real effluents. Mechanistic models may also be developed based on external convective mass transfer and intra-particle diffusion limitations to understand the kinetics of dye adsorption on pillared clays. These may also require robust and accurate multicomponent adsorption isotherms generated over a wide range of pH, temperature, and nature of adsorbent material.

The accumulated body of knowledge on the preparation of various metal pillared clays in literature provides the basis for investigating different combinations of clays and metals to produce applications-specific layered structured materials. The development and applications for such materials have to be based on sufficient theory that includes robust isotherms, mechanistic kinetic models, and molecular level understanding, e.g., through DFT simulations. It is recommended that the various pillared clay materials may extensively demonstrated for real effluents and in continuous columns as well so that their potential in industrial applications may be understood.

References

1. Aguiar JE, Bezerra BTC, Siqueira ACA, Barrera D, Sapag K, Azevedo DCS, Lucena SMP, Silva IJ (2014) Improvement in the adsorption of anionic and cationic dyes from aqueous solutions: a comparative study using aluminium pillared clays and activated carbon. Sep Sci Technol 49(5):741–751. https://doi.org/10.1080/01496395.2013.862720
2. Akazdam S, Chafi M, Yassine W, Gourich B (2017) Removal of acid orange 7 dye from aqueous solution using the exchange resin amberlite FPA-98 as an efficient adsorbent: kinetics, isotherms, and thermodynamics study, p 20
3. Akitt JW, Farthing A (1981) Aluminium-27 nuclear magnetic resonance studies of the hydrolysis of Aluminium III. Part 2. Gel-permeation chromatography. Dalton Trans 7:1606–1608
4. Akitt JW, Farthing A (1981) Aluminium-27 nuclear magnetic resonance studies of the hydrolysis of Aluminium (III). Part 4. Hydrolysis using sodium carbonate. Dalton Trans 7:1617–1623
5. Akitt JW, Farthing A (1981) Aluminium-27 nuclear magnetic resonance studies of the hydrolysis of Aluminium III. Part 5. Slow hydrolyse using aluminium metal. Dalton Trans 7:1624–1628
6. Anirudhan TS, Bringle CD, Radhakrishnan PG (2012) Heavy metal interactions with phosphatic clay: kinetic and equilibrium studies. Chem Eng J 200–202(August):149–157. https://doi.org/10.1016/j.cej.2012.06.024
7. Ararem A, Bouras O, Arbaoui F (2011) Adsorption of caesium from aqueous solution on binary mixture of iron pillared layered montmorillonite and goethite. Chem Eng J 172(1):230–236. https://doi.org/10.1016/j.cej.2011.05.095
8. Baloyi J, Ntho T, Moma J (2018) Synthesis and application of pillared clay heterogeneous catalysts for wastewater treatment: a review. RSC Adv 8(10):5197–5211. https://doi.org/10.1039/C7RA12924F
9. Hadjltaief B, Haithem PD, Costa ME, Galvez, and Mourad Ben Zina. (2013) Influence of operational parameters in the heterogeneous photo-fenton discoloration of wastewaters in the presence of an iron-pillared clay. Ind Eng Chem Res 52(47):16656–16665. https://doi.org/10.1021/ie4018258
10. Bergaya F, Aouad A, Mandalia T (2006) Chapter 7.5 pillared clays and clay minerals. Dev Clay Sci 1:393–421. https://doi.org/10.1016/S1572-4352(05)01012-3. (Elsevier)
11. Bergaya F, Lagaly G (2006) Chapter 1 general introduction: clays, clay minerals, and clay science. Dev Clay Sci 1:1–18. https://doi.org/10.1016/S1572-4352(05)01001-9. (Elsevier)
12. Bhattacharyya KG, Gupta SS (2006) Adsorption of Chromium(VI) from water by clays. Ind Eng Chem Res 45(21):7232–7240. https://doi.org/10.1021/ie060586j
13. Bouberka Z, Kacha S, Kameche M, Elmaleh S, Derriche Z (2005) Sorption study of an acid dye from an aqueous solutions using modified clays. J Hazard Mater 119(1–3):117–124. https://doi.org/10.1016/j.jhazmat.2004.11.026
14. Brigatti MF, Galan E, Theng BKG (2006) Chapter 2 structures and mineralogy of clay minerals. Dev Clay Sci 1:19–86. https://doi.org/10.1016/S1572-4352(05)01002-0. (Elsevier)
15. Butman MF, Ovchinnikov NL, Karasev NS, Kapinos AN, Belozerov AG, Kochkina NE (2017) Adsorption of anion and cation dyes onto pillared montmorillonite. Prot Met Phys Chem Surf 53(4):632–638. https://doi.org/10.1134/S2070205117020083

16. Chauhan M, Saini VK, Suthar S (2020a) Enhancement in selective adsorption and removal efficiency of natural clay by intercalation of Zr-pillars into its layered nanostructure. J Clean Prod 258(June):120686. https://doi.org/10.1016/j.jclepro.2020.120686

17. Chauhan M, Saini VK, Suthar S (2020b) Ti-pillared montmorillonite clay for adsorptive removal of amoxicillin, imipramine, diclofenac-sodium, and paracetamol from water. J Hazard Mater 399(November):122832. https://doi.org/10.1016/j.jhazmat.2020.122832

18. Chen F, Zhao E, Kim T, Wang J, Hableel G, Reardon PJT, Ananthakrishna SJ et al (2017) Organosilica nanoparticles with an intrinsic secondary amine: an efficient and reusable adsorbent for dyes. ACS Appl Mater Interfaces 9(18):15566–15576. https://doi.org/10.1021/acsami.7b04181

19. Chigwada G, Wang D, Wilkie CA (2006) Polystyrene nanocomposites based on quinolinium and pyridinium surfactants. Polym Degrad Stab 91(4):848–855. https://doi.org/10.1016/j.polymdegradstab.2005.06.016

20. Coelho AV, Poncelet G (1991) Gallium, aluminium and mixed gallium-aluminium pillared montmorillonite preparation and characterization, p 12

21. Cool P, Vansant EF (1998) Pillared clays: preparation, characterization and applications. In: Synthesis, 1:265–88. Molecular Sieves. Springer, Berlin. https://doi.org/10.1007/3-540-696 15-6_9

22. Darmawan A, Fuad K, Azmiyawati C (2019) Synthesis of chromium pillared clay for adsorption of methylene blue. In: IOP conference series: materials science and engineering vol 509 (May), p 012003. https://doi.org/10.1088/1757-899X/509/1/012003

23. De León MA, Castiglioni J, Bussi J, Sergio M (2008) Catalytic activity of an iron-pillared montmorillonitic clay mineral in heterogeneous photo-fenton process. Catal Today 133–135(April):600–605. https://doi.org/10.1016/j.cattod.2007.12.130

24. Dhahri M, Frini-Srasra N, Srasra E (2016) The effect of preparation method on textural and structural properties of alumina-pillared interstratified illite-smectite. Surf Eng Appl Electrochem 52(6):524–530. https://doi.org/10.3103/S1068375516060053

25. Dongyun D, Xiaorong Z, Xiaohua L (2005) Comparison of conventional and microwave-assisted synthesis and characteristics of aluminum-pillared rectorite. J Wuhan Univ Technol-Mater Sci Ed 20(2):53–56. https://doi.org/10.1007/BF02838488

26. Dvininov E, Popovici E, Pode R, Cocheci L, Barvinschi P, Nica V (2009) Synthesis and characterization of TiO$_2$-pillared Romanian clay and their application for azoic dyes photodegradation. J Hazard Mater 167(1–3):1050–1056. https://doi.org/10.1016/j.jhazmat.2009.01.105

27. Fernández-Pérez A, Marbán G (2020) Visible light spectroscopic analysis of methylene blue in water; what comes after dimer? ACS Omega 5(46):29801–29815. https://doi.org/10.1021/acsomega.0c03830

28. Fetter G, Hernández V, V Rodríguez, M.A Valenzuela, V.H Lara, and P Bosch. (2003) Effect of microwave irradiation time on the synthesis of zirconia-pillared clays. Mater Lett 57(5–6):1220–1223. https://doi.org/10.1016/S0167-577X(02)00961-8

29. Fournier A (2008) Interactions of pure aluminium hydrolytic species—keggin polyoxocations—and hydroxide with biologically relevant molecules. The Nottingham Trent University

30. Georgescu A-M, Nardou F, Zichil V, Nistor ID (2018) Adsorption of Lead(II) ions from aqueous solutions onto Cr-pillared clays. Appl Clay Sci 152(February):44–50. https://doi.org/10.1016/j.clay.2017.10.031

31. Gil A, Assis FCC, Albeniz S, Korili SA (2011) Removal of dyes from wastewaters by adsorption on pillared clays. Chem Eng J 168(3):1032–1040. https://doi.org/10.1016/j.cej.2011.01.078

32. Gil A, Korili SA, Trujillano R, Vicente MA (2011) A review on characterization of pillared clays by specific techniques. Appl Clay Sci 53(2):97–105. https://doi.org/10.1016/j.clay.2010.09.018

33. Gil A, Vicente MA (2020) Progress and perspectives on pillared clays applied in energetic and environmental remediation processes. Curr Opin Green Sustain Chem 21(February):56–63. https://doi.org/10.1016/j.cogsc.2019.12.004

34. González B, Pérez A, Trujillano R, Gil A, Vicente M (2017) Microwave-assisted pillaring of a montmorillonite with Al-polycations in concentrated media. Materials 10(8):886. https://doi.org/10.3390/ma10080886

35. González B, Trujillano R, Vicente MA, Rives V, de Faria EH, Ciuffi KJ, Korili SA, Gil A (2017) Doped Ti-pillared clays as effective adsorbents—application to methylene blue and trimethoprim removal. Environ Chem 14(5):267. https://doi.org/10.1071/EN16192

36. Guerra DL, Lemos VP, Angélica RS, Airoldi C (2008) The modified clay performance in adsorption process of Pb2+ ions from aqueous phase—thermodynamic study. Colloids Surf, A 322(1–3):79–86. https://doi.org/10.1016/j.colsurfa.2008.02.024

37. Gupta SS, Bhattacharyya KG (2005) Interaction of metal ions with clays: I. A case study with Pb(II). Appl Clay Sci 30(3–4):199–208. https://doi.org/10.1016/j.clay.2005.03.008

38. Hao YF, Yan LG, Yu HQ, Yang K, Yu SJ, Shan RR, Du B (2014) Comparative study on adsorption of basic and acid dyes by hydroxy-aluminum pillared bentonite. J Mol Liq 199(November):202–7. https://doi.org/10.1016/j.molliq.2014.09.005

39. Hasegawa N, Okamoto H, Kato M, Usuki A (2000) 'Preparation and mechanical properties of polypropylene–clay hybrids based on modified polypropylene and organophilic clay, 78:1918–22

40. He H, Ma L, Zhu J, Frost RL, Theng BKG, Bergaya F (2014) Synthesis of organoclays: a critical review and some unresolved issues. Appl Clay Sci 100(October):22–28. https://doi.org/10.1016/j.clay.2014.02.008

41. Herrera NN, Letoffe J-M, Reymond J-P, Bourgeat-Lami E (2005) Silylation of laponite clay particles with monofunctional and trifunctional vinyl alkoxysilanes. J Mater Chem 15(8):863. https://doi.org/10.1039/b415618h

42. Hou M-F, Ma C-X, Zhang W-D, Tang X-Y, Fan Y-N, Wan H-F (2011) Removal of rhodamine B using iron-pillared bentonite. J Hazard Mater 186(2–3):1118–1123. https://doi.org/10.1016/j.jhazmat.2010.11.110

43. Huang R, Hu C, Yang B, Zhao J (2016) Zirconium-immobilized bentonite for the removal of methyl orange (MO) from aqueous solutions. Desalination Water Treat 57(23):10646–54. https://doi.org/10.1080/19443994.2015.1040849

44. Issaoui O, Amor HB, Ismail M, Jeday MR (2017) Preparation of Al-pillared clay and application of methylene blue adsorption. In: 2017 International conference on green energy conversion systems (GECS), pp 1–4. IEEE, Hammamet, Tunisia. https://doi.org/10.1109/GECS.2017.8066219

45. Aziz K, Bakhtyar DM, Salh SK, Bertier P (2019) The high efficiency of anionic dye removal using Ce-Al13/pillared clay from darbandikhan natural clay. Molecules 24(15):2720. https://doi.org/10.3390/molecules24152720

46. Katdare SP, Ramaswamy V, Ramaswamy AV (1997) Intercalation of Al oligomers into Ca2+-montmorillonite using ultrasonics. J Mater Chem 7(11):2197–2199. https://doi.org/10.1039/a705001a

47. Klapyta Z, Fujita T, Iyi N (2001) Adsorption of dodecyl- and octadecyltrimethylammonium ions on a smectite and synthetic micas. Appl Clay Sci 19(1–6):5–10. https://doi.org/10.1016/S0169-1317(01)00059-X

48. Kon'kova TV., Alekhina MB, Mikhailichenko AI, Kandelaki GI, Morozov AN (2014) Adsorption properties of pillared clays. Prot Metals Phys Chem Surf 50(3):326–30. https://doi.org/10.1134/S2070205114030083

49. Lellis B, Fávaro-Polonio CZ, Pamphile JA, Polonio JC (2019) Effects of textile dyes on health and the environment and bioremediation potential of living organisms. Biotechnol Res Innov 3(2):275–290. https://doi.org/10.1016/j.biori.2019.09.001

50. Lenoble V, Bouras O, Deluchat V, Serpaud B, Bollinger J-C (2002) Arsenic adsorption onto pillared clays and iron oxides. J Colloid Interface Sci 255(1):52–58. https://doi.org/10.1006/jcis.2002.8646

51. Liu J, Zhang G (2014) Recent advances in synthesis and applications of clay-based photocatalysts: a review. Phys Chem Chem Phys 16(18):8178–8192. https://doi.org/10.1039/C3CP54146K

52. Luengo C, Puccia V, Avena M (2011) Arsenate adsorption and desorption kinetics on a Fe(III)-modified montmorillonite. J Hazard Mater 186(2–3):1713–1719. https://doi.org/10.1016/j.jha zmat.2010.12.074
53. Market Report and Forecast 2021–2026 (2021). EMR 2021. https://www.expertmarketres earch.com/reports/india-dyes-and-pigments-market
54. Matthes W, Madsen FW, Kahr G (1999) Sorption of heavy-metal cations by Al and Zr-hydroxy-intercalated and pillared bentonite. Clays Clay Miner 47(5):13
55. Mishra T, Parida KM, Rao SB (1996) Transition metal oxide pillared clay. J Colloid Interface Sci 183(1):176–183. https://doi.org/10.1006/jcis.1996.0532
56. Mishra T, Mahato DK (2016) A comparative study on enhanced Arsenic(V) and Arsenic(III) removal by iron oxide and manganese oxide pillared clays from ground water. J Environ Chem Eng 4(1):1224–1230. https://doi.org/10.1016/j.jece.2016.01.022
57. Mishra T, Parida K (1997) Transition-metal oxide pillared clays. J Mater Chem 7(1):147–152. https://doi.org/10.1039/a603797f
58. Mishra T, Parida K (1998) Transition metal pillared clay 4. A comparative study of textural, acidic and catalytic properties of chromia pillared montmorillonite and acid activated montmorillonite. Appl Catal A 166(1):123–133. https://doi.org/10.1016/S0926-860X(97)002 47-0
59. Mokaya R (2000a) Novel layered materials: non-phosphates, pp 1610–17. Academic Press.
60. Mokaya R (2000b) 'Novel layered materials: non-phosphates', pp 1610–17. Academic Press.
61. Muthukkumaran A, Aravamudan K (2017) Combined homogeneous surface diffusion model—design of experiments approach to optimize dye adsorption considering both equilibrium and kinetic aspects. J Environ Manage 204(December):424–435. https://doi.org/10.1016/j.jen vman.2017.09.010
62. Najafi H, Farajfaed S, Zolgharnian S, Mirak SHM, Asasian-Kolur N, Sharifian S (2021) A comprehensive study on modified-pillared clays as an adsorbent in wastewater treatment processes. Process Saf Environ Prot 147(March):8–36. https://doi.org/10.1016/j.psep.2020. 09.028
63. Ninago MD, López OV, Gabriela Passaretti M, Fernanda Horst M, Lassalle VL, Ramos IC, Di Santo R, Ciolino AE, Villar MA (2017) Mild microwave-assisted synthesis of aluminum-pillared bentonites: thermal behavior and potential applications. J Therm Anal Calorim 129(3):1517–1531. https://doi.org/10.1007/s10973-017-6304-6
64. Olaya A, Moreno S, Molina R (2009) Synthesis of pillared clays with aluminum by means of concentrated suspensions and microwave radiation. Catal Commun 10(5):697–701. https:// doi.org/10.1016/j.catcom.2008.11.029
65. Padak B, Brunetti M, Lewis A, Wilcox J (2006) Mercury binding on activated carbon. Environ Prog 25(4):319–326. https://doi.org/10.1002/ep.10165
66. Paiva LB, de, Ana Rita Morales, and Francisco R. Valenzuela Díaz. (2008) Organoclays: properties, preparation and applications. Appl Clay Sci 42(1–2):8–24. https://doi.org/10.1016/ j.clay.2008.02.006
67. Pandey P, Saini VK (2019) Pillared interlayered clays: sustainable materials for pollution abatement. Environ Chem Lett 17(2):721–727. https://doi.org/10.1007/s10311-018-00826-0
68. Patel H, Somani R, Bajaj H, Jasra R (2007) Preparation and characterization of phosphonium montmorillonite with enhanced thermal stability. Appl Clay Sci 35(3–4):194–200. https://doi. org/10.1016/j.clay.2006.09.012
69. Pinnavaia TJ, Tzou MS, Landau SD (1985) New chromia pillared clay catalysts. J Am Chem Soc 107(16):4783–4785. https://doi.org/10.1021/ja00302a033
70. Qin Z, Yuan P, Yang S, Liu D, He H, Zhu J (2014) Silylation of Al13-intercalated mont-morillonite with trimethylchlorosilane and their adsorption for orange II. Appl Clay Sci 99(September):229–236. https://doi.org/10.1016/j.clay.2014.06.038
71. Schoonheydt RA, Pinnavaia T, Lagaly G, Gangas N (1999) Pillared clays and pillared layered solids, vol 71, pp 2367–71. https://doi.org/10.1351/pac199971122367
72. Ramesh A, Hasegawa H, Maki T, Ueda K (2007) Adsorption of inorganic and organic arsenic from aqueous solutions by polymeric Al/Fe modified montmorillonite. Sep Purif Technol 56(1):90–100. https://doi.org/10.1016/j.seppur.2007.01.025

73. Kacha S, Ouali MS, Elmale S (1997) Dye abatement of textile industry wastewater with bentonite and aluminium salts, p 233
74. Sahel K, Bouhent M, Belkhadem F, Ferchichi M, Dappozze F, Guillard C, Figueras F (2014) Photocatalytic degradation of anionic and cationic dyes over TiO_2 P_25, and Ti-pillared clays and Ag-doped Ti-pillared clays. Appl Clay Sci 95(June):205–210. https://doi.org/10.1016/j. clay.2014.04.014
75. Sanabria NR, Molina R, Moreno S (2009) Effect of ultrasound on the structural and textural properties of Al–Fe pillared clays in a concentrated medium. Catal Lett 130(3–4):664–671. https://doi.org/10.1007/s10562-009-9956-4
76. Shalaby AA (2020) Determination of acid dissociation constants of alizarin red s, methyl orange, bromothymol blue and bromophenol blue using a digital camera. RSC Adv 6
77. Shen Y-H (2001) Preparations of organobentonite using nonionic surfactants. Chemosphere 44(5):989–995. https://doi.org/10.1016/S0045-6535(00)00564-6
78. Shokrollahi A, Firoozbakht F (2016) Determination of the acidity constants of neutral red and bromocresol green by solution scanometric method and comparison with spectrophoto-metric results. Beni-Suef Univ J Basic Appl Sci 5(1):13–20. https://doi.org/10.1016/j.bjbas. 2016.02.003
79. Tepmatee P, Siriphannon P (2013) Effect of preparation method on structure and adsorption capacity of aluminum pillared montmorillonite. Mater Res Bull 48(11):4856–4866. https://doi. org/10.1016/j.materresbull.2013.06.066
80. Thomas WJ, Crittenden BD (1998) Adsorption technology and design. Oxford, Boston, Butterworth-Heinemann
81. Tomul F (2011) Synthesis, characterization, and adsorption properties of Fe/Cr-pillared bentonites. Ind Eng Chem Res 50(12):7228–7240. https://doi.org/10.1021/ie102073v
82. Tomul F (2011) Effect of ultrasound on the structural and textural properties of copper-impregnated cerium-modified zirconium-pillared bentonite. Appl Surf Sci 258(5):1836–1848. https://doi.org/10.1016/j.apsusc.2011.10.056
83. Vinod VP, Anirudhan TS (2002) Sorption of tannic acid on zirconium pillared clay. J Chem Technol Biotechnol 77(1):92–101. https://doi.org/10.1002/jctb.530
84. Worch E (2012a) Adsorption technology in water treatment: fundamentals. Processes, and modeling. Berlin, Boston, De Gruyter
85. Worch E (2012b) Adsorption technology in water treatment: fundamentals. In: Processes, and modeling. Berlin, Boston, De Gruyter
86. Yaseen DA, Scholz M (2019) Textile dye wastewater characteristics and constituents of synthetic effluents: a critical review. Int J Environ Sci Technol 16(2):1193–1226. https://doi. org/10.1007/s13762-018-2130-z
87. Yilmaz N, Yapar S (2004) Adsorption properties of tetradecyl- and hexadecyl trimethy-lammonium bentonites. Appl Clay Sci 27(3–4):223–228. https://doi.org/10.1016/j.clay.2004. 08.001
88. Zermane F, Bouras O, Baudu M, Basly J-P (2010) Cooperative Coadsorption of 4-nitrophenol and basic yellow 28 dye onto an iron organo-inorgano pillared montmorillonite clay. J Colloid Interface Sci 350(1):315–319. https://doi.org/10.1016/j.jcis.2010.06.040
89. Zhao L, Wang J, Zhang P, Gu Q, Gao C (2018) Absorption of heavy metal ions by alginate. In: Bioactive seaweeds for food applications, pp 255–68. Elsevier. https://doi.org/10.1016/B978-0-12-813312-5.00013-3
90. Zhu J, He H, Zhu L, Wen X, Deng F (2005) Characterization of organic phases in the interlayer of montmorillonite using FTIR and 13C NMR. J Colloid Interface Sci 286(1):239–244. https:// doi.org/10.1016/j.jcis.2004.12.048

Versatile Fabrication and Use of Polyurethane in Textile Wastewater Dye Removal via Adsorption and Degradation

Muhammad Iqhrammullah, Rahmi, Hery Suyanto, Kana Puspita, Haya Fathana, and Syahrun Nur Abdulmadjid

Abstract Polyurethane is known for its wide applicability with easy fabrication along with adjustable characteristics. Furthermore, polyurethane-based materials have been reported to carry effective removal of various synthetic dyes. This chapter reviews the use of polyurethane based on updated reports on dye removal. The development of polyurethane and how green routes synthesis are researched are also highlighted. Modifications of polyurethane have been conducted via grafting, composite, and functionalization. Currently, biobased polyols employed in the fabrication of polyurethane are β-cyclodextrin, castor oil, palm oil, and *Moringa oleifera* gum. Zr-based metal organic framework and carbonaceous materials are notably promising fillers for adsorptive removal of dyes. Incorporation of nanoparticles into polyurethane matrix provides a photocatalytic feature to the material, hence, working synergistically with the adsorption. Polyurethane has a significant role as a support

M. Iqhrammullah (✉) · H. Fathana
Graduate School of Mathematics and Applied Sciences, Universtias Syiah Kuala, Banda Aceh 23111, Indonesia
e-mail: m.iqhram@oia.unsyiah.ac.id

M. Iqhrammullah · Rahmi · H. Fathana
Department of Chemistry, Faculty of Mathematics and Natural Sciences, Universitas Syiah Kuala, Banda Aceh 23111, Indonesia

H. Suyanto
Faculty of Mathematics and Natural Sciences, Udayana University, Kampus Bukit Jimbaran, 80361 Bali, Indonesia

K. Puspita
Department of Chemistry Education, Faculty of Education and Teacher Training, Universitas Syiah Kuala, Banda Aceh 23111, Indonesia

S. N. Abdulmadjid (✉)
Department of Physics, Faculty of Mathematics and Natural Sciences, Universitas Syiah Kuala, Banda Aceh 23111, Indonesia
e-mail: syahrun_madjid@unsyiah.ac.id

© The Author(s), under exclusive license to Springer Nature Singapore Pte Ltd. 2022   179
A. Khadir and S. S. Muthu (eds.), *Polymer Technology in Dye-containing Wastewater*,
Sustainable Textiles: Production, Processing, Manufacturing & Chemistry,
https://doi.org/10.1007/978-981-19-1516-1_7

material for immobilization of fungal culture that could perform enzymatic degradation against dyes. In conclusion, polyurethane could be applied in the wastewater treatment of textile industry by considering the removal efficiency, stability, reusability, and cost-efficiency.

Keywords Polyurethane · Dye · Adsorption · Photocatalyst · Nanoparticle · Composite

1   Introduction

A uniquely versatile material, polyurethane, has contributed to the significant development of polymer application [73]. The nomenclature itself is different from that of mostly used, where the polymer is often named after its monomer. Polyurethane is named based on the urethane linkage formed through a condensation between a polyol and isocyanate. The simplicity of fabricating this material has led its use in various forms (Fig. 1), along with adjustable properties [2]. Applications of polyurethane include biomedical applications, building, construction, spaceship, automotive, textiles, pollution control, and so on [17, 72]. Polyurethane is a common ingredient for coating [88], heat insulation [66], sound insulation [86], and dyeing [37].

Application of polyurethane in pollutant removal has been well studied with significant research progress and development keep on going. Based on current reports, polyurethane has been employed to remove pollutants such as heavy metals [21, 34], NH_3-N [55], oil spills [58], dyes [35, 48, 87], and other organic pollutants [57]. One of the concerning pollutants is synthetic dye from textile wastewater stem from its toxic, mutagenic, and carcinogenic properties. Furthermore, the release of dye pollutant in environment leads to higher biochemical and chemical oxygen demand as well as hampering the photosynthesis of aquatic organisms [44]. Herein, we will discuss the role of polyurethane in the current updates of textile wastewater

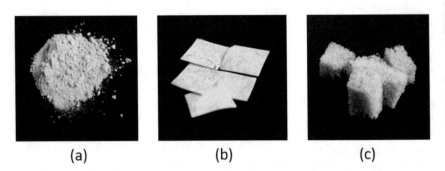

(a)   (b)   (c)

Fig. 1 Different forms of polyurethane; powder (**a**), film membrane (**b**), and foam (**c**)

treatment. We will also discuss the early and current development of polyurethane, and how green approaches have been researched in the fabrication of polyurethane.

2 Early Discovery and Development of Polyurethane

The first discovery of polyurethane was recorded in 1937 by a German professor, Ottor Bayer, along with his colleagues, by successfully reacting aliphatic diisocyanate with glycol. This discovery had successfully arisen Germany's reputation in the field of polymer sciences [81]. In late 1938, Rinke and colleagues from the US had obtained a patent on their research in synthesizing polyurethane by reacting aliphatic 1,8-octane diisocyanate and 1,4-butanadiol [77]. The polyurethane product obtained from the aforementioned reaction has low viscosity allowing its conversion into fibers. Owing to fibrous structure-related wide applicability of polyurethane, it consequently attracted higher market demand [26].

As of today, the market demand for polyurethane keeps increasing, in which its application has reached industrial sectors such as automotive, air transportation, packaging, electronic, coating, textile, toys, and construction materials [40, 61]. It is due to the fact that polyurethane can be easily synthesized using simple materials, which could be obtained from polyol and diisocyanate reaction. Other than previously stated, polyols used commonly in polyurethane synthesis are polyether polyol [60], tris(polyoxypropylene ether)propane [31], and polypropylene glycol [56]. As for the diisocyanate sources, for the most part, they are 4,4'-diphenylmethane diisocyanate (MDI) [15, 60], toluene diisocyanate (TDI) [20, 31], isophorone diisocyanate [18], and hexamethylene diisocyanate [42]. Diisocyanates with cyclic aromatic structure, such as MDI and TDI, are very reactive—requiring no catalyst in the reaction [34].

Several additives might be used in the fabrication to adjust the properties of polyurethane product which include catalyst, chain extender, plasticizing, and others. To obtain polyurethane in a foam shape, a blowing agent should be added into the prepolymer mixture [7, 15, 39]. The characteristics of polyurethane, however, do not only depend on the type of materials or additives used but also the composition [9]. For instance, to enhance the rigidity of polyurethane, the composition of the diisocyanate could be increased [16]. Other factors affecting polyurethane include the polymerization temperature, curing temperature, stirring, pressure, and density [13].

Unfortunately, those aforementioned materials are mostly non-renewable. For instance, 1,4-butanadiol which is a derivative product from petrochemical industry [78]. An additive, ethylene glycol, which was also used in polyurethane fabrication [23], is also derived from petrochemical route [85]. Due to its versatility in the prepolymer material choice, researches have been focused on utilizing bio-sourced polyols in producing polyurethane. This way, polyurethane could be produced from renewable materials alongside with reducing the solid waste from biomass. The green approach in synthesizing polyurethane also includes the replacement of aromatic cyclic isocyanate in order to produce less to non-toxic polyurethane.

3 Green Approaches in Polyurethane Synthesis

Acting as the replacement of petrochemical route-derived polyols, bio-sourced polyols could reach a market demand in 2021 up to 4.7 billion US dollars [26]. Biopolyols themselves have been produced on an industrial scale by several famous chemical industries: Shell Chemical, Dow Chemical, and Bayer Material Science. Significant progress of green approach in polyurethane synthesis could be attributed to the fact that biopolyol could be easily derived from biomass. Of which, the most abundant biopolymer, cellulose, after purification processes [11], could be used to fabricate polyurethane. Other reported biopolyols include chitosan [15], starch [4], soy bean oil [1, 43], sunflower seed oil [19], and castor oil [22, 33]. Even ball-milled crude algae could be directly used as the biopolyol with a reaction mechanism presented (Fig. 2).

Polymerization through the formation urethane linkage could improve the properties, either chemical or physical, of the carbohydrate biopolymers. Of which is cellulose, reported improved in terms of its mechanical strength and hydrophobicity [27, 30]. Similarly, starch-based polyurethane also has improvement in hydrophobicity albeit the biopolyol source is hydrophilic [82]. Moreover, functional group richness in carbohydrates could be attributed to the high pollutant uptake, including that of membrane [71], powder [45], and foam [41] shapes. The structure of carbohydrate macromolecule could also contribute to the removal of pollutant, notably shown by cyclodextrin that possesses a toroidal shape with dual water affinity. The inner part of the toroidal shape is hydrophobic, meanwhile the outer part is hydrophobic, which is suitable for separating textile wastewater in the presence of oil [69].

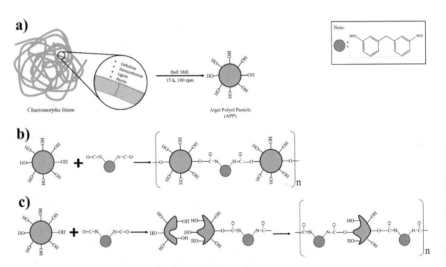

Fig. 2 Reaction between ball-milled algal particle with MDI forming a polyurethane membrane. Reproduced with the permission from Elsevier citing [55]

Some reports employed the vegetable oil as biopolyol feeds to prepare polyurethane for heat and sound insulation and its application as superconductor [10, 25]. For adsorbent materials, the researches have produced the polyurethane from castor oil [39, 54] and palm oil [6]. In this case, however, vegetable oils could not be directly used as polyol, except for castor oil and Lesquerella seed oil due to their possession of reactive OH groups [14, 33]. To add the reactive OH groups to the fatty acids in a vegetable oil, several reaction routes could be conducted namely ozonolysis [65], hydroformylation [29], epoxidation followed by oxirene-ring opening [62], transesterification [83], metathesis [64], and thiol-ene coupling reaction [63]. Castor oil could also be treated with the aforementioned reaction routes in order to produce more reactive OH groups to yield easier polymerization (especially when aliphatic isocyanate is used).

Notably, a polyurethane foam derived from castor oil and prepared using thiol-ene coupling reaction had a rigid characteristic with mechanical strength and thermal stability similar to that prepared using commercial polyols. On the other hand, the advanced ozonolysis does not require catalyst, by skipping the dehydrogenation through the addition of ethylene glycol or glycerol altogether with O_3 gases yielding ester alcohol with terminal OH groups [65]. Among the stated reaction routes, only transesterification yields by-products di- and triglyceride, which could act as natural plasticizing agents [22]. Therefore, transesterification is suitable to be employed in the fabrication of flexible polyurethane.

A greener approach in the current advance of polyurethane synthesis is the isocyanate-free route, obtained via cyclocarbonates and diamines reaction. Hence, modification of fatty acid triglycerides should be carried out in order to obtain the cyclocarbonate through epoxidation followed by slow coupling using CO_2. Since this process is carried out with CO_2, it contributes to the eco-friendliness of polyurethane synthesis owing to CO_2 abundance and carbon footprint reduction potential [28, 36]. The reaction has been reported in the preparation of sunflower seed-based polyol [19]. In that reported study, the occurrence of aminolysis was observed (Fig. 3). The use of diamines with secondary amines was reported resulting in lower urethane conversion rate with unobservable aminolysis [8, 19].

4 Application in Textile Wastewater

The application of polyurethane materials in textile wastewater treatment was studied employing artificial samples containing dyes (Table 1). Examples of cationic dyes used in the studies are crystal violet, rhodamine B, methylene blue, ethidium, malachite green, and methyl violet. As for anionic dyes, they could be methyl orange, eosin Y, and remazol brilliant blue R. Xylidine and acid black 1 are categorized as neutral dyes. Of which, methylene blue is the most commonly used model in dye removal investigations. Several studies used the dyes in combination; cationic–cationic [48, 59, 87], cationic–anionic [53, 80], and cationic–anionic–neutral [51].

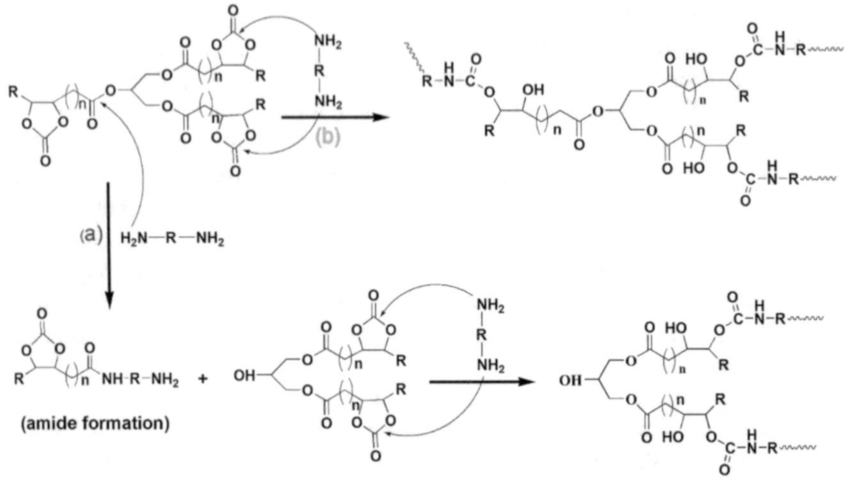

Fig. 3 Schematic mechanism of urethane linkage formation from a sunflower seed oil-derived cyclocarbonate and a diamine. The reaction induces aminolysis forming an amide (**a**) or directly forms urethane linkage (**b**). Reproduced with permission from Elsevier citing [19]

By using the combination, ones could draw a conclusion whether their fabricated materials are specific to a certain type of dye. A study used a combination of cationic dye methylene blue and anionic dye amaranth red found that amaranth red persisted after the treatment [38]. Similarly, the removal was only effective against methylene blue in the presence of methyl orange in the solution, as shown in Fig. 4 [52]. This is in line with studies using methyl orange alone, where low removal efficiency was reported [6, 35]. Nonetheless, a study using graphene nanosheet-loaded polyurethane foam yielded higher adsorption capacity against anionic eosin [53]. By employing photocatalytic degradation, a group of researchers has achieved excellent removal for both methyl orange and methylene blue [80]. Another research group also performed a study in cationic–anionic–neutral combination of dyes and obtained satisfying removal efficiency using *Peniophora laxitexta* immobilized in polyurethane foam [51].

Since polyurethane is not built from specific monomers, its removal ability could be dependent on the type of polyol and isocyanate and, more importantly, the modification. As shown in Table 1, most of the polyurethanes were commercial. There are, however, studies using biopolyols such as cellulose acetate [35], β-cyclodextrin [59], *Moringa oleifera* gum [70], castor oil [46], and palm oil-derived monoester [6]. Of which, cellulose acetate and β-cyclodextrin are known for their high pollutant uptake capacity attributed to functional group richness and molecular geometry [47, 79]. Additionally, *Moringa oleifera* gum contributed to the antibacterial properties of the polyurethane foam [70].

Polymer grafting is common in developing polyurethane because it can add more functional group and assist filler loading, where polydopamine [87] and octadecylamine [80] are some of the examples. Functionalization could also be conducted for

Table 1 Polyurethane-based material used in textile wastewater treatment with various removal methods

References	Material		Characteristics	Dye	Removal method and mechanism
	Polyurethane base	Modification			
[87]	Commercial polyurethane foam	Grafted with polydopamine and loaded with Zr-MOFs through *in-situ* hydrothermal	Foam with open cell structure possessing rich O-containing functional group with good stability and reusability	Crystal violet, rhodamine B	Batch adsorption dependent on chemisorption
[80]	Commercial polyurethane foam	Grafted with octadecylamine and loaded with molybdenum disulfide and Fe_3O_4 via absorption	Foam with superhydrophobicity, robust stability in extreme environments, and great oil/water separation ability	Methyl orange, methylene blue	Using a reactor to separate the oil from water by flowing the wastewater through the modified foam while photodegrading the pollutant
[76]	Ethylene oxide capped polyether triol and diphenylmethane diisocyanate	Loaded with nano-based freeze-dried lignin	The size of filler particles contributes to improved charged surface and polyurethane coating	Methylene blue	Batch adsorption
[53]	Commercial polyurethane foam (prepolymer)	Loaded with graphene nanosheets	Highly durable, flexible, and hydrophilic	Methylene blue, ethidium bromide, eosin Y	Batch adsorption dependent on electrostatic interaction and π–π stacking
[52]	Commercial polyurethane foam	Loaded with disk-derived activated carbon black via crosslinking using sodium alginate in $CaCl_2$ solution	Superhydrophilic and higher mechanical properties	Methylene blue	Batch adsorption dependent on hydrogen bonding, electrostatic interaction, and π–π stacking, as well as mechanical entrapment

(continued)

Table 1 (continued)

References	Material		Characteristics	Dye	Removal method and mechanism
	Polyurethane base	Modification			
[51]	Commercial polyurethane foam	*Peniophora laxitexta* immobilization in a polyurethane foam-sandwiched lignocellulosic medium	The medium and polyurethane foam was colonized by *P. laxitexta*	Xylidine, malachite green, remazol brilliant blue R	Adsorption and laccase-based enzymatic degradation
[38]	Commercial polyurethane foam	Functionalized using N-(3, 4-dihydroxyphenethyl) acrylamide	Smooth surface and contains C = C double bond functional groups excellent mechanical stability, outstanding reusability	Methylene blue	Batch adsorption using continuous and automatic setup
[32]	Commercial foam	Functionalized using NaOH and loaded with ZnO nanoparticles	UV and visible light photocatalytic activities	Acid black 1	Batch adsorption and photodegradation under UV and sunlight
[35]	Cellulose acetate and MDI	Cellulose acetate was crosslinked using MDI forming a urethane linkage	MDI-reduced agglomeration with porous structure and O- and N-containing functional groups	Methyl orange	Batch adsorption dependent on chemisorption

(continued)

Table 1 (continued)

References	Material		Characteristics	Dye	Removal method and mechanism
	Polyurethane base	Modification			
[70]	Moringa oleifera gum and MDI	Loaded with ash	Antibacterial activities. Rougher and more amorphous. Rich with N- and O-containing functional groups	Malachite green	Batch adsorption dependent on chemisorption and intraparticle diffusion
[59]	Hydroxypropyl-β-cyclodextrin and MDI	Hydroxypropyl-β-cyclodextrin (HPβCD), HPβCD-conjugated magnetic nanoparticles (HPMN) and HPβCD-conjugated magnetic nanoparticles with polyurethane networks	Nanoparticle-induced rough structure. HMDI reduces the removal efficiency	Crystal violet, methyl violet	Batch adsorption. The adsorption is positively affected by adsorbent dose and contact time and negatively affected by HMDI/HPβCD ratio and dye concentration
[48]	Commercial PUF	Loaded with carbon quantum dots and TiO_2 nanoparticle	Photocatalytic properties with more than 30% of high-energy crystal planes and good reusability	Methylene blue, rhodamine B	Adsorption and photodegradation under 350 W xenon lamp
[46]	Castor oil based	Zirconium-based metal organic frameworks loaded on polyurethane foam	Flake-like surface structure, O- and N-containing functional group, pore diameter $= 5.68$ nm	Rhodamine B	Batch adsorption dependent on Lewis acid–base interaction, hydrogen bonds, and electrostatic interaction

(continued)

Table 1 (continued)

References	Material		Characteristics	Dye	Removal method and mechanism
	Polyurethane base	Modification			
[6]	Palm-based monoester and 4,4-methylene diphenyl diisocyanate	Polyethylene glycol as the chain extender	Foam with hollow structure, strong mechanical properties (tensile strength: 7.42 MPa) and O- and N- containing functional groups	Methyl orange	Batch adsorption

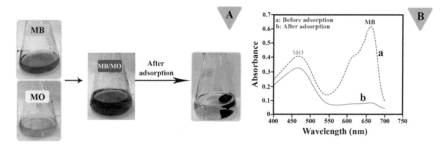

Fig. 4 Adsorption of methylene blue (MB) and methyl orange (MO) on disk-derived activated carbon black-modified polyurethane foams. Depicted in visual appearance (**a**) and UV–Vis spectra (**b**). Reproduced with permission from Elsevier citing [52]

the same objective, where some reports used N-(3,4-dihydroxyohenethyl acrylamide [38] and NaOH [32]. Through functionalization, polyurethane materials are enriched by O- and N-containing functional groups that are responsible to entrap the pollutant on the adsorbent surface [55, 67].

Addition of fillers is found in many reports as a mean of improving the pollutant uptake as well as the overall properties of the material. The fillers include metal organic framework [46, 87], nanoparticles [32, 48, 59, 80], lignin [76], carbonaceous materials [52, 53], and ash [70]. Embedment of nanoparticles into polyurethane matrix allow the photocatalytic degradation of dyes [32, 48, 80], in which combination of adsorption and photodegradation is preferred [68]. Other than the aforementioned purpose, fillers are loaded to polyurethane matrix to improve hydrophobicity, which is required to separate oil from water [80].

5 Removal Mechanism

Adsorption is the underlying mechanism of dye removal using polyurethane, including that with photocatalytic and enzymatic degradations. The entrapment of pollutants onto adsorbent surface relies on several factors. O- and N-containing functional groups along with $C = C$ double bond functional groups contribute to higher intermolecular interaction [12]. The interaction might be in a form of Lewis acid–base interaction, hydrogen bond, electrostatic interaction, π-π stacking, or in combination. Furthermore, physical properties such as pore size, surface area, amorphicity, and hydrophilicity also affect the adsorption performance by enhancing the diffusion of aqueous pollutants [5, 84]. In wastewater, where various dyes are present, interaction could also occur between adsorbates. The illustration of adsorbate interacting with the active site of an adsorbent surface could be observed in Fig. 5.

Adsorption parameters, namely contact time, pH, temperature, and ion strength, also play a role in pollutant uptake. Contact time is dependent on the diffusion aqueous medium in reaching the surface of the entire adsorbent, highly correlated with the

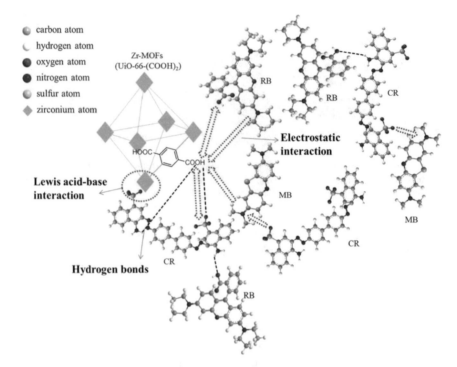

Fig. 5 Schematic mechanism of rhodamine B (RB), methylene blue (MB), and Congo red (CR) adsorptions onto polyurethane foam loaded with Zr-Metal Organic Framework (Zr-MOF). Reproduced with the permission from Elsevier citing [46]

hydrophilicity of the material [33, 54]. Meanwhile, the presence of non-pollutant ions could inhibit the adsorption as they compete with the ionic dyes. Adsorbent surface is charged dependently to the pH level. When the surface charge and ion charge are the same, the repulsion might occur, and vice versa [3, 68]. In the case of temperature, thermodynamic studies could be carried out to determine whether adsorption requires higher energy for the interaction to occur [53]. In terms of the kinetics, studies concluded that the adsorption is governed by chemisorption based on the experimental data fitting with that from the pseudo-second-order kinetic model [52, 53, 70].

As previously stated, modification using photocatalyst nanoparticles could enhance the removal efficiency by providing photodegradation against the pollutant which synergistically works with the adsorption [68]. In this light, the photodegradation helps to self-cleanse the adsorbed pollutants, thus, making the active site to be re-unoccupied. A challenge in the application of photocatalyst nanoparticles is their powder form which is hard to separate at the end of the treatment. Moreover, the nanoparticle has a tendency to form agglomerate reducing the contact surface [49]. Embedment of nanoparticle in polyurethane matrix could prevent the agglomeration which consequently could elevate the pollutant removal performance [32, 80]. It is worth mentioning that coupling TiO_2 nanoparticles with carbon quantum dots results in higher photocatalytic activity [48]. This is ascribed to the incorporation of the new energy states around the Fermi level close to the lowest conduction band by the C 2p state of carbon quantum dots [50].

An approach of utilizing microbiota to degrade the dyes from textile wastewater has been reported [51]. The typical microorganisms used for this purpose are fungi producing ligninolytic enzymes, including laccases and some peroxidases [75]. The immobilized culture was conducted on lignocellulosic medium consisting of cellulose spent casings, soybean hulls, or wheat straw [51] because they are cheap and effective biosorbents [3, 24]. For practical use, the immobilized culture is placed on supports that could perform physical adsorption such as polyurethane foam. This strategy does not only yield higher removal efficiency but also higher continuity and regeneration since the support suppress the apparent broth viscosity and allow more oxygen supply associated with improved rheological properties [74].

6   Insights for Practical Application

For practical application of the polyurethane-based materials in dyes removal, there are several aspects need to be fulfilled. The first is removal efficiency; adjusting the material ratio, contact time, and dye concentration could exhibit some improvement [59]. Secondly, the material should perform a robust dyes uptake in various conditions including the working pH range and the presence of interfering matrix. Separation of water and oil from the wastewater effluent could be a help in this regard [48, 52, 80]. Since textile wastewater contains more than one dye, investigation on the mixed solution is required, as reported by several studies [51, 53, 80].

To maintain the excellent performance of the material in removing dyes from the wastewater, the stability of the materials is important. Some studies have shown materials with good stability after performing the removal [38, 46]. Stability of material using a living organism or easily destructible ingredient requires further studies in this regard [51, 76]. Reusability of the material is a significant aspect that has been contributed by polyurethane-based material [51, 52, 70]. Under the same idea, either the fabrication of the material or the removal processes needs to be cost-efficient. For instance, the cost of removing 1 g MB at high concentration range with maximum adsorption capacity of 386.49 mg/g is estimated to be 8.91 Euro [52].

7   Conclusions

Versatility of polyurethane fabrication enables its usage in wastewater treatment in various ways. Adsorption becomes the most underlying mechanism of polyurethane for the dye removal, regardless of the method used. Adsorptive dye removal could be coupled with photocatalytic and enzymatic degradation. Modifications via grafting, filler addition, and functionalization of polyurethane could improve the dye uptake. Future researches are expected to develop polyurethane via green synthesis routes.

Acknowledgements Authors wish to honor the passing of Prof. Ir. Marlina, M.Si from the Department of Chemistry, Faculty of Mathematics and Natural Sciences, Universitas Syiah Kuala, Banda Aceh, Indonesia. She had made a significant contribution in the preparation of this chapter. The first author wishes to thank his mother, sister, and friends (Rayyan, Naufal, Rini, Nenden, Nanas, Nizam, Andhika, Iqbal, Valdi, and others) for the support during the making of this chapter.

References

1. Acik G, Kamaci M, Altinkok C, Karabulut HRF, Tasdelen MA (2018) Synthesis and properties of soybean oil-based biodegradable polyurethane films. Prog Org Coat 123:261–266. https://doi.org/10.1016/j.porgcoat.2018.07.020
2. Akindoyo JO, Beg MDH, Ghazali S, Islam MR, Jeyaratnam N, Yuvaraj AR (2016) Polyurethane types, synthesis and applications-a review. RSC Adv. https://doi.org/10.1039/c6ra14525f
3. Alni A, Puspita K, Zulfikar MA (2019) Biosorbent from Chinese cabbage (Brassica pekinensia L.) for phenol contaminated waste water treatment. Key Eng Mater 811:71–79. https://doi.org/10.4028/www.scientific.net/KEM.811.71
4. Ashjari HR, Dorraji MSS, Fakhrzadeh V, Eslami H, Rasoulifard MH, Rastgouy-Houjaghan M, Gholizadeh P, Kafil HS (2018) Starch-based polyurethane/CuO nanocomposite foam: antibacterial effects for infection control. Int J Biol Macromol 111:1076–1082. https://doi.org/10.1016/j.ijbiomac.2018.01.137
5. Ayotte P, Smith RS, Stevenson KP, Dohnálek Z, Kimmel GA, Kay BD (2001) Effect of porosity on the adsorption, desorption, trapping, and release of volatile gases by amorphous solid water. J Geophy Res Planets 106(E12):33387–33392. https://doi.org/10.1029/2000JE001362
6. Badri KH, Ismail FH, Shakir ASA, Mohamad S, Hamuzan HA, Hassan NS (2018) Polyurethane membrane as an adsorbent for methyl orange and ethyl violet dyes Malaysian. J Anal Sci 22(6):1040–1047. https://doi.org/10.17576/mjas-2018-2206-14
7. Badruddoza AZM, Bhattarai B, Suri RPS (2017) Environmentally friendly β-Cyclodextrin– Ionic liquid polyurethane-modified magnetic sorbent for the removal of PFOA, PFOS, and Cr(VI) from water. ACS Sustain Chem Eng 5(10):9223–9232. https://doi.org/10.1021/acssuschemeng.7b02186
8. Boyer A, Cloutet E, Tassaing T, Gadenne B, Alfos C, Cramail H (2010) Solubility in CO2 and carbonation studies of epoxidized fatty acid diesters: towards novel precursors for polyurethane synthesis. Green Chem 12(12):2205. https://doi.org/10.1039/c0gc00371a
9. Carriço CS, Fraga T, Carvalho VE, Pasa VMD (2017) Polyurethane foams for thermal insulation uses produced from castor oil and crude glycerol biopolyols. Molecules. https://doi.org/10.3390/molecules22071091
10. Carriço CS, Fraga T, Pasa VMD (2016) Production and characterization of polyurethane foams from a simple mixture of castor oil, crude glycerol and untreated lignin as bio-based polyols. Eur Polymer J. https://doi.org/10.1016/j.eurpolymj.2016.10.012
11. Chandra Mohan C, Harini K, Vajiha Aafrin B, Lalitha Priya U, Maria Jenita P, Babuskin S, Karthikeyan S, Sudarshan K, Renuka V, Sukumar M (2018) Extraction and characterization of polysaccharides from tamarind seeds, rice mill residue, okra waste and sugarcane bagasse for its Bio-thermoplastic properties. Carbohyd Poly 186:394–401. https://doi.org/10.1016/J.CARBPOL.2018.01.057
12. Chen Q, Zheng J, Wen L, Yang C, Zhang L (2019) A multi-functional-group modified cellulose for enhanced heavy metal cadmium adsorption: Performance and quantum chemical mechanism. Chemosphere 224:509–518. https://doi.org/10.1016/j.chemosphere.2019.02.138

13. Choong CE, Lee G, Jang M, Park CM, Ibrahim S (2019) Fabrication of seashell-incorporated polyurethane for sustainable remediation of Fe(II)-contaminated acidic wastewater. J Polym Environ 27(2):309–317. https://doi.org/10.1007/s10924-018-1339-8
14. Contreras J, Valdés O, Mirabal-Gallardo Y, de la Torre AF, Navarrete J, Lisperguer J, Durán-Lara EF, Santos LS, Nachtigall FM, Cabrera-Barjas G, Abril D (2020) Development of eco-friendly polyurethane foams based on Lesquerella fendleri (A. Grey) oil-based polyol. Eur Poly J 128:109606. https://doi.org/10.1016/j.eurpolymj.2020.109606
15. da Rosa Schio R, da Rosa BC, Gonçalves JO, Pinto LAA, Mallmann ES, Dotto GL (2019) Synthesis of a bio–based polyurethane/chitosan composite foam using ricinoleic acid for the adsorption of Food Red 17 dye. Int J Biol Macromol 121:373–380. https://doi.org/10.1016/j.ijbiomac.2018.09.186
16. Darmadi D, Irfan M, Iqhramullah M, Marlina Marlina MRL (2018) Synthesis of chitosan modified polyurethane foam for adsorption of mercury (II) ions. J Bahan Alam Terbarukan 7(1):18–27. https://doi.org/10.15294/jbat.v7i1.13614
17. Das A, Mahanwar P (2020) A brief discussion on advances in polyurethane applications. Adv Ind Eng Poly Res 3(3):93–101. https://doi.org/10.1016/j.aiepr.2020.07.002
18. Dave VJ, Patel HS (2017) Synthesis and characterization of interpenetrating polymer networks from transesterified castor oil based polyurethane and polystyrene. J Saudi Chem Soc 21(1):18–24. https://doi.org/10.1016/J.JSCS.2013.08.001
19. Doley S, Dolui SK (2018) Solvent and catalyst-free synthesis of sunflower oil based polyurethane through non-isocyanate route and its coatings properties. Eur Polymer J 102:161–168. https://doi.org/10.1016/j.eurpolymj.2018.03.030
20. Fang C-H, Liu P-I, Chung L-C, Shao H, Ho C-H, Chen R-S, Fan H-T, Liang T-M, Chang M-C, Horng R-Y (2016) A flexible and hydrophobic polyurethane elastomer used as binder for the activated carbon electrode in capacitive deionization. Desalination 399:34–39. https://doi.org/10.1016/J.DESAL.2016.08.005
21. Fang Y, Liu X, Wu X, Tao X, Fei W (2021) Electrospun polyurethane/phytic acid nanofibrous membrane for high efficient removal of heavy metal ions. Environ Technol 42(7):1053–1060. https://doi.org/10.1080/09593330.2019.1652695
22. Fernández Rojas M, Pacheco Miranda L, Martinez Ramirez A, Pradilla Quintero K, Bernard F, Einloft S, Carreño Díaz LA (2017) New biocomposites based on castor oil polyurethane foams and ionic liquids for CO2 capture. Fluid Phase Equilib 452:103–112. https://doi.org/10.1016/J.FLUID.2017.08.026
23. Firdaus FE (2014) Synthesis and characterization of soy-based polyurethane foam with utilization of ethylene glycol in polyol. Makara J Technol 18(1). http://journal.ui.ac.id/technology/journal/article/view/2937
24. Fitri RA, Wirakusuma A, Fahrina A, Bilad MR, Arahman N (2019) Adsorption performance of low-cost java plum leaves and guava fruits as natural adsorbents for removal of free fatty acids from coconut oil. Int J Eng 32(10):1372–1378. https://doi.org/10.5829/ije.2019.32.10a.06
25. Gaidukova G, Ivdre A, Fridrihsone A, Verovkins A, Cabulis U, Gaidukovs S (2017) Polyurethane rigid foams obtained from polyols containing bio-based and recycled components and functional additives. Ind Crops Prod 102:133–143. https://doi.org/10.1016/J.INDCROP.2017.03.024
26. Gama N, Ferreira A, Barros-Timmons A (2018) Polyurethane foams: past, present, and future. Materials 11(10):1841. https://doi.org/10.3390/ma11101841
27. Garces IT, Aslanzadeh S, Boluk Y, Ayranci C (2018) Cellulose nanocrystals (CNC) reinforced shape memory polyurethane ribbons for future biomedical applications and design. J Thermoplast Compos Mater 089270571880633. https://doi.org/10.1177/0892705718806334
28. Gennen S, Grignard B, Jérôme C, Detrembleur C (2019) CO$_2$ -sourced non-isocyanate poly(Urethane)s with pH-sensitive Imine linkages. Adv Synth Catal 361(2):355–365. https://doi.org/10.1002/adsc.201801230
29. Guo A, Demydov D, Zhang W, Petrovic ZS (2002) Polyols and polyurethanes from hydroformylation of soybean oil. J Polym Environ 10(1):49–52. https://doi.org/10.1023/A:1021022123733

30. Hadjadj A, Jbara O, Tara A, Gilliot M, Malek F, Maafi EM, Tighzert L (2016) Effects of cellulose fiber content on physical properties of polyurethane based composites. Compos Struct 135:217–223. https://doi.org/10.1016/j.compstruct.2015.09.043

31. Hussein FB, Abu-Zahra NH (2016) Synthesis, characterization and performance of polyurethane foam nanocomposite for arsenic removal from drinking water. J Water Proc Eng 13:1–5. https://doi.org/10.1016/j.jwpe.2016.07.005

32. Inderyas A, Bhatti IA, Ashar A, Ashraf M, Ghani A, Yousaf M, Mohsin M, Ahmad M, Rafique S, Masood N, Iqbal M (2020) Synthesis of immobilized ZnO over polyurethane and photocatalytic activity evaluation for the degradation of azo dye under UV and solar light irardiation. Mater Res Express 7(2):025033. https://doi.org/10.1088/2053-1591/ab715f

33. Iqhrammullah M, Marlina, Hedwig R, Karnadi I, Kurniawan KH, Olaiya NG, Mohamad Haafiz MK, Abdul Khalil HPS, Abdulmadjid SN (2020) Filler-modified castor oil-based polyurethane foam for the removal of aqueous heavy metals detected using laser-induced breakdown spectroscopy (LIBS) technique. Polymers 12(4):903. https://doi.org/10.3390/polym12040903

34. Iqhrammullah M, Marlina M, Khalil HPSA, Kurniawan KH, Suyanto H, Hedwig R, Karnadi I, Olaiya NG, Abdullah CK, Abdulmadjid SN (2020) Characterization and performance evaluation of cellulose acetate-polyurethane film for lead II ion removal. Polymers 12(6):1317. https://doi.org/10.3390/polym12061317

35. Iqhrammullah M, Marlina, Nur S (2020) Adsorption behaviour of hazardous dye (methyl orange) on cellulose-acetate polyurethane sheets. IOP Conf Series Mater Sci Eng 845:012035. https://doi.org/10.1088/1757-899X/845/1/012035

36. Jazi ME, Al-Mohanna T, Aghabozorgi F (2016) Synthesis and applications of isocyanate free polyurethane materials. Global J Sci Front Res B Chem 16(3):1–20. https://pdfs.semanticscholar.org/d887/5723ff1ea02ade527129bd0bace17347fbbc.pdf

37. Jia J, Yin Y, Liu W, Li X, Wang C (2020) Novel colored polyurethane nanoparticle for recyclable dyeing polyester fabric. J Clean Prod 265:121601. https://doi.org/10.1016/j.jclepro.2020.121601

38. Jin L, Gao Y, Yin J, Zhang X, He C, Wei Q, Liu X, Liang F, Zhao W, Zhao C (2020) Functionalized polyurethane sponge based on dopamine derivative for facile and instantaneous clean-up of cationic dyes in a large scale. J Hazard Mater 400:123203. https://doi.org/10.1016/j.jhazmat.2020.123203

39. Khan TA, Nazir M, Khan EA, Riaz U (2015) Multiwalled carbon nanotube–polyurethane (MWCNT/PU) composite adsorbent for safranin T and Pb(II) removal from aqueous solution: batch and fixed-bed studies. J Mol Liq 212:467–479. https://doi.org/10.1016/j.molliq.2015.09.036

40. Kim J, Kumar R, Bandodkar AJ, Wang J (2017) Advanced materials for printed wearable electrochemical devices: a review. Adv Electr Mater 3(1):1600260. https://doi.org/10.1002/aelm.201600260

41. Kumari S, Chauhan GS, Ahn J (2016) Novel cellulose nanowhiskers-based polyurethane foam for rapid and persistent removal of methylene blue from its aqueous solutions. Chem Eng J 304:728–736. https://doi.org/10.1016/j.cej.2016.07.008

42. Kupeta AJK, Naidoo EB, Ofomaja AE (2018) Kinetics and equilibrium study of 2-nitrophenol adsorption onto polyurethane cross-linked pine cone biomass. J Clean Prod 179:191–209. https://doi.org/10.1016/j.jclepro.2018.01.034

43. Lee A, Deng Y (2015) Green polyurethane from lignin and soybean oil through non-isocyanate reactions. Eur Polymer J 63:67–73. https://doi.org/10.1016/j.eurpolymj.2014.11.023

44. Lellis B, Fávaro-Polonio CZ, Pamphile JA, Polonio JC (2019) Effects of textile dyes on health and the environment and bioremediation potential of living organisms. Biotechnol Res Innov 3(2):275–290. https://doi.org/10.1016/j.biori.2019.09.001

45. Li G, Chai K, Zhou L, Tong Z, Ji H (2019) Easy fabrication of aromatic-rich cellulose-urethane polymer for preferential adsorption of acetophenone over 1-phenylethanol. Carbohyd Polym 206:716–725. https://doi.org/10.1016/j.carbpol.2018.11.057

46. Li J, Gong J-L, Zeng G-M, Zhang P, Song B, Cao W-C, Liu H-Y, Huan S-Y (2018) Zirconium-based metal organic frameworks loaded on polyurethane foam membrane for simultaneous removal of dyes with different charges. J Colloid Interface Sci 527:267–279. https://doi.org/10.1016/j.jcis.2018.05.028

47. Liu Q, Zhou Y, Lu J, Zhou Y (2020) Novel cyclodextrin-based adsorbents for removing pollutants from wastewater: a critical review. Chemosphere 241:125043. https://doi.org/10.1016/j.chemosphere.2019.125043

48. Liu T, Sun S, Zhou L, Li P, Su Z, Wei G (2019) Polyurethane-supported graphene oxide foam functionalized with carbon dots and TiO_2 particles for photocatalytic degradation of dyes. Appl Sci 9(2):293. https://doi.org/10.3390/app9020293

49. Lubis S, Maulana I, Masyithah (2018) Synthesis and characterization of TiO_2/α-Fe_2O_3 composite using hematite from iron sand for photodegradation removal of dye. J Natural 18(1):38–43. https://doi.org/10.24815/jn.v18i1.8649

50. Mahmood A, Shi G, Wang Z, Rao Z, Xiao W, Xie X, Sun J (2021) Carbon quantum dots-TiO_2 nanocomposite as an efficient photocatalyst for the photodegradation of aromatic ring-containing mixed VOCs: An experimental and DFT studies of adsorption and electronic structure of the interface. J Hazard Mater 401:123402. https://doi.org/10.1016/j.jhazmat.2020.123402

51. Majul L, Wirth S, Levin L (2020) High dye removal capacity of Peniophora laxitexta immobilized in a combined support based on polyurethane foam and lignocellulosic substrates. Environ Technol 1–12. https://doi.org/10.1080/09593330.2020.1801851

52. Mallakpour S, Behranvand V (2021) Polyurethane sponge modified by alginate and activated carbon with abilities of oil absorption, and selective cationic and anionic dyes clean-up. J Clean Prod 312:127513. https://doi.org/10.1016/j.jclepro.2021.127513

53. Manabe S, Adavan Kiliyankil V, Takiguchi S, Kumashiro T, Fugetsu B, Sakata I (2021) Graphene nanosheets homogeneously incorporated in polyurethane sponge for the elimination of water-soluble organic dyes. J Colloid Interface Sci 584:816–826. https://doi.org/10.1016/j.jcis.2020.10.012

54. Marlina M, Iqhrammullah M, Darmadi, Mustafa I, Rahmi M (2019) The application of chitosan modified polyurethane foam adsorbent. RASĀYAN J Chem 12(2):494–501. https://doi.org/10.31788/RJC.2019.1225080

55. Marlina M, Iqhrammullah M, Saleha S, Fathurrahmi, Maulina FP, Idroes R (2020) Polyurethane film prepared from ball-milled algal polyol particle and activated carbon filler for NH_3–N removal. Heliyon 6(8):e04590. https://doi.org/10.1016/j.heliyon.2020.e04590

56. Mohammadi A, Lakouraj MM, Barikani M (2014) Preparation and characterization of p-tert-butyl thiacalix[4]arene imbedded flexible polyurethane foam: an efficient novel cationic dye adsorbent. React Funct Polym 83:14–23. https://doi.org/10.1016/j.reactfunctpolym.2014.07.003

57. Mohammadi Y, Faghihi K (2020) A new magnetic β-cyclodextrin polyurethane nanocomposite for the removal of organic pollutants in wastewater. Iran Polym J 29(10):933–942. https://doi.org/10.1007/s13726-020-00852-2

58. Mohammadpour R, Mir Mohamad Sadeghi G (2020) Effect of liquefied lignin content on synthesis of bio-based polyurethane foam for oil adsorption application. J Poly Environ 28(3):892–905. https://doi.org/10.1007/s10924-019-01650-5

59. Nasiri S, Alizadeh N (2019) Synthesis and adsorption behavior of hydroxypropyl-β-cyclodextrin–polyurethane magnetic nanoconjugates for crystal and methyl violet dyes removal from aqueous solutions. RSC Adv 9(42):24603–24616. https://doi.org/10.1039/C9RA03335A

60. Nikkhah AA, Zilouei H, Asadinezhad A, Keshavarz A (2015) Removal of oil from water using polyurethane foam modified with nanoclay. Chem Eng J 262:278–285. https://doi.org/10.1016/j.cej.2014.09.077

61. Panda SS, Panda BP, Nayak SK, Mohanty S (2018) A review on waterborne thermosetting polyurethane coatings based on castor oil: synthesis, characterization, and application. Polym Plast Technol Eng 57(6):500–522. https://doi.org/10.1080/03602559.2016.1275681

62. Pantone V, Annese C, Fusco C, Fini P, Nacci A, Russo A, D'Accolti L (2017) One-pot conversion of epoxidized soybean oil (ESO) into soy-based polyurethanes by $MoCl_2O_2$ catalysis. Molecules 22(2):333. https://doi.org/10.3390/molecules22020333

63. Patil CK, Jirimali HD, Paradeshi JS, Chaudhari BL, Alagi PK, Hong SC, Gite VV (2018) Synthesis of biobased polyols using algae oil for multifunctional polyurethane coatings. Green Mater 6(4):165–177. https://doi.org/10.1680/jgrma.18.00046

64. Pillai PKS, Li S, Bouzidi L, Narine SS (2018) Polyurethane foams from chlorinated and non-chlorinated metathesis modified canola oil polyols. J Appl Polym Sci 135(33):46616. https://doi.org/10.1002/app.46616

65. Purwanto E, Riadi L, Tamara NI, Mellisha Ika K (2014) The optimization of ozonolysis reaction for synthesis of biopolyol from used palm cooking oil. ASEAN J Chem Eng 14(1):1–12

66. Rabello LG, da Conceição C, Ribeiro R (2021) A novel vermiculite/vegetable polyurethane resin-composite for thermal insulation eco-brick production. Compos B Eng 221:109035. https://doi.org/10.1016/j.compositesb.2021.109035

67. Rahmi R, Iqhrammullah M, Audina U, Husin H, Fathana H (2021) Adsorptive removal of Cd (II) using oil palm empty fruit bunch-based charcoal/chitosan-EDTA film composite. Sustain Chem Phar 21:100449. https://doi.org/10.1016/j.scp.2021.100449

68. Rahmi R, Lubis S, Az-Zahra N, Puspita K, Iqhrammullah M (2021) Synergetic photocatalytic and adsorptive removals of metanil yellow using TiO_2/grass-derived cellulose/chitosan (TiO_2/GC/CH) film composite. Int J Eng 34(8):1827–1836. https://doi.org/10.5829/ije.2021.34.08b.03

69. Rajput KN, Patel KC, Trivedi UB (2016) β -cyclodextrin production by cyclodextrin glucanotransferase from an alkaliphile microbacterium terrae KNR 9 using different starch substrates. Biotechnol Res Int 2016:1–7. https://doi.org/10.1155/2016/2034359

70. Ranote S, Kumar D, Kumari S, Kumar R, Chauhan GS, Joshi V (2019) Green synthesis of Moringa oleifera gum-based bifunctional polyurethane foam braced with ash for rapid and efficient dye removal. Chem Eng J 361:1586–1596. https://doi.org/10.1016/j.cej.2018.10.194

71. Riaz T, Ahmad A, Saleemi S, Adrees M, Jamshed F, Hai AM, Jamil T (2016) Synthesis and characterization of polyurethane-cellulose acetate blend membrane for chromium (VI) removal. Carbohyd Polym 153:582–591. https://doi.org/10.1016/j.carbpol.2016.08.011

72. Savelyev Y, Rudenko A, Robota L, Koval E, Savelyeva O, Markovskaya L, Veselov V (2009) Novel polymer materials for protecting crew and structural elements of orbital station against microorganisms attack throughout long-term operation. Acta Astronaut 64(1):36–40. https://doi.org/10.1016/j.actaastro.2008.06.013

73. Selvasembian R, Gwenzi W, Chaukura N, Mthembu S (2021) Recent advances in the polyurethane-based adsorbents for the decontamination of hazardous wastewater pollutants. J Hazard Mater 417:125960. https://doi.org/10.1016/j.jhazmat.2021.125960

74. Sen SK, Raut S, Bandyopadhyay P, Raut S (2016) Fungal decolouration and degradation of azo dyes: a review. Fungal Biol Rev 30(3):112–133. https://doi.org/10.1016/j.fbr.2016.06.003

75. Senthivelan T, Kanagaraj J, Panda RC (2016) Recent trends in fungal laccase for various industrial applications: an eco-friendly approach-a review. Biotechnol Bioprocess Eng 21(1):19–38. https://doi.org/10.1007/s12257-015-0278-7

76. Seto C, Chang BP, Tzoganakis C, Mekonnen TH (2021) Lignin derived nano-biocarbon and its deposition on polyurethane foam for wastewater dye adsorption. Int J Biol Macromol 185:629–643. https://doi.org/10.1016/j.ijbiomac.2021.06.185

77. Sharmin E, Zafar F (2012) Polyurethane: An Introduction. In: Zafar, F., Sharmin, E., editors. Polyurethane [Internet]. London: IntechOpen; 2012 [cited 2022 Mar 18]. https://www.intechopen.com/chapters/38589/10.5772/51663

78. Silva RGC, Ferreira TF, Borges ÉR (2020) Identification of potential technologies for 1, 4-Butanediol production using prospecting methodology. J Chem Technol Biotechnol 95(12):3057–3070. https://doi.org/10.1002/jctb.6518

79. Suhas, Gupta VK, Carrott PJM, Singh R, Chaudhary M, Kushwaha S (2016) Cellulose: a review as natural, modified and activated carbon adsorbent. Biores Technol 216:1066–1076. https://doi.org/10.1016/j.biortech.2016.05.106

80. Sui S, Quan H, Hu Y, Hou M, Guo S (2021) A strategy of heterogeneous polyurethane-based sponge for water purification: Combination of superhydrophobicity and photocatalysis to conduct oil/water separation and dyes degradation. J Colloid Interface Sci 589:275–285. https://doi.org/10.1016/j.jcis.2020.12.122

81. Szycher M (2013) Szycher's handbook of polyurethanes 2nd ed.

82. Tai NL, Adhikari R, Shanks R, Halley P, Adhikari B (2018) Flexible starch-polyurethane films: effect of mixed macrodiol polyurethane ionomers on physicochemical characteristics and hydrophobicity. Carbohyd Polym 197:312–325. https://doi.org/10.1016/j.carbpol.2018.06.019

83. Wong YC, Tan YP, Taufiq YH, Ramli I, Tee HS (2015) Biodiesel production via transesterification of palm oil by using CaO–CeO$_2$ mixed oxide catalysts. Fuel 162:288–293. https://doi.org/10.1016/j.fuel.2015.09.012

84. Yamamoto K, Shiono T, Yoshimura R, Matsui Y, Yoneda M (2018) Influence of hydrophilicity on adsorption of caffeine onto montmorillonite. Adsorpt Sci Technol 36(3–4):967–981. https://doi.org/10.1177/0263617417735480

85. Yue H, Zhao Y, Ma X, Gong J (2012) Ethylene glycol: properties, synthesis, and applications. Chem Soc Rev 41(11):4218. https://doi.org/10.1039/c2cs15359a

86. Zhang Z, Jiang H, Li R, Gao S, Wang Q, Wang G, Ouyang X, Wei H (2021) High-damping polyurethane/hollow glass microspheres sound insulation materials: preparation and characterization. J Appl Polym Sci 138(10):49970. https://doi.org/10.1002/app.49970

87. Zhao J, Xu L, Su Y, Yu H, Liu H, Qian S, Zheng W, Zhao Y (2021) Zr-MOFs loaded on polyurethane foam by polydopamine for enhanced dye adsorption. J Environ Sci 101:177–188. https://doi.org/10.1016/j.jes.2020.08.021

88. Zou C, Zhang H, Qiao L, Wang X, Wang F (2020) Near neutral waterborne cationic polyurethane from CO$_2$ -polyol, a compatible binder to aqueous conducting polyaniline for eco-friendly anti-corrosion purposes. Green Chem 22(22):7823–7831. https://doi.org/10.1039/D0GC02592E

Application of Polymer/Carbon Nanocomposite for Organic Wastewater Treatment

Adane Adugna Ayalew

Abstract Over the last years, numerous investigations on the utilization of carbon-based materials for environmental sustainability remediation have been studied. Modifying the carbon particles by using nanopolymeric materials has been underlined as a novel group of economically kind for the application of dyes and pigment removal in textile wastewater. Polymers provide typical properties with high surface area that improve processability and selectivity to remove various contaminants. Thus, the emerging of polymeric materials onto carbon precursors is an innovative way to encounter the environmental water standard by removing water pollutants. This review emphasized the actual and latest accomplishments in production, physiochemical analysis, and usage of polymeric/carbon nanocomposites for the application of textile effluent management. The surface modification of carbon materials by using polymeric materials enhanced the physical and chemical of the properties of composite materials. Besides, the production of new nanocomposite materials from carbon nanoparticles and polymer materials is also expressed. At specific, higher consideration has been focused on the current progresses with different types of carbon and polymer nanocomposites with the main application for the elimination of several pollutants including dyes and pigments. Besides, the upcoming potential and some problems are also predicted to touch the optimum routine of the various adsorbent materials.

Keywords Nanocarbon · Nanocomposites · Bio-polymer · Dyes · Wastewater

1 Introduction

Nowadays, the standard level of water and wastewater is become declining because of the fast development of industries that release water sources. The use of synthetic chemical products including dyes, pigments, pharmaceuticals, synthetic detergents, fertilizers, food additives, and agrochemicals increases through time in developing

A. A. Ayalew (✉)
Faculty of Chemical and Food Engineering, Bahir Dar Institute of Technology, Bahir Dar University, Bahir Dar, Ethiopia

© The Author(s), under exclusive license to Springer Nature Singapore Pte Ltd. 2022
A. Khadir and S. S. Muthu (eds.), *Polymer Technology in Dye-containing Wastewater*,
Sustainable Textiles: Production, Processing, Manufacturing & Chemistry,
https://doi.org/10.1007/978-981-19-1516-1_8

and developed countries [24, 29]. The community uses such products at a fast rate simultaneously and consequently, and huge pollutants are generated during the operation process and in waste disposal [61]. Moreover, the effluent waste contains undesirable compound interaction such as carbon powder release toxic nature and upsets the ecology [43].

The textile industry production is one of the leading parts that consumes a huge quantity of water and uses more than 500 different complexes and additives [34]. The discharge of dyestuffs wastes concentration is found mostly in the range of 10–200 mg/L which encloses large salty, oil suspension, suspended solids, and detergents, which outcomes in strong coloured, salt, and basic effluents [8, 69, 75]. Due to the increasing usage of textile dyestuffs and pigments, the elimination of pigments and dyes colour from textile effluents is becoming the main ecological challenge. It is so difficult to remove the extremely minor amounts of dyestuffs which present at textile effluents especially, dyestuffs found at 1.5 mg/L and below this concentration are very noticeable and undesirable [48, 60]. Through the production process, the pigment colour and the fragment of toxic aromatic compounds are moved into water bodies. These effluents are producing a vast volume of polycyclic or heterocyclic organic molecules present with the effluent are dangers to social and animals well-being [11]. On the other way, the aromatic molecules have one or more rings and are more established and also have a complex configuration such as dyes that can stay in the ecosystem without disintegrating for an extensive time. The different aromatic compounds within difficult configurations have been existing in textile wastewater effluents, producing hard effluent wastewater plant because of higher initial dyestuff concentration, high colour, more stability, and non-degradable easily [96]. Currently, above 100,000 various kinds of dyestuffs and pigments arrangements are prepared and different dyestuff configurations are produced and currently, greater than half a million-ton dyes are utilized for textile industries [73, 92]. Among synthetic dyestuffs, 10% are lost in the waste during the application process [10]. In recent years, dyes in textile industries have been developed. Synthetic dyes are miscible in water and cause significant health and environmental problems during mixed into water bodies [58].

In recent years, there is rising consciousness of the significance of existing environmental outgoing practices proficient in eliminating contaminants such as dyestuffs from water bodies and in the same way protecting the health of affected populations [21]. So, serious consideration should be compensated to manage the wastewater before releasing to the atmosphere [47, 59]. In concept, a diversity of chemical, physical, and biological approaches can be utilized in the elimination of pigment and dyestuffs encompass with the effluents, for instance, ion exchange, electrochemical, sedimentation, oxidation, flocculation, and ultrafiltration methods are mainly conducted in effluents treatment [52, 94]. Nevertheless, all these techniques can only be operative and financially advantageous at higher dye concentrations. Besides this, they have drawbacks in high consumption of energy [15, 97, 104]. Adsorption process is currently an attractive and good-looking way for the elimination of pigments enclosed effluents because of its inexpensive and simple operational techniques [17, 25, 78, 100]. Currently, because of the distinctive structural, physical, mechanical, chemical and morphological characteristics of carbon-based nanocomposites including hardness, simple modification, good temperature, and chemical resistance, and conductivity carbon/polymer composite attracted high attention to research in the elimination of pigments, and dyes from textile wastewater [89, 90].

Recently, many investigations paid attention to carbon/polymer nanocomposite which can be used to remove dyestuffs from textile effluents by using several approaches including membranes, adsorptions, and photocatalytic [63].

Currently, there are four basic types of carbon nano-sized particles including graphene oxide nanoparticles, carbon nanofibers nanoparticles, graphene nanoparticles, and reduced graphene oxide nanoparticles. These classes of nano-sized carbon materials have been received wonderful consideration for applying in textile effluent treatment because of their superior adsorption properties [9, 67].

Thus, the literature reported that nanomaterial derived from carbon sources through surface modification by a polymer can be used for organic effluent treatment [72]. This chapter focuses on the invention and a prospective application of polymer/carbon nanocomposites through particular consideration towards the removal of organic contaminants mainly from textile wastewater effluents.

2 Carbon Nanocomposite Materials

Nanocarbon composite nanomaterials mainly contain carbon crystalline elements, which are used as packing components in the composite. These nanomaterials comprise solid chemically attached carbon particles and commonly show strong material characteristics.

The entire pattern of carbon nanomaterial is displayed from Fig. 1. Recent most developing carbon-based nanomaterials are encompassed graphene oxide nanoparticles, carbon nanofibers nanoparticles, graphene nanoparticles, and reduced graphene oxide nanoparticles.

The graphene carbon is the single sheet of the carbon atom and the highest solid kind of material in normal situations. Nanocarbon nanoparticles in the graphene molecules are arranged by replication their atoms to form a hexagonal structure. Graphene oxide nonparties can be prepared through the oxidation of graphite

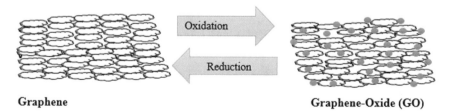

Fig. 1 Structure of graphene and graphene oxide nanomaterials through oxidation and reduction reaction

nanoparticles. When graphite molecules are dispersed in a basic solution, graphene oxide is produced. It is clearly shown that graphene oxide is directly formed through the oxidation of graphite molecules. Here, the key distinction among the two carbon nanoparticles (graphene oxide and graphite) is the interlayer distance between the entity nuclear sheets. Since the graphene oxides are more chemically reactive with water molecules while graphite oxide has a multi-layer during graphene oxide dispersion whereas graphene oxide contains a monolayer. Furthermore, reduced graphite oxide is formed through the deoxygenation of graphene oxide after being exposed to oxidation and reduction reactions. When reduction reaction is applied to graphene oxide, the conductivity property of graphene is became restored [5, 20]. Carbon nanotubes have a number of cavities that contain double graphene oxide nanoparticles films. These layers are arranged in a rod-shaped hexagonal frame structure.

Carbon nanotube particles encompass the different types of coating and shapes including single-walled carbon nanotubes (CNTs), multi-walled (MWCNTs) [6, 54]. Graphene oxide sheets in carbon nanofibers have been layout in the form of shaft structures [12, 62]. Hence, the entire structure and configurations of nanofiber particles can be analysed through angle diffraction between the graphene oxide particles sheet and fibres axis. However, for MWNCTs, the angle is zero [37, 46].

2.1   Category of Carbon Nanomaterials

Nanotechnology displayed a great approach with the area of water and effluent treatment for the elimination and offering a sustainable method to secure water resources. Figure 2 shows the cataloguing of carbon nanomaterials including graphene oxide nanoparticles, carbon nanofibers nanoparticles, graphene nanoparticles, and reduced graphene oxide nanoparticles [18, 66]. These nanomaterials have excellent and distinctive characteristics to utilize effectively in composites with other materials or individuals. But, regarding industrial effluent treatment, these molecules have

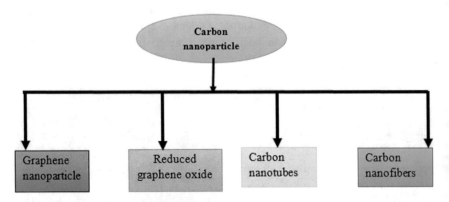

Fig. 2 Classification of carbon-based nanomaterials

basic practical blocks including the difficulty of recovery, formation of accumulation particles, producing unfavourable effluences on ecology and also to the environment. Thus, the favourable method toward sustain the utilization of these nanoparticles materials is to emerge new nanocomposite materials by taking the benefit of composite materials properties [98].

2.2 Physiochemical Properties of Carbon Nanocomposites

The preparation and physicochemical analysis of nanomaterials provide a new approach to resolve the problem faced by the exclusion of pigments, dyes, and colorants in industrial effluents. This can be used to minimize consumption of energy as well as raw material expenses [36]. Through controlling and monitoring the movement of molecules at nanometer scale molecular diffusion over permeability particles is a developing investigation point. Since now a day nanomaterial contains abroad prospective use, principally in the area of wastewater treatment. Because of wonderful hydrophobicity, frictionless interior, carbon nanocomposites has been aided distinctive mass transfer with liquid molecules interface according to the molecular configuration [30].

The electrostatic charges have also an impact during distribution on carbon nanocomposites in various ions and water separation effects. It has been investigated that through altering the configurations on the carbon nanocomposites, the carbon nanocomposites could be monitored to enable the water intake though completely obstructing the transfer of molecules via the stations [13]. Carbon nanocomposites have been confirmed as favourable adsorbents for eliminating numerous environmental contaminants because of their exceptional properties such as higher pore structure, good thermal stability, higher specific surface areas, and chemical resistance [82]. Furthermore, their adsorption abilities and intake can be increased thru leading some functional groups towards their exterior surfaces such as polymers, carboxyl group, and hydroxyl [19, 40].

Graphene contains great potential for use in water transport and adsorbent of water pollutants. Due to its tremendously great mechanical strength, presence of good active site and porosity, graphene can be used effectively for water purification and effluent treatment [40, 81]. Graphene has π-bonds, and graphene oxide structures have plentiful oxygen-promising functional groups. Both have very great specific surface areas, which enhance for adsorption of dyes and pigment from textile water effluent [88].

Overall, there are five likely interaction systems during removal of various pollutants from effluents by using graphene nanoparticles, such as hydrogen bonds, π bonds, hydrophobic effects, electrostatic interactions and covalent bonding [103]. Graphene oxide and its nanocomposites exhibited pretty favourable towards for complete elimination of dyestuffs and pigments. Moreover, the sorption process and the adsorption process for cationic and anionic dyes primarily occur by covalent bonding and electrostatic interaction [98]. Graphene oxide showed a high adsorption competence for both positively and negatively charge dyes; however, small attraction is occurred for negatively charged dyes because of the presence of high repulse interaction between the neighbouring particle layer. On the other way, graphene nanocomposite materials are mainly good nanoadsorbnet for the removal of negatively charged molecules because of the presence of higher chemical interaction between the adsorbate and adsorbent surface [53].

3 Polymeric Nanocomposites

Composite materials are molded by the mixture of two or above materials with various physicochemical properties [45]. These nanomaterials are described by possession detached within the identical configuration. The component materials of the composites are categorized into two classifications such as support and matrix. Matrix materials are accountable for keeping the locations of the reinforcement materials, by adjacent and assisting them whereas reinforcement materials add to their mechanical strength, thermal stability, chemical resistance, and higher adsorbent efficiency that deliver novel properties to the matrix. As a result, combined materials have mutual physiochemical properties of both materials i.e. matrix and reinforcement. But, the characteristics of the composites may differ completely from the original materials that compose them [44, 76, 86].

Polymeric materials are confirmed with exceptional properties such as the comfort of preparation, inexpensive, high flexibility, biocompatibility, strong machinability, and lightweight. However, its use is occasionally restricted by its small units, low resistance, and low conductivity related to other materials such as ceramics or metals. One substitute to advance these physicochemical properties is by introducing small amounts of nanofillers within different shapes and configurations. This reinforcement has been supported for many periods to improve chemical, thermal, optical, and electrical properties and hence would be an outstanding adsorbent material for the removal of dyes and pigment molecules [3, 22].

A known amount of organic and inorganic nanofillers have been mixed with different kinds of polymers to obtain polymer nanocomposites (PNCs) deliberates with definite properties for the application of pollutant removal as adsorbents in textile wastewater. Among organic nanofillers are carbon-based polymers composite nanofiller including graphene, carbon fibres, and graphite, which are responsible for the adsorption process during the elimination of enormous contaminants from textile industries effluents.

4 Polymer/carbon Nanocomposites (PCNCs)

Carbon has a distinctive electrical structure and is competent to form weak bonds with various polymer materials. Carbon can occur in several atom forms, and carbon nanomaterials can be prepared and were used for environmental application for adsorbents in the removal of pollutants. At the nanoscale level, carbon has unique features relative to other adsorbent materials, including strong physical properties, high conductivity, good temperature solidity, and good chemical resistance. Based on the carbon geometrical structure and shape, carbon nanomaterials are categorized into several kinds. Up to now, graphene nanoparticles, graphite, carbon nanofibers nanoparticles, reduced graphene oxide nanoparticles, and carbon nanotube nanoparticles are the famous type of carbon nanoparticles adsorbent materials and they have been effectively used for the removal of organic pigment and dyes.

Recently, polymer-carbon nanocomposite materials are looking into several purposes, principally in the treatment of organic dyestuffs from textile industry effluents where the polymer/carbon nanocomposite obsessed outstanding physiochemical properties. Polymer/carbon nanocomposite has incredible physiochemical distinguishes that is constructive for the removal of pigments and dyes.

4.1 Polymer-Carbon Nanotube Nanocomposites

Nanocarbon nanotubes are contained several tubes which are made of carbon molecules with cylindrical bulky particles inclosing hexagonal configurations. Carbon nanotubes have an incredible explicit character such as higher specific particle area, porosity, high selectivity, structural diversity and are favourable for textile water treatment. In recent times, carbon nanotubes are broadly used as a low-cost adsorbent for the treatment of the industry's effluents, one of which is the textile industry effluent [79].

Today, the composite research of carbon nanotube with polymer has fascinated lot of notice in order to advance the carbon nanotube exterior features and also to improve the removal of dyestuffs and pigments. The surface modification and coating of carbon nanotubes by using polymeric materials are a superficial method to develop a new nanoadsorbent material with a higher sorption capacity of dyestuffs and pigments.

Recently, carbon nanotubes were combined by a known polymer called poly (sodium-polystyrene sulfonate) compound. Explicitly, carbon nanotube was first layered by using dopamine chemicals. The mixture of the solution is then polymerized in order to produce a composite of polydopamine/carbon nanotube. The resulted polymer/carbon nanotube composite nanomaterial has displayed incredible disperse in the aqueous solution that would improve the exclusion of dye molecules such as methylene blue. The maximum adsorption capacity for methylene blue removal from textile industrial effluent was obtained to be 174.0 mg/at 25.0 min contact time [90]. The other polymer is called polyaniline, which modified the carbon nanotube surface to improve the removal capacity of the alizarin yellow dyestuff. The nanocomposite was synthesized by using the situ-polymerization method [89]. The optimum sorption value for removal of alizarin dyestuff was found to be 884.80 mg/g. The adsorption records were followed the Langmuir isotherm models and also the sorption kinetics was confirmed as a pseudo-second-order. Moreover, the composite energy changes exhibited the sorption process for removal of alizarin yellow dyestuff was followed a spontaneous and also exothermic. Thus, this polymer/carbon nanotube nanocomposite can be used as a prospective novel adsorbent material for the elimination of organic effluents.

In polymer/carbon nanotube nanocomposite adsorbent preparation, most researchers have concentrated on expanding biopolymers. Chitosan is one of the known biopolymer compounds that can be used as composite biomaterials with carbon nanotube as presented in Table 1. Recently, chitosan has become the best solid nanomaterial for textile industry effluent removal because of its extensive accessibility, non-toxicity, and biodegradability. Moreover, chitosan has an abundant number of active sites on its entire surface, which is accountable for enhancing to the removal of contaminants [85]. Besides that, chitosan could also aid as excellent for carbon nanotechnology that is incapable toward dissolve efficiently at an aqueous solution hence that clarifies chitosan's main function normally used for polymer/carbon nanotubes nanocomposites [28, 56]. Similarly, Abbasi [1] synthesized nanocarbon nanotube-chitosan composite material for the removal of most commonly used in the textile industry namely direct blue 71 and reactive blue 19. From result showed that the adsorption value towards the nanocomposite for direct blue 71 was found as 61.35 mg/g and for reactive blue 19 was 97.08 mg/g.

Correspondingly, Zhu et al. [102] prepared a composite material from carbon nanotube/chitosan using cadmium sulphide, hydrogen peroxide as surface modifiers to enhance the removal efficiency of methylene orange dyestuff. The result has shown that the dyestuff was completely removed from the wastewater. In addition to these, Zhu et al. [101] produced a nanocomposite material from graphite carbon nanotube/chitosan to eliminate chromium (III) from wastewater. The specific surface area of the pore nanocomposite particle was obtained as 39.2 m^2, and the adsorption capacity was outstanding and was found 261.9 mg/g with the adsorption pH value was 6.2. Further, biopolymer materials are integrated with other biopolymer sources including natural starches [57]. The authors synthesized a nanocomposite adsorbent from a biopolymer compound such as carbon nanotube/starches/ vitamin C is used for the eliminating of methylene orange dyes.

The adsorption capacity for methyl orange removal on prepared nanocomposite was obtained as 11.11 mg/g at a 2.0 pH value. Similarly, Chang et al. [16] studied the sorption efficiency of the two most types of dyes such as methylene orange and methylene blue using the nanocomposite adsorbent prepared from MWCNTs/starch/iron oxide nanocomposite. In this composite, the starch primary was coated towards the external surface of the composites. In this investigation, the biopolymer particles were effectively attached and dispersed with the help of iron oxide nanoparticles. The result showed that the sorption capacity for methyl orange and methylene blue was found to be 135.5 and 94.2 mg/g consistently.

Correspondingly Yan et al. [93] synthesized a composite from guar-gum/carbon nanotube for the elimination of two common dyestuffs such as methyl orange dye and neutral red dye from wastewater effluents. Guar-gum is hydrophilic and the addition of this towards magnetic iron oxide/carbon nanotubes resulted in a better spreading

Table 1 Removal of organic effluents using carbon/polymer nanocomposite through adsorption

Composite	Waste dye	Surface area (m^2/g)	Adsorbent dosage (g/L)	pH	Adsorption capacity (mg/g)	Isotherm	References
CNTs/chitosan/SiO$_2$	Direct Blue71	–	1.0	6.8	61.35	Langmuir	[1]
	Reactive Blue19	–	1.0	2.0	97.08	Langmuir	
Magnetic graphite carbon/chitosan	Congo Red dye	39.2	1.0	6.3	261.8	Langmuir	[101]
Fe$_3$O$_4$/polyester/MWCNT	Orange (II) dye	–	0.1	6.2	67.57	Langmuir	[27]
Polypyrrole/CNTs-CoFe$_2$O$_4$	Methylene blue	150.41	1.0	6.0–9.0	138.0	Langmuir	[50]
	Methyl orange	150.41	1.0	3.0–9.0	116	Langmuir	
Guargum/CNTs/iron oxide	Methylene blue	–	–	–	57.87	Langmuir	[93]
Starch/MWCNTs/iron oxide	Methyl orange	155.36	0.5	–	135.8	–	[16]
	Methyl blue	155.36	0.5	–	94.1	–	
MWCNTs/Ferro ferric oxide/ionic based Polyether	Sunset orange	–	0.1	6.2	85.47	Langmuir	[27]
Cadmium sulfide/MWCNTs/Chitosan	Methyl orang	–	0.1	–	–	–	[102]
Starch/MWCNTs/vitamin C	Acid fuschskin	–	1.0	2.0–9.0	132.15	Langmuir	[57]

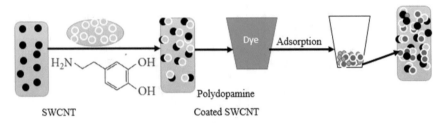

Fig. 3 Illustration of MB removal by using Polydopamine/SWCNT nanocomposite

of carbon nanotubes in solution. In this report, the maximum adsorption competence for neutral red and methylene blue was found as 89.8 and 61.9 mg/g respectively [35].

On one side, carbon nanotubes are contributed to the polymer composite in order to improve the stability of the material during the adsorption process. Such activities were investigated by Gao et al. [27] and Li et al. [49]. The removal of the process of methylene blue by using a Polydopamine and single-wall carbon nanotube (SWCNT) can be observed from Fig. 3. In this case, the polymer offers enhanced SWCNT material to be more stable and to adsorb dye molecules from textile industry effluent [2].

From the study, Gao et al. [27] detached to the negative charge dye species such as safranin-O, orange-II (OII), and Amaranth dyes. The authors investigated the adsorption capacity of each dye by using magnetic/polyester/MWCNTs composite. The result showed that the adsorption value for safranin-O was 65.57 mg/g, orange (II) was 85.47 mg/g and finally for Amaranth dyes was found to be 47.4 mg/g. The optimum pH value for all dye removal was obtained as 6.2. Similarly, Li et al. [49] prepared CNT-CoFe$_2$O$_4$/Polypyrrole composite to remove some dyestuff including Alexa fluor, methyl orange, and methylene blue dyes from industrial effluents. The result showed that the prepared composite material has a capacity to adsorbed dye molecules effectively. The adsorption result for Alexa fluor was found 132.15 mg/g, for methyl orange was found as 116.0 mg/g, and finally for methylene blue was obtained to be 137.0 mg/g. The prepared nanocomposite material has magnetic, adsorptive catalytic properties and this has higher adsorption capacity than the individual components.

4.2 Graphene-Polymer Nanocomposite

Graphene is a nanomaterial of carbon and contains a two-dimensional arrangement with a monolayer of the carbon molecule, which in atoms are bonded together in a repeatedly way to form hexagonal structure. Graphene nanomaterials have remarkable physical and chemical possessions which create them a smart material for wastewater treatment applications [38]. Recently, graphene/composite has been attracted numerous research attention as novel nanoadsorbents because such composites have more than one geometrical dimension configuration which offers advanced porosity, outstanding ionic conduction, good thermal stability, higher material strength, and higher chemical resistance. Graphene/polymer nanocomposite materials have been involved for different applications including for ultra-filtration membrane development. Thus, graphene-based nanocomposite materials have been reported as tremendous nominees and potential for the elimination of dyes and pigments from industrial effluents. Graphene/polymer nanocomposite materials are getting substantial rank in textile water treatment because of the combination of such kinds of hybrid nanomaterials. Hence, the utilization of graphene and polymeric materials as single composite nanomaterial is offering a promising novel adsorbent class of material containing high selectivity, higher porosity, higher adsorption efficiency, and higher mechanical strain.

Polymer/graphene nanocomposites can be prepared through several methods depending on the requisite belongings in the objective adsorbent shape and structure.

Nanographene oxide nanoparticles have two-dimension hexagonal matrixes of covalently bonded carbon molecules and have a porosity area of particles nearly 2600 m^2/g. Moreover, they have distinctive physicochemical properties. The presence of higher particles surface area, good electrical conductivity, higher material strength, electrostatic pilling property, thermal stability, and other physicochemical properties allowed graphene oxide for textile wastewater treatment [91]. Moreover, graphene oxide enclosed by epoxide and hydroxide on the bottom and top of the entire surface, furthers the carboxyl groups disseminated in arbitrary mode at the boundaries of graphene oxide pages [77].

Polymer/graphene nanocomposite has incredible properties. In the latest investigation, graphene nanoparticles were hybrid with the sodium salt of alginic acid called sodium alginate to synthesis a new adsorbent material. The result revealed that the prepared nanocomposite material contains a number of pore and active site which is a tremendous separation capacity with higher permeability. Membrane filtration was developed from the hybrid of higher temperate resistance type of polymer called poly (2-diethyl aminoethyl methacrylate) and graphene oxide nanoparticles for removal of dyes and pigments. The result of composite material has a uniform and controlled pores distribution on the membrane surface. The composite of the polymer/graphene nanoparticles is modified by reacting with the CO_2 in water system. In this manner, protonated is occurred and also when argon molecule is fed to the system, deprotonated is followed. This surface modification could improve the morphology of the composite material to speed up the separation process of the membrane [23]. Moreover, surface modification can be used to develop and fabricate porous polymeric nanocomposite membranes with strong and high removal efficiency [99]. In the carbon/polymer nanocomposite, the composite of polymer with graphene oxide nanoparticles has a promising application for the removal of numerous dyestuffs effluents.

4.3   Polymer/graphite Composite

Activated carbon is a shapeless dense nanomaterial, which comprise the graphene layer. Activated carbon can be obtained in different forms such as powder and fibres. Recently, nanocrabon materials can be used for textile industry effluent waste management due to their low-cost, easily preparation from agricultural biomass, better particle surface area, and high porosity. In spite of those benefits, activated carbon in the form of fine particles is challenging to eliminate from the treated aqueous media, due to the existence of the extremely small ground particles. Besides, their particles become combined and directed toward the reduction adsorption capacity. So, there is a need for a new method of using activated carbons as a prospective adsorbent by composite formation with polymer, which has enhanced to be a favourable method for textile wastewater treatment. Currently, nanocomposite material was synthesized from activated carbon and cellulose triacetate through the evaporation/precipitation techniques for exclusion of dyes and pigments pollutants [7].

Similarly, the synthetic polymer of polypyrrole was hybrid with activated carbon to produce a novel composite material used for elimination of Pb (II) molecules. The composite of carbon/polymer was performed in a carbonization manner with the presence of a chemical activator. The supported polymeric material altered the structure and morphology of the composite to have numerous functions in the adsorption process.

4.4   Carbon-Polymer Nanofibers Nanocomposite

Recently, nanocarbon arrangement particles have been obtained attention for the removal of dyestuffs and other organic effluents because of the presence of good material characteristics and comprehensive sort of prospective applications. From the list of nanocarbon nanomaterials including graphene, carbon nanotubes, as well as carbon nanofibers worked additional prospective nanomaterials for the elimination of dyestuffs and pigments.

Carbon nanofibers contain a distinct nanostructure either dual or single coat and also have the capacity to form cores, pores and hollow. The composite of carbon nanofiber nanoparticles with polymer could gain a higher chemical resistance, high surface area and high thermal stability. These nano-sized particles permit the carbon nanofiber to use for polymer reinforcement nanocomposites. When the polymer is combined with the nanofiber carbon, the nanocomposite material's strength becomes more developed than the individual components. The performance of nanocomposite materials is considerably enhanced mostly in distribution and strength. Surface modification of carbon nanofiber materials to advance their compatibility was attained through integrating them with a polymer coating and blending.

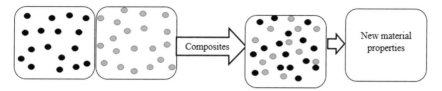

Fig. 4 Diagrammatic representation of the combining of carbon nanoparticles with polymer matrix

When carbon nanomaterials are combined with polymer matrix, the composite considerably changes the property such as mechanical [33], optical [51], structural, and electrical [65] conserving the mass of polymer as shown in Fig. 4. Polymer/carbon nanocomposite surface interaction is considered to be contributory in the determination of interfacial polymers [26], and this leads to the macroscopic properties of polymer nanocomposite materials [4, 71].

4.5 Polymer/Graphene Oxide (GO) Nanocomposite

Similar to carbon nanotubes, nanographene oxide nanoparticles are combined with different polymeric nanomaterials to prepare nanocomposite new adsorbent materials or filtration processes. Graphene oxide nanoparticles have been combined with polyvinyl alcohol (PVA) to produce membrane nanocomposite. Similar to carbon nanotubes, the fabricated mixed matrix membrane is comprised of graphene oxide nanoparticles for the application of nanofiltration process. The other study focused that the preparation of nanocomposite material was conducted by using natural and synthetic biopolymer with graphene oxide nanoparticles. Currently, there are different biopolymers materials used to produce composite materials including chitin, gum, and chitosan which ferric oxide was also mixed as it contributes to having the composite material a magnetic property. From all types of biopolymers investigations, the graphene/chitosan nanocomposite has been properly known. Table 2 shows the adsorption capacity of some composite material from graphene oxide nanoparticle and chitosan materials used for effluent removal. The main usage of chitosan particles in the composites is principally to improve the dispersion of particles within the solution environment [103].

The degree of dyestuffs removal using graphene oxide/polymer is varying because of the different structures and ions present in different dyes. Sheshmani et al. [74] synthesized a new composite adsorbent by using graphene oxide nanoparticles and chitosan polymer through a solvothermal technique to remove acid orange 7 (AO7). The highest adsorption capacity for the removal of acid orange was found to be 42.7 mg/g at 3.0 pH.

Similarly, Jiang et al. [39] syntheses a nanocomposite material using chitosan polymer and graphene oxide nanoparticles for the elimination of methylene orange dye from the solution. This investigation revealed that the optimum methylene blue

Table 2 List of polymer/graphene nanocomposite for removal of dyestuff and pigments

Nanocomposite materials	Waste dyes	Particle surface area (m^2/g)	Adsorbent dose (g/L)	pH	Adsorption capacity	Isotherm	Reference
Chitosan/graphene oxide	Methylene blue	–	0.5	4.0	398.1	Langmuir	[39]
Chitosan/graphene oxide	Rhodamine 6G	–	0.2	–	288	–	[32]
	Methyl violet dye	–	1.0	10	17.6	Langmuir	–
	Alizarin yellow	–	1.0	6.0	14.2	Langmuir	–
Gum/GO/PVA/Fe$_2$O$_3$	Congo red	–	1.0	2–8	94	Langmuir	[70]
	Crystal violet	–	1.0	5.0–8.0	101.74	Langmuir	–
Graphene oxide/polyacrylamide (PAM)	Methylene blue	–	0.2	–	292.9		[95]
RGO/Fe$_3$O$_4$/chitosan	Methylene blue	54.7	0.1	9.0	249.2	Langmuir	[84]
Chitosan/graphene oxide	Congo red	1.81	1.0	5.0	175.9	Langmuir	[41]
Inulin/GO	Methylene blue	35	0.5	–	790	Langmuir	[68]
	Rhodamine 6G	–	0.5	–	980	–	
	Orange II	–	0.5	–	77	–	
	Acid fuschkin	–	0.5	–	200	–	
Xylan/graphene oxide	Methylene blue	42	0.5	–	678	–	
	Rhodamine 6G	–	0.5	–	640	–	

(continued)

Table 2 (continued)

Nanocomposite materials	Waste dyes	Particle surface area (m^2/g)	Adsorbent dose (g/L)	pH	Adsorption capacity	Isotherm	Reference
	Orange II	–	0.5	–	90	–	
	Acid fuschkin	–	0.5	–	276	–	
Chitosan/Fe_3O_4/graphene	Acid orange7 dyes	–	0.1	3.7	42.7	Frundlich	[74]
Chitin/graphene oxide	Remazol black dyes	–	1.0	4.0	70	Freunlidch	[31]
K-carrageenan	Methylene blue	37	0.5	–	750	–	[68]

removal capacity was obtained to be 398.1 mg/g, which is tremendous removal efficiency, at the optimal pH value of 4.0. Moreover, the prepared nanocomposites of chitosan/graphene were used as antibacterial properties that exhibited a greater inactivation proportion of E. coli type of bacteria at 99%. Similarly, Kamal et al. [41] studied the removal efficiency of Congo red dye by using prepared nanocomposite.

According to the authors' report, the maximum adsorption capacity was reached 176 mg/g, which is greater than the polymer itself with no midfield by graphene oxide particles. Similarly, Van Hoa et al. [84] also investigated by adding Fe_3O_2 into the composite chitosan/reduced graphene oxide through the hydrothermal process used for the elimination of methylene blue dye ions. The authors found that the removal capacity for methylene blue removal towards the nanocomposite was significantly high and found to be 249.3 mg/g with an optimum pH value of 6.0. Similar adsorbent nanomaterial was examined by Gul et al. [32] to remove alizarin yellow dyestuffs and methyl violet dyes from effluent wastes. The authors found the maximum adsorption capacity for alizarin yellow was 14.2 mg/g at a pH of 10.0 and for methyl violet was to be 17.6 mg/g at pH of 6.0. Moreover, some studies have also been performed on synthesized nanocomposite materials by using graphene with plant gums as described by Qi et al. [68, 70]. Sahraei et al. [70] synthesis of new bio-sorbent composite materials containing a plentifully accessible gum, tragacanth/graphene oxide for the elimination of two common dyes namely Congo red and crystal violet dyes. The composites of these materials were modified by using some tragacanth surfactants and co-polymers including 2-acrylamido-2-methyl-1-propanesulfonic acid, 1-viniyal imidazole (IV). The functionalized tragacanth was later assorted using Fe_2O_3 and polyvinyl alcohol through the sol–gel system in reaction boric acid and acetone. The result of magnetic composite adsorbent has shown a remarkable adsorption capacity for the removal of crystal violet and Congo red. The adsorption capacity for Congo red and crystal violet dyes by using this magnetic nanocomposite was found at 94.0 mg/g and 10.74 mg/g correspondingly. On the other way, Qi et al. [68] investigated using abundant renewal polymeric materials including (inculin, xylan, and K-carrageenan) combined with graphene oxide nanoparticles in order to produce new adsorbent materials. The authors have confirmed that the biopolymers were dispersed in the graphene oxide layer and are promising for effective removal of dyestuffs and pigments.

Even though the particles surface pore of the nanocomposite of polysaccharide/graphene oxide was smaller when compared to the biopolymer/graphene oxide nanocomposite that has been informed, the removal capacity for methylene blue, Alexa Fluor, orange II (OII), and rhodamine 6G (R6G) dyes was greater, with an absorption capacity of around seven times than of the pure graphene oxide. The outcomes were endorsed for a greater percentage of enormous holes and sites which permit the contaminated molecules to move toward the active site of the adsorbent surface.

In addition to natural polymers, the synthetic polymer has been offered a potential application in material surface modification. Yang et al. [95] examined for adsorption of methylene blue dyes and rhodamine 6G dyes by using a hydrogel prepared from polyacrylamide/graphene oxide nanoparticles composite. The result showed that the adsorption potential for both rhodamine 6G dyes and methylene blue at pH of 7.0 was found as 288 and 292.2 mg/g respectively, which is significantly greater than other biopolymer composites. Furthermore, the different nature of both constituents, graphene oxide and polymer also directs to boosted adsorption competence as defined by González et al. [31] who syntheses graphene oxide nanoparticles and chitosan polymer composites. The authors investigated the effluences of several ratios of graphene oxide nanoparticles onto chitosan nanocomposite for the elimination of neutral red and remazol black dyes from textile industry effluents. According to their study, it is proposed that the addition of graphene oxide exhibited minor elimination of dyes and pigments as compared to the chitin alone. However, the existence of graphene oxide graphene oxide affords mechanical struggle to the chitin.

5 Mechanisms of Adsorption

To absorb any molecules using an adsorption system, knowing the adoption mechanism is significant. The adsorption occurrence is depending on intrinsic features of adsorbents; therefore, the major adsorbent mechanism is mainly depending on the adsorbent's nature and properties. So, studying the composite adsorbent property on the capacity of adsorption is an essential value [42].

Flow rate and retention time are major factors for contribution in determination adsorption mechanism during the functional adsorption process. Azhar and Farshi [64] prepared a nanocomposite by using polyaniline and starch with montmorillonite modification for removal of reactive dye from waste effluents. The authors confirm that the removal efficiency of reactive dyes depends on the charges of their mutual attraction between the nanocomposite adsorbent surface and the dye molecules.

Similarly, Wang et al. [87] investigated the adsorption process and removal mechanisms of dyes molecules with the adsorbent surface. The authors decided that from the investigational data and thermodynamically study, the sorption mechanism for removal of dyes was an exothermic process in nature. According to the authors' explanation, the infrared analysis showed that the adsorption mechanism for removal of methyl violet molecules depends on the particles' electron attraction between the adsorbate and adsorbent surface [87].

Polymer nanocomposites are recognized as having extreme adsorption performance because of the occurrence of nanoparticles with great specific surface area in the polymer background/matrix.

As the literature described, there are several adsorption and pollutant removal mechanisms reported. The adsorption mechanism for removal of contaminants effluents may depend on the adsorbent and adsorbate type, origins, structure, and charges of ions.

Figure 5 illustrates the type of adsorption mechanisms including chemisorption and physisorption. The chemisorption adsorption mechanisms encompass the chemical reaction between the surface adsorbent and the adsorbate molecule such as

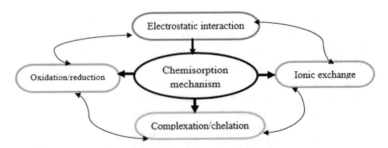

Fig. 5 Graphic representation of various adsorption mechanisms encompasses in chemisorptions

charge interaction, reduction or oxidation, hydrogen bonding, and Lewis acid inter-action. Chemisorption mechanisms are typically attended through an exothermic way, where reaction energy is released when pollutants are attached to the surface of the adsorbent. In such a mechanism, the entropy value can be reduced because the adsorption process is spontaneous. This type of adsorption mechanism commonly displayed a monolayer could develop the contaminates on the active site of the adsorbent [83]. Chemisorption commonly follows kinetics called pseudo-second-order for the adsorption process mechanism and also the sorption system usually as a function of adsorbate concentrations, the adsorbent structure, and adsorption operational parameters [80] [55]. While in the physisorption adsorption mechanism, adsorption happens among the adsorbent surface and adsorbate molecules through a weak boding namely Vander Waals forces.

Figure 6 displays the physical adsorption mechanism (right side). As clearly shown, the adsorbent surface has no functional modification relative to the chemisorptions exterior while in chemisorptions (left side) there is a number of active sites on the surface due to the presence of some functional groups and modification. So, adsorbate molecules move towards the entire surface via weak Vander Walls forces.

The adsorption and adsorption rate are significantly fast. Physisorption mechanisms typically involve kinetics called pseudo-second-order adsorption the sorption

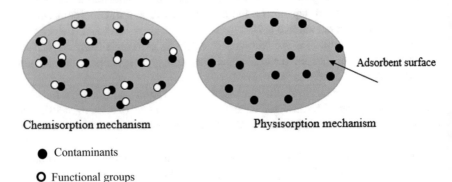

Fig. 6 Graphic representation of chemisorptions and physisorption adsorption process

process is principal a function of the adsorbents configurations, a charge of adsorbent and adsorbate molecules, and adsorbate concentration [14]. Thus, materials which have porous with a larger particle pore area and pore volume are good nanoadsorbent material for physical adsorption systems. Thermodynamically, the physical adsorption does not involve maximum activation energy to precede adsorption.

6 Future Perspectives

This chapter mainly addressed the advanced application of nanocomposite materials derived from nanocarbon and polymers essentially use for the elimination of dyes and organic pigments from industry effluents. As the request for freshwater has significantly increased in the past periods, several approaches were being engaged to treat the effluent wastewater. It is found that adsorption become more attracted among other treatment techniques in removal of dyes and pigments from textile wastewater. The enormous range of polymer/carbon for dyes and pigments elimination has been used at the laboratory level, and the result showed an encouraging outcome. Most of the prepared polymer/carbon nanocomposite adsorbent materials are testing still on a lab-scale by using the stock solution and their willingness to be commercialized is in its early stages. Moreover, research should be focused to test the carbon/polymer nanocomposite adsorbent materials' performance by using actual effluents rather than batch experiments.

7 Conclusion

The chapter is basically summarized the broad application of carbon/polymer nanocomposite materials adsorbent for the removal of dyestuffs from textile industry effluents. Moreover, the utilization of carbon nanomaterials including graphene oxide nanoparticles, carbon nanofibers nanoparticles, graphene nanoparticles, and reduced graphene oxide nanoparticles have practically applied in the removal of dyes and pigments. The utilization of nanocarbon materials with other nanocomposite has a wonderful benefit in wastewater treatment, specifically in textile effluents. Recently, carbon/polymer nanocomposites have emerged with inherent physicochemical properties than the individual components. Previous study shows that carbon nanoparticle composite materials have larger and unique physiochemical assets, such as high specific surface area, good mechanical strength, high thermal stability and greater adsorption capacity. Thus, due to the incessant investigation on the emerging of novel composite, there will be an expected that polymer/carbon nanocomposite material will be delivered for longstanding in the application of industry level.

Acknowledgements The author would like to thank those organizations, and individuals whom he has used previously published materials. The author apologizes to all people, and organizations whose contribution to the field may have been reviewed by the fault or not adequately acknowledged.

Declaration of Competing Interest The author declares that he has no known opposing for economic interests or personal relationships that could have seemed to inspire the work reported in this work.

References

1. Abbasi M (2017) Synthesis and characterization of magnetic nanocomposite of chitosan/SiO$_2$/carbon nanotubes and its application for dyes removal. J Clean Prod 145:105–113. https://doi.org/10.1016/j.jclepro.2017.01.046
2. Abdullah N, Tajuddin MH, Yusof N (2018) Carbon-based polymer nanocomposites for dye and pigment removal. Carbon Based Poly Nanocompos Environ Energy Appl. https://doi.org/10.1016/B978-0-12-813574-7.00013-7
3. Afreen S, Omar RA, Talreja N, Chauhan D, Ashfaq M (2018) Carbon-based nanostructured materials for energy and environmental remediation applications. Nanotechnol Life Sci 369–392. https://doi.org/10.1007/978-3-030-02369-0_17
4. Agboola O, Fayomi OSI, Ayodeji A, Ayeni AO, Alagbe EE, Sanni SE, Okoro EE, Moropeng L, Sadiku R, Kupolati KW, Oni BA (2021) A review on polymer nanocomposites and their effective applications in membranes and adsorbents for water treatment and gas separation. Membranes 11:139. https://doi.org/10.3390/membranes11020139
5. Aghigh A, Alizadeh V, Wong HY, Islam MS, Amin N, Zaman M (2015) Recent advances in utilization of graphene for filtration and desalination of water: a review. Desalination 365:389–397. https://doi.org/10.1016/j.desal.2015.03.024
6. Ajayan PM, Schadler LS, Giannaris C, Rubio A (2000) Single-walled carbon nanotube-polymer composites: strength and weakness. Adv Mater 10:750–753. https://doi.org/10.1002/(SICI)1521-4095(200005)12:10%3c750::AID-ADMA750%3e3.0.CO;2-6
7. Akharame MO, Fatoki OS, Opeolu BO, Olorunfemi DI, Oputu OU (2018) Polymeric nanocomposites (Pncs) for wastewater remediation: an overview. Polym Plast Technol Eng 57:1801–1827. https://doi.org/10.1080/03602559.2018.1434666
8. Al-Anzi BS, Siang OC (2017) Recent developments of carbon based nanomaterials and membranes for oily wastewater treatment. RSC Adv 7:20981–20994. https://doi.org/10.1039/c7ra02501g
9. Al-Saleh MH, Sundararaj U (2011) Review of the mechanical properties of carbon nanofiber/polymer composites. Compos A Appl Sci Manuf 42:2126–2142. https://doi.org/10.1016/j.compositesa.2011.08.005
10. Alhassani HA, Rauf MA, Ashraf SS (2007) Efficient microbial degradation of Toluidine Blue dye by Brevibacillus sp. Dyes Pigm 75:395–400. https://doi.org/10.1016/j.dyepig.2006.06.019
11. Amini M, Arami M, Mahmoodi NM, Akbari A (2011) Dye removal from colored textile wastewater using acrylic grafted nanomembrane. Desalination 267:107–113. https://doi.org/10.1016/j.desal.2010.09.014
12. Ao C, Yuan W, Zhao J, He X, Zhang X, Li Q, Xia T, Zhang W, Lu C (2017) Super-hydrophilic graphene oxide@electrospun cellulose nanofiber hybrid membrane for high-efficiency oil/water separation. Carbohyd Polym 175:216–222. https://doi.org/10.1016/j.carbpol.2017.07.085
13. Banerjee S, Murad S, Puri IK (2007) Preferential ion and water intake using charged carbon nanotubes. Chem Phys Lett 434:292–296. https://doi.org/10.1016/j.cplett.2006.12.025

14. Berber MR (2020) Current advances of polymer composites for water treatment and desalination. J Chem 2020:1–19. https://doi.org/10.1155/2020/7608423

15. Cazetta AL, Vargas AMM, Nogami EM, Kunita MH, Guilherme MR, Martins AC, Silva TL, Moraes JCG, Almeida VC (2011) NaOH-activated carbon of high surface area produced from coconut shell: kinetics and equilibrium studies from the methylene blue adsorption. Chem Eng J 174:117–125. https://doi.org/10.1016/j.cej.2011.08.058

16. Chang PR, Zheng P, Liu B, Anderson DP, Yu J, Ma X (2011) Characterization of magnetic soluble starch-functionalized carbon nanotubes and its application for the adsorption of the dyes. J Hazard Mater 186:2144–2150. https://doi.org/10.1016/j.jhazmat.2010.12.119

17. Chen C, Wang X (2006) Adsorption of Ni(II) from aqueous solution using Oxidized multiwall carbon nanotubes. Ind Eng Chem Res 45:9144–9149. https://doi.org/10.1021/ie060791z

18. Chen D, Feng H, Li J (2012) Graphene oxide: preparation, functionalization, and electrochemical applications. Chem Rev 112:6027–6053. https://doi.org/10.1021/cr300115g

19. Chen J, Wang Y, Huang Y, Xu K, Li N, Wen Q, Zhou Y (2015) Magnetic multiwall carbon nanotubes modified with dual hydroxy functional ionic liquid for the solid-phase extraction of protein. Analyst 140:3474–3483. https://doi.org/10.1039/c5an00201j

20. Das R, Ali ME, Hamid SBA, Ramakrishna S, Chowdhury ZZ (2014) Carbon nanotube membranes for water purification: a bright future in water desalination. Desalination 336:97–109. https://doi.org/10.1016/j.desal.2013.12.026

21. De Gisi S, Lofrano G, Grassi M, Notarnicola M (2016) Characteristics and adsorption capacities of low-cost sorbents for wastewater treatment: a review. Sustain Mater Technol 9:10–40. https://doi.org/10.1016/j.susmat.2016.06.002

22. De Volder MFL, Tawfick SH, Baughman RH, Hart AJ (2013) Carbon nanotubes: present and future commercial applications. Science 339:535–539. https://doi.org/10.1126/science.1222453

23. Dong L, Fan W, Tong X, Zhang H, Chen M, Zhao Y (2018) A CO2-responsive graphene oxide/polymer composite nanofiltration membrane for water purification. J Mater Chem A 6:6785–6791. https://doi.org/10.1039/c8ta00623g

24. Errais E, Duplay J, Darragi F, M'Rabet I, Aubert A, Huber F, Morvan G (2011) Efficient anionic dye adsorption on natural untreated clay: kinetic study and thermodynamic parameters. Desalination 275:74–81. https://doi.org/10.1016/j.desal.2011.02.031

25. Fan L, Luo C, Sun M, Li X, Lu F, Qiu H (2012) Preparation of novel magnetic chitosan/graphene oxide composite as effective adsorbents toward methylene blue. Biores Technol 114:703–706. https://doi.org/10.1016/j.biortech.2012.02.067

26. Forrest JA, Dalnoki-Veress K (2001) The glass transition in thin polymer films. Adv Coll Interface Sci 94:167–195. https://doi.org/10.1016/S0001-8686(01)00060-4

27. Gao H, Zhao S, Cheng X, Wang X, Zheng L (2013) Removal of anionic azo dyes from aqueous solution using magnetic polymer multi-wall carbon nanotube nanocomposite as adsorbent. Chem Eng J 223:84–90. https://doi.org/10.1016/j.cej.2013.03.004

28. Geckeler KE, Premkumar T (2011) Carbon nanotubes: are they dispersed or dissolved in liquids? Nanoscale Res Lett 6:1–3. https://doi.org/10.1186/1556-276X-6-136

29. Geng Z, Lin Y, Yu X, Shen Q, Ma L, Li Z, Pan N, Wang X (2012) Highly efficient dye adsorption and removal: a functional hybrid of reduced graphene oxide-Fe 3O 4 nanoparticles as an easily regenerative adsorbent. J Mater Chem 22:3527–3535. https://doi.org/10.1039/c2jm15544c

30. Goh PS, Ismail AF, Ng BC (2013) Carbon nanotubes for desalination: performance evaluation and current hurdles. Desalination 308:2–14. https://doi.org/10.1016/j.desal.2012.07.040

31. González JA, Villanueva ME, Piehl LL, Copello GJ (2015) Development of a chitin/graphene oxide hybrid composite for the removal of pollutant dyes: adsorption and desorption study. Chem Eng J 280:41–48. https://doi.org/10.1016/j.cej.2015.05.112

32. Gul K, Sohni S, Waqar M, Ahmad F, Norulaini NAN, Mohd Omar AK (2016) Functionalization of magnetic chitosan with graphene oxide for removal of cationic and anionic dyes from aqueous solution. Carbohyd Poly 152, 520–531. https://doi.org/10.1016/j.carbpol.2016.06.045

33. Gusev AA (2006) Micromechanical mechanism of reinforcement and losses in filled rubbers. Macromolecules 39:5960–5962. https://doi.org/10.1021/ma061308z

34. Hessel C, Allegre C, Maisseu M, Charbit F, Moulin P (2007) Guidelines and legislation for dye house effluents. J Environ Manage 83:171–180. https://doi.org/10.1016/j.jenvman.2006. 02.012

35. Hu L, Gao S, Ding X, Wang D, Jiang J, Jin J, Jiang L (2015) Photothermal-responsive single-walled carbon nanotube-based ultrathin membranes for on/off switchable separation of oil-in-water nanoemulsions. ACS Nano 9:4835–4842. https://doi.org/10.1021/nn5062854

36. Humplik T, Lee J, O'Hern SC, Fellman BA, Baig MA, Hassan SF, Atieh MA, Rahman F, Laoui T, Karnik R, Wang EN (2011) Nanostructured materials for water desalination. Nanotechnology 22:292001. https://doi.org/10.1088/0957-4484/22/29/292001

37. Ihsanullah (2019) Carbon nanotube membranes for water purification: developments, challenges, and prospects for the future. Sep Purif Technol 209:307–337. https://doi.org/10.1016/ j.seppur.2018.07.043

38. Jayakaran P, Nirmala GS, Govindarajan L (2019) Qualitative and Quantitative Analysis of Graphene-Based Adsorbents in Wastewater Treatment. Int J Chem Eng 2019. https://doi.org/ 10.1155/2019/9872502

39. Jiang Y, Gong JL, Zeng GM, Ou XM, Chang YN, Deng CH, Zhang J, Liu HY, Huang SY (2016) Magnetic chitosan-graphene oxide composite for anti-microbial and dye removal applications. Int J Biol Macromol 82:702–710. https://doi.org/10.1016/j.ijbiomac.2015.11.021

40. Joshi RK, Alwarappan S, Yoshimura M, Sahajwalla V, Nishina Y (2015) Graphene oxide: the new membrane material. Appl Mater Today 1:1–12. https://doi.org/10.1016/j.apmt.2015. 06.002

41. Kamal MA, Bibi S, Bokhari SW, Siddique AH, Yasin T (2017) Synthesis and adsorptive characteristics of novel chitosan/graphene oxide nanocomposite for dye uptake. React Funct Polym 110:21–29. https://doi.org/10.1016/j.reactfunctpolym.2016.11.002

42. Karimi S, Tavakkoli Yaraki M, Karri RR (2019) A comprehensive review of the adsorption mechanisms and factors influencing the adsorption process from the perspective of bioethanol dehydration. Renew Sustain Energy Rev 107:535–553. https://doi.org/10.1016/j.rser.2019. 03.025

43. Kelly FJ, Fussell JC (2015) Air pollution and public health: emerging hazards and improved understanding of risk. Environ Geochem Health 37:631–649. https://doi.org/10.1007/s10653-015-9720-1

44. Khare P, Singh A, Verma S, Bhati A, Sonker AK, Tripathi KM, Sonkar SK (2018) Sunlight-induced selective photocatalytic degradation of methylene blue in bacterial culture by pollutant soot derived nontoxic graphene nanosheets. ACS Sustain Chem Eng 6:579–589. https://doi.org/10.1021/acssuschemeng.7b02929

45. Khezri K, Mahdavi H (2016) Polystyrene-silica aerogel nanocomposites by in situ simultaneous reverse and normal initiation technique for ATRP. Microporous Mesoporous Mater 228:132–140. https://doi.org/10.1016/j.micromeso.2016.03.022

46. Klein KL, Melechko AV, McKnight TE, Retterer ST, Rack PD, Fowlkes JD, Joy DC, Simpson ML (2008) Surface characterization and functionalization of carbon nanofibers. J Appl Phys 103:3. https://doi.org/10.1063/1.2840049

47. Kousha M, Daneshvar E, Sohrabi MS, Jokar M, Bhatnagar A (2012) Adsorption of acid orange II dye by raw and chemically modified brown macroalga Stoechospermum marginatum. Chem Eng J 192:67–76. https://doi.org/10.1016/j.cej.2012.03.057

48. Kudaibergenov S, Koetz J, Nuraje N (2018) Nanostructured hydrophobic polyampholytes: self-assembly, stimuli-sensitivity, and application. Adv Compos Hybrid Mater 1:649–684. https://doi.org/10.1007/s42114-018-0059-9

49. Li F, Yu Z, Shi H, Yang Q, Chen Q, Pan Y, Zeng G, Yan L (2017) A Mussel-inspired method to fabricate reduced graphene oxide/g-C_3N_4 composites membranes for catalytic decomposition and oil-in-water emulsion separation. Chem Eng J 322:33–45. https://doi.org/10.1016/j.cej. 2017.03.145

50. Li X, Lu H, Zhang Y, He F (2017) Efficient removal of organic pollutants from aqueous media using newly synthesized polypyrrole/CNTs-CoFe$_2$O$_4$ magnetic nanocomposites. Chem Eng J 316:893–902. https://doi.org/10.1016/j.cej.2017.02.037

51. Li Y, Tao P, Viswanath A, Benicewicz BC, Schadler LS (2013) Bimodal surface ligand engineering: the key to tunable nanocomposites. Langmuir 29:1211–1220. https://doi.org/10.1021/la3036192

52. Liang J, Ning X, Kong M, Liu D, Wang G, Cai H, Sun J, Zhang Y, Lu X, Yuan Y (2017) Elimination and ecotoxicity evaluation of phthalic acid esters from textile-dyeing wastewater. Environ Pollut 231:115–122. https://doi.org/10.1016/j.envpol.2017.08.006

53. Liu F, Chung S, Oh G, Seo TS (2012) Three-dimensional graphene oxide nanostructure for fast and efficient water-soluble dye removal. ACS Appl Mater Interfaces 4:922–927. https://doi.org/10.1021/am201590z

54. Liu J, Li X, Jia W, Ding M, Zhang Y, Ren S (2016) Separation of emulsified oil from oily wastewater by functionalized multiwalled carbon nanotubes. J Dispers Sci Technol 37:1294–1302. https://doi.org/10.1080/01932691.2015.1090320

55. Liu Y, Liu X, Dong W, Zhang L, Kong Q, Wang W (2017) Efficient adsorption of sulfamethazine onto modified activated carbon: a plausible adsorption mechanism. Sci Rep 7:1–12. https://doi.org/10.1038/s41598-017-12805-6

56. Ma PC, Siddiqui NA, Marom G, Kim JK (2010) Dispersion and functionalization of carbon nanotubes for polymer-based nanocomposites: a review. Compos A Appl Sci Manuf 41:1345–1367. https://doi.org/10.1016/j.compositesa.2010.07.003

57. Mallakpour S, Rashidimoghadam S (2017) Starch/MWCNT-vitamin C nanocomposites: electrical, thermal properties and their utilization for removal of methyl orange. Carbohyd Polym 169:23–32. https://doi.org/10.1016/j.carbpol.2017.03.081

58. Mishra P, Soni R (2016) Analysis of dyeing and printing waste water of balotara textile industries. Int J Chem Sci 14:1929–1938

59. Mondal T, Bhowmick AK, Krishnamoorti R (2013) Synthesis and characterization of bi-functionalized graphene and expanded graphite using n-butyl lithium and their use for efficient water soluble dye adsorption. J Mater Chem A 1:8144–8153. https://doi.org/10.1039/c3ta11212h

60. Mu B, Wang A (2016) Adsorption of dyes onto palygorskite and its composites: a review. J Environ Chem Eng 4:1274–1294. https://doi.org/10.1016/j.jece.2016.01.036

61. Náray-Szabó G, Mika LT (2018) Conservative evolution and industrial metabolism in Green chemistry. Green Chem 20:2171–2191. https://doi.org/10.1039/c8gc00514a

62. Ngo Q, Yamada T, Suzuki M, Ominami Y, Cassell AM, Li J, Meyyappan M, Yang CY (2007) Structural and electrical characterization of carbon nanofibers for interconnect via applications. IEEE Trans Nanotechnol 6:688–695. https://doi.org/10.1109/TNANO.2007.907400

63. Noamani S, Niroomand S, Rastgar M, Sadrzadeh M (2019) Carbon-based polymer nanocomposite membranes for oily wastewater treatment. Npj Clean Water 2:1–14. https://doi.org/10.1038/s41545-019-0044-z

64. Olad A, Azhar FF (2014) Eco-friendly biopolymer/clay/conducting polymer nanocomposite: characterization and its application in reactive dye removal. Fibers Poly 15:1321–1329. https://doi.org/10.1007/s12221-014-1321-6

65. Ounaies Z, Park C, Wise KE, Siochi EJ, Harrison JS (2003) Electrical properties of single wall carbon nanotube reinforced polyimide composites. Compos Sci Technol 63:1637–1646. https://doi.org/10.1016/S0266-3538(03)00067-8

66. Padaki M, Surya Murali R, Abdullah MS, Misdan N, Moslehyani A, Kassim MA, Hilal N, Ismail AF (2015) Membrane technology enhancement in oil-water separation a review. Desalination 357:197–207. https://doi.org/10.1016/j.desal.2014.11.023

67. Peng Y, Yu Z, Li F, Chen Q, Yin D, Min X (2018) A novel reduced graphene oxide-based composite membrane prepared via a facile deposition method for multifunctional applications: oil/water separation and cationic dyes removal. Sep Purif Technol 200:130–140. https://doi.org/10.1016/j.seppur.2018.01.059

68. Qi Y, Yang M, Xu W, He S, Men Y (2017) Natural polysaccharides-modified graphene oxide for adsorption of organic dyes from aqueous solutions. J Colloid Interf Sci 486:84–96. https://doi.org/10.1016/j.jcis.2016.09.058

69. Rayaroth MP, Aravind UK, Aravindakumar CT (2017) Ultrasound based AOP for emerging pollutants: from degradation to mechanism. Environ Sci Pollut Res 24:6261–6269. https://doi.org/10.1007/s11356-016-6606-4

70. Sahraei R, Sekhavat Pour Z, Ghaemy M (2017) Novel magnetic bio-sorbent hydrogel beads based on modified gum tragacanth/graphene oxide: removal of heavy metals and dyes from water. J Clean Prod 142:2973–2984. https://doi.org/10.1016/j.jclepro.2016.10.170

71. Senses E, Faraone A, Akcora P (2016) Microscopic chain motion in polymer nanocomposites with dynamically asymmetric interphases. Sci Rep 6:1–11. https://doi.org/10.1038/srep29326

72. Seyed Arabi SM, Lalehloo RS, Olyai MRTB, Ali GAM, Sadegh H (2019) Removal of congo red azo dye from aqueous solution by ZnO nanoparticles loaded on multiwall carbon nanotubes. Physica E 106:150–155. https://doi.org/10.1016/j.physe.2018.10.030

73. Shahid M, Islam S-U, Mohammad F (2013) Recent advancements in natural dye applications: a review. J Cleaner Prod 53:310–331. https://doi.org/10.1016/j.jclepro.2013.03.031

74. Sheshmani S, Ashori A, Hasanzadeh S (2014) Removal of acid orange 7 from aqueous solution using magnetic graphene/chitosan: a promising nano-adsorbent. Int J Biol Macromol 68:218–224. https://doi.org/10.1016/j.ijbiomac.2014.04.057

75. Shi Z, Zhang W, Zhang F, Liu X, Wang D, Jin J, Jiang L (2013) Ultrafast separation of emulsified oil/water mixtures by ultrathin free-standing single-walled carbon nanotube network films. Adv Mater 25:2422–2427. https://doi.org/10.1002/adma.201204873

76. Singh A, Khare P, Verma S, Bhati A, Sonker AK, Tripathi KM, Sonkar SK (2017) Pollutant soot for pollutant dye degradation: soluble graphene nanosheets for visible light induced photodegradation of methylene blue. ACS Sustain Chem Eng 5:8860–8869. https://doi.org/10.1021/acssuschemeng.7b01645

77. Stankovich S, Piner RD, Nguyen SBT, Ruoff RS (2006) Synthesis and exfoliation of isocyanate-treated graphene oxide nanoplatelets. Carbon 44:3342–3347. https://doi.org/10.1016/j.carbon.2006.06.004

78. Sun Y, Yang S, Sheng G, Guo Z, Wang X (2012) The removal of U(VI) from aqueous solution by oxidized multiwalled carbon nanotubes. J Environ Radioact 105:40–47. https://doi.org/10.1016/j.jenvrad.2011.10.009

79. Tan KB, Vakili M, Horri BA, Poh PE, Abdullah AZ, Salamatinia B (2015) Adsorption of dyes by nanomaterials: recent developments and adsorption mechanisms. Sep Purif Technol 150:229–242. https://doi.org/10.1016/j.seppur.2015.07.009

80. Tan KL, Hameed BH (2017) Insight into the adsorption kinetics models for the removal of contaminants from aqueous solutions. J Taiwan Inst Chem Eng 74:25–48. https://doi.org/10.1016/j.jtice.2017.01.024

81. Thines RK, Mubarak NM, Nizamuddin S, Sahu JN, Abdullah EC, Ganesan P (2017) Application potential of carbon nanomaterials in water and wastewater treatment: a review. J Taiwan Inst Chem Eng 72:116–133. https://doi.org/10.1016/j.jtice.2017.01.018

82. Thommes M, Cychosz KA (2014) Physical adsorption characterization of nanoporous materials: progress and challenges. Adsorption 20:233–250. https://doi.org/10.1007/s10450-014-9606-z

83. Tran HN, Chao HP (2018) Adsorption and desorption of potentially toxic metals on modified biosorbents through new green grafting process. Environ Sci Pollut Res 25:12808–12820. https://doi.org/10.1007/s11356-018-1295-9

84. Van Hoa N, Khong TT, Thi Hoang Quyen T, Si Trung T (2016) One-step facile synthesis of mesoporous graphene/Fe$_3$O$_4$/chitosan nanocomposite and its adsorption capacity for a textile dye. J Water Process Eng 9:170–178. https://doi.org/10.1016/j.jwpe.2015.12.005

85. Wan Ngah WS, Teong LC, Hanafiah MAKM (2011) Adsorption of dyes and heavy metal ions by chitosan composites: a review. Carbohyd Polym 83:1446–1456. https://doi.org/10.1016/j.carbpol.2010.11.004

86. Wang H, Ma H, Zheng W, An D, Na C (2014) Multifunctional and recollectable carbon nanotube ponytails for water purification. ACS Appl Mater Interfaces 6:9426–9434. https://doi.org/10.1021/am501810f

87. Wang M, Gu Q, Luo Y, Bukhvalov D, Ma X, Zhu L, Li G, Luo Z (2019) Understanding mechanism of adsorption in the decolorization of aqueous methyl violet (6B) solution by okra polysaccharides: experiment and theory. ACS Omega 4:17880–17889. https://doi.org/10.1021/acsomega.9b02768

88. Wu K, Yu J, Jiang X (2018) Multi-walled carbon nanotubes modified by polyaniline for the removal of alizarin yellow R from aqueous solutions. Adsorpt Sci Technol 36:198–214. https://doi.org/10.1177/0263617416687564

89. Wu Z, Zhang C, Peng K, Wang Q, Wang Z (2018) Hydrophilic/underwater superoleophobic graphene oxide membrane intercalated by TiO_2 nanotubes for oil/water separation. Front Environ Sci Eng 12:1–10. https://doi.org/10.1007/s11783-018-1042-y

90. Xie Y, He C, Liu L, Mao L, Wang K, Huang Q, Liu M, Wan Q, Deng F, Huang H, Zhang X, Wei Y (2015) Carbon nanotube based polymer nanocomposites: biomimic preparation and organic dye adsorption applications. RSC Adv 5:2503–82512. https://doi.org/10.1039/c5ra15626b

91. Yakout AA, Shaker MA (2016) Dodecyl sulphate functionalized magnetic graphene oxide nanosorbent for the investigation of fast and efficient removal of aqueous malachite green. J Taiwan Inst Chem Eng 63:81–88. https://doi.org/10.1016/j.jtice.2016.03.027

92. Yamjala K, Nainar MS, Ramisetti NR (2016) Methods for the analysis of azo dyes employed in food industry-a review. Food Chem 192:813–824. https://doi.org/10.1016/j.foodchem.2015.07.085

93. Yan L, Chang PR, Zheng P, Ma X (2012) Characterization of magnetic guar gum-grafted carbon nanotubes and the adsorption of the dyes. Carbohyd Polym 87:1919–1924. https://doi.org/10.1016/j.carbpol.2011.09.086

94. Yang K, Wei W, Qi L, Wu W, Jing Q, Lin D (2014) Are engineered nanomaterials superior adsorbents for removal and pre-concentration of heavy metal cations from water? RSC Adv 4:46122–46125. https://doi.org/10.1039/c4ra09375e

95. Yang Y, Song S, Zhao Z (2017) Graphene oxide (GO)/polyacrylamide (PAM) composite hydrogels as efficient cationic dye adsorbents. Colloids Surf A 513:315–324. https://doi.org/10.1016/j.colsurfa.2016.10.060

96. Yaseen DA, Scholz M (2019) Textile dye wastewater characteristics and constituents of synthetic effluents: a critical review. Int J Environ Sci Technol 16:1193–1226. https://doi.org/10.1007/s13762-018-2130-z

97. Zhang S, Zeng M, Li J, Li J, Xu J, Wang X (2014) Porous magnetic carbon sheets from biomass as an adsorbent for the fast removal of organic pollutants from aqueous solution. J Mater Chem A 3:2754–2764. https://doi.org/10.1039/c3ta14604a

98. Zhang Y, Wu B, Xu H, Liu H, Wang M, He Y, Pan B (2016) Nanomaterials-enabled water and wastewater treatment. NanoImpact 3:22–39. https://doi.org/10.1016/j.impact.2016.09.004

99. Zhao J, Chen H, Ye H, Zhang B, Xu L (2019) Poly(dimethylsiloxane)/graphene oxide composite sponge: a robust and reusable adsorbent for efficient oil/water separation. Soft Matter 15:9224–9232. https://doi.org/10.1039/c9sm01984g

100. Zhao Q, Huang H, Li F (2011) Phosphorescent heavy-metal complexes for bioimaging. Chem Soc Rev 40:2508–2524. https://doi.org/10.1039/c0cs00114g

101. Zhu H, Fu Y, Jiang R, Yao J, Liu L, Chen Y, Xiao L, Zeng G (2013) Preparation, characterization and adsorption properties of chitosan modified magnetic graphitized multi-walled carbon nanotubes for highly effective removal of a carcinogenic dye from aqueous solution. Appl Surf Sci 285:865–873. https://doi.org/10.1016/j.apsusc.2013.09.003

102. Zhu HY, Yao J, Jiang R, Fu YQ, Wu YH, Zeng GM (2014) Enhanced decolorization of azo dye solution by cadmium sulfide/multi-walled carbon nanotubes/polymer composite in combination with hydrogen peroxide under simulated solar light irradiation. Ceram Int 40:3769–3777. https://doi.org/10.1016/j.ceramint.2013.09.043

103. Zhu Y, Murali S, Cai W, Li X, Suk JW, Potts JR, Ruoff RS (2010) Graphene and graphene oxide: synthesis, properties, and applications. Adv Mater 22:3906–3924. https://doi.org/10.1002/adma.201001068

104. Zou B, Chen K, Wang Y, Niu C, Zhou S (2015) Amino-functionalized magnetic magnesium silicate double-shelled hollow microspheres for enhanced removal of lead ions. RSC Adv 5:22973–22979. https://doi.org/10.1039/c5ra01373a

"Environmental Issues Concerned with Poly (Vinyl Alcohol) (PVA) in Textile Wastewater"

Muhammad Hamad Zeeshan, Umm E. Ruman, Gaohong He, Aneela Sabir, Muhammad Shafiq, and Muhammad Zubair

Abstract Polyvinyl Alcohol (PVA), a synthetic polymer with whitish powdery appearance is usually appraised as biodegradable replica of natural polymers. PVA has been widely used in industrial sectors. A massive amount of PVA goes to textile industry on daily basis due to its use as sizing agent and dye adsorbent. So, the untreated textile wastewater (containing high concentration of PVA and other toxic compounds) when directly divulged into clean water bodies, it adversely affects the environment. This chapter provides the brief review of PVA in textile wastewater. The main points are (1) uses of PVA in textile, PVA degradation pathways in wastewater and number of pretreatment methods that have been adopted for its degradation followed by biodegradation (which significantly reduces its BOD, and COD values), (2) PVA containing wastewater give birth to environmental dilemmas by disturbing aquatic life, by changing soil conditions and by deleteriously affecting human health. Furthermore, some reliable and innoxious alternatives to PVA are also referenced.

Keywords Poly (vinyl alcohol) · PVA degradation · Sizing agent · Advanced oxidation processes · PVA-based composites · PVA environmental affect

1 Introduction

Polyvinyl Alcohol (PVA) is a synthetic polymer that has been produced by the complete or biased hydroxylation of Polyvinyl Acetate. Hermann and Haehnal got credit by first time synthesizing the PVA solution by saponifying the polyvinyl ester

M. H. Zeeshan (✉) · G. He
State Key Laboratory of Fine Chemicals, R&D Center of Membrane Science and Technology, School of Chemical Engineering, Dalian University of Technology, Dalian 116024, China
e-mail: hamad.xeeshan@mail.dlut.edu.cn

U. E. Ruman · M. Zubair
Department of Chemistry, University of Gujrat, Gujrat 57200, Pakistan

A. Sabir · M. Shafiq
Department of Polymer and Textile Engineering, University of the Punjab, Lahore 54590, Pakistan

with sodium hydroxide in 1924 [3]. On industrial scale, PVA is synthesized by an ongoing process in which the acetate groups of polyvinyl acetate undergo hydrolysis in the presence of methanol and aqueous sodium hydroxide [4]. PVA is recognized by the general formula $[CH_2CH(OH)_n]$. Figure 1 depicts the structure of polyvinyl alcohol. It is considered as biodegradable replica of natural polymers that has broad applications at commercial level [18]. Tables 1 and 2 enlist some important physical and chemical properties of polyvinyl alcohol.

Number of mechanical properties of PVA depend on two factors, i.e. vinyl acetate polymer chain and the degree of hydrolysis under various acidic and alkaline conditions. It has been observed that variation of both these factors produce PVA products that significantly vary in their different properties such as molecular weight, flexibility, solubility, adhesiveness and tensile strength. The grade of PVA can be

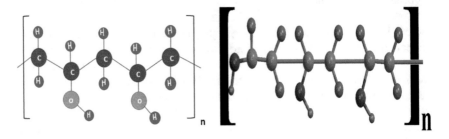

Fig. 1 Structure of polyvinyl alcohol

Table 1 Physical properties of PVA

Physical property	Explanation	References
Appearance	Whitish granular powder/odorless	[3]
Density	1.19–1.31 g/cm³	[3]
Specific gravity(solid)	1.27–1.31	[12]
Specific gravity of 10% soln	1.02	[12]
Solubility	Soluble in water and mixed solvents	[12]
Glass transition temperature(Tg)	75–85	[37]
Boiling point	228 °C	[3]
Melting point	180–190 °C	[36]
Heat of Fusion(by diluent method)	1.64 kcal/mol	[36]
Heat of Fusion(by copolymer method)	0.56 kcal/mol	[36]
Flammability	Burns like paper	[12]

Table 2 Chemical properties of PVA

Chemical property	Explanation	References
Molecular weight	30 000–200 000 Da	[6]
Structural formula	-(-CH$_2$CHOCOCH$_3$-)-m, -(-CH$_2$CHOH-)-n	[6]
Empirical formula	(C$_2$H$_4$O)$_n$	[6]
Degradation	Degrades slowly at 100 °C Degrades rapidly at 200 °C	[34]

characterized by its molecular weight and the degree of hydrolysis, which in turn affect the pH, viscosity, melting point and refractive index of the PVA [15].

2 Application of PVA

PVA has been widely used at industrial level, i.e. in paper coating, in stiffening rubber hoses and rollers, in construction for strengthening the concrete, in fishing industry, in 3D printing and the most importantly in sizing process of textile industries for strengthening of yarn and modification of cellulosic fabric. The large number of applications of PVA are due to its high tensile strength, water solubility, high thermal stability and superior flexibility [57]. PVA also has been used in drug delivery system due to its drug compatibility, water solubility and amusing swelling properties [35].

PVA has a crucial role as the sizing agent in textile industry. Sizing process is shown in Fig. 2. PVA can be used as adhesive as well as sizing agent in the textile sizing and in the production of non-woven fabrics. In order to strengthen the fabric, the sizing agents are placed on the fabric before its weaving the process of textile (warp) sizing indicates that by using PVA as sizing agent number of barriers can be overcome in synthesizing finishes [57]. PVA has magical adhesive properties,

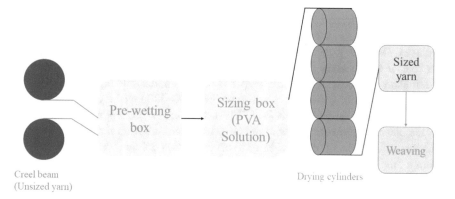

Fig. 2 Sizing process in textile industries

Table 3 Poly (vinyl alcohol)-based amalgams for the removal of various dyes

Adsorbent	Dye name	References
PVA	Methylene blue	[49]
PVA	Bromothymol blue	[1]
PVA/Poly (Acrylic acid)/Polydopamine membranes	Methylene blue	[53]
Carboxymethyl starch-g- poly (vinyl imidazole)/PVA	Crystal violet	[28]
PVA/Carboxymethyl starch-g- poly (vinyl imidazole)	Congo red	[28]
PVA/coconut coir dust	Malachite green	[16]
PVA/Triethylenetetramine/ Glutaraldehyde	Direct red 80	[26]
PVA/Glutaraldehyde/β-cyclodextrin	Reactive dye	[20]
PVA/alginate single network hydrogel	Methylene blue	[24]
PVA/alginate double network hydrogel	Methylene blue	[24]

and in warp sizing, adequate adhesion of the sizing agent to fiber is one of the main demands. The sufficient adhesion property is, the sizing agent will adhere to the warp yarn in better way and hence will increase its strength [50].

Another important use of PVA in the textile wastewater treatment is the removal of dyes from the textile wastewater because colored water is unacceptable for drinking, municipal as well as for agricultural purposes [40]. Poly (vinyl alcohol) and its deployed amalgams have been acknowledged as efficient adsorbents in the latest researches. Moreover, modification of PVA by cross-linking it with other polymers has been done and modified PVA has been appreciated as dye adsorbent [33]. Table 3 enlists poly (vinyl alcohol) based amalgams for the removal of some important dyes in textile waste water.

3 Degradation of PVA

PVA is present in textile wastewater, as it is applied for the removal of dyes in wastewater as well as it comes from the desizing process occurring in the textile industry. Once the weaving process has completed and the sizing agent has finalized its role, it is necessary to remove this sizing agent as it creates inconvenience in the finishing process. As a result, PVA has become intractable organic pollutant of textile wastewater. According to an estimation, 200–300 kg of PVA enters into the wastewater plant on daily basis [41]. When this PVA-containing wastewater enters in the clean water bodies without effective treatment it covers the water surface in the form of large areas of foam [45]. The accumulation of foam on water surface will hamper the activities of aquatic life. Furthermore, poorly degradable PVA increases the movements of heavy metals deposited in the grounds of streams, lakes and oceans which are hazardous to human as well as aquatic life [7].

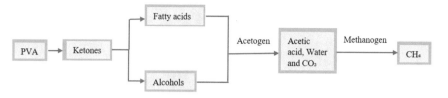

Fig. 3 Degradation pathways of PVA

The huge amount of rejected PVA is becoming a significant issue. Literature provides the information about the degradation pathway of PVA in wastewater as shown in Fig. 3. During the degradation process the PVA degrades into smaller molecules such as ketones, which then undergo hydrolysis and produce fatty acids and alcohols. The fatty acids and alcohols continue their degradation and produce acetic acid, water and carbon dioxide. Methanogens, in turn act on these products and completely degrades PVA [52]. During the thermal degradation different concentrations of acetic acid, 2-butenal, acetone, acetaldehyde, ethanol, water and CO_2 are observed at different temperatures. However the end products of degradation are usually H_2O, CO_2 and methane which are harmless [48].

Number of studies have been done on the degradation of PVA and many methods have been developed regarding the rapid degradation of PVA in textile wastewater [25]. Figure 4 illustrate thermal degradation of PVA at different temperatures. Figure 5 depicts some potential PVA degradation methods. PVA biodegradation is the conventional method used for the degradation of PVA in wastewater but PVA has poor biodegradability as well as different microbial species has different tendencies and specifications, so before subjecting the PVA saturated textile wastewater to biodegradation, pretreatment is necessary. Pretreatment involves different method, sometimes only the pretreatment method is sufficient for PVA degradation [43].

These Potential degradation processes include photocatalytic degradation (which efficiently degrades PVA into CO_2 and H_2O with the help of mineralization process), Ionizing radiation induced degradation (which includes electron beam radiation and the γ-radiations is suitable for wastewater containing low initial concentration of PVA), chemical oxidation (which has been conveniently applied for the removal of PVA from textile wastewater by maintaining the temperature of the reactor above 200 °C) and electro-coagulation (the process can be carried out at low temperature and low voltage, however, the degradation rate has been found different for different electrode pairs) [10, 11, 46, 51].

Chemical coagulation such as certain organic and inorganic coagulants has been considered efficient for the pretreatment of PVA-containing wastewater under different conditions. Adsorption process (carbon adsorption is considered the most effective and affordable with specification to powdered activated carbon, which has PVA removal efficiency above 90%), fenton process (almost 40% of the Dissolved Organic Carbon of PVA is removed within 2 h, however when the fenton process is photochemically accelerated, the DOC removal efficiency is amplified to 90%) and

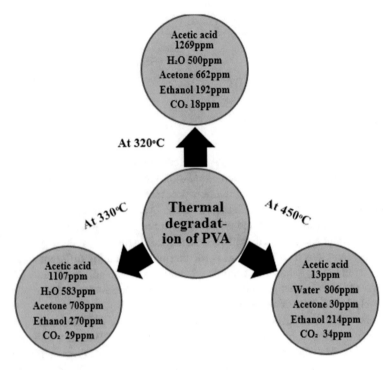

Fig. 4 Thermal degradation of PVA at different temperatures

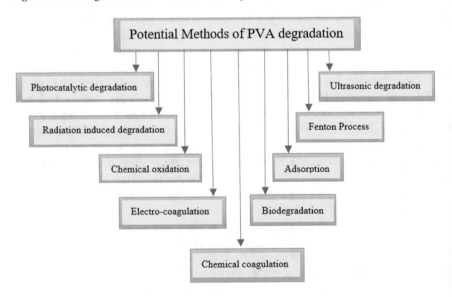

Fig. 5 PVA degradation processes in textile wastewater

Ultrasonic degradation (use of ultrasonic energy for polymer degradation is considered effective, lowest degradation is observed at higher frequency due to the negation of physical effects) [5, 7, 31, 32].

4 Environmental Effect of PVA

Extensive environmental problems that are interlinked with textile industry are those which result from the untreated textile wastewater containing toxic chemicals and harmful dyes [23]. Figure 6 depicts some environmental effects of PVA. Textile wastewater containing PVA, other sizing agents and dyes, when directly exposed to the water bodies like rivers and seas pollutes the clean water which results in aquatic life deterioration and worsening of crops [42]. PVA present in the textile wastewater also has health hazards, it has been found as carcinogenic, mutagenic and has deleterious effect on central nervous system [22].

4.1 Effect of PVA-Containing Wastewater on Aquatic Life

PVA and other effluents present in textile wastewater bring out substantial effect on aquatic life. It has been found that one gram PVA is correspondent to 0.016 g of BOD_5 and 1.76 g of COD and the ratio of BOD_5/COD appear to be 0.011. The treatment of PVA-containing wastewater is necessary and if untreated wastewater is exposed to the natural water bodies, it creates the foamy seal on the surface of

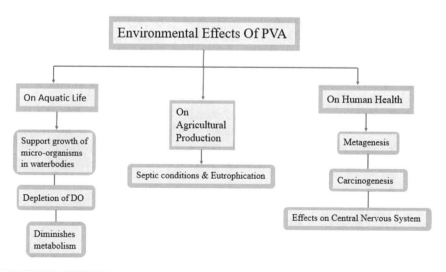

Fig. 6 Environmental effects of PVA

these waterbodies. Due to the presence of this sponge on surface the interaction of aquatic life (both aquatic plants and animals) to the outer atmosphere is minimized and hence results in the devastation of aquatic life [54].

High concentration of BOD and COD results in very supportive environment for the growth of microorganisms. These microorganisms sustain themselves by decomposing organic matter using dissolved oxygen (DO). Hence, when there is high concentration of BOD and COD, the depletion of DO will increase which is necessary for the maintenance of aquatic life, most commonly for fish and the aerobic organisms. The depleted DO effects the reproduction and metabolism of aquatic organism [27]. Moreover, the removal of DO has been reported to have devastative effects on aquatic biodiversity by diminishing metabolism and reducing the water ability to carry water oxygen [14].

4.2 Effect of Containing PVA Wastewater on Agricultural Production

Implementation of wastewater to crops and plants has been found out to be favorable because it contains important quantities of nitrogen, phosphorus and few other micronutrients. However, the presence of some toxic substances in wastewater is offering serious challenges to the physico-chemical properties of the exposed soil [2]. According to an estimation, annually 32,500 ha agricultural land is irrigated with the help of 30% of 962,335 million gallons of untreated wastewater in Pakistan. The untreated textile wastewater containing PVA and other dyes possesses some important properties such as high COD, BOD and some other solids dissolved in it [39].

Very limited work is reported on the effects of PVA (present in untreated textile wastewater) on the agricultural production. However, it has been discovered that when PVA-containing wastewater is used for irrigation it results in septic conditions as well as have adverse effects on the growth of plants and crops due to change in pH [29]. PVA and other organic species present in the wastewater also results in eutrophication [17].

Untreated textile wastewater effluent contains many harmful dyes and heavy metals. Hence, those crops and vegetables which are grown in lands irrigated by textile effluents are badly affected in both quality and quantity. Heavy metals accumulate in the seedlings and inhibit the process of seed germination [30].

4.3 Effect of PVA-Containing Wastewater on Human Health

Textile industry is considered the most polluting among the whole industrial zone. Untreated wastewater effluents from the textile industries are degenerating the fresh water bodies. The PVA, dyes and other heavy metals present in wastewater found out to be carcinogenic in their origin and bring about certain abnormalities in fetus, if the contaminated water is used for long time. Some dyes are harmful to such an extent that these can cause DNA damage [8].

During the singe and printing process the textile wastewater discharge includes different chemicals such as starch, gum, benzene and most commonly PVA. Number of pollutants arise from this discharge such as resins, waxes, fats and starch. When this PVA-containing water is used for drinking purposes it causes many health hazards such as mutagenesis, carcinogenesis and effects central nervous system [19].

However, many clinical studies indicated that there are very limited issues that arise from the exposure to PVA. Very less number of cases of human carcinogenesis have been reported [15]. Highly concentrated solution of PVA acts as irritant for eyes and skin. According to some studies it amplifies the medical conditions but most recent studies indicated that its carcinogenic and mutagenic effects are not justified for humans and PVA is still in the way to be classified as carcinogenic [9].

5 PVA Alternatives

PVA present in wastewater is birthing number of environmental issues related to aquatic ecosystems, agriculture and health. Proper pretreatment of PVA in wastewater can significantly reduce the adverse effects of PVA. AOP (advanced oxidation process) has been certified as successful methods to boost up the biodegradability of PVA [55]. AOPs are developed to produce •OH. The hydroxyl radicals produced by the AOPs oxidize the PVA to small molecules [47]. It has been estimated that in the presence of oxidants, decomposition of PVA molecules in wastewater is pseudo first order [13].

There is a serious need to find out the sizing agent that can substitute PVA as it results in destruction of environment. Keratin, that is obtained from chicken feathers has been proved as potential sizing agent on polyester and can be easily degraded in sludge [38]. Modification of starch has been done by manifesting it with acrylates and methacrylates and being used as successful and harmless sizing agent [56]. Some biopolymers such as cyclodextrans and chitosan are also studied as sizing agents with very little pollution [21, 44].

6 Conclusion

PVA has been used by the textile industries as major sizing agent and as PVA composites for removal of dyes since long time. The treatment of PVA is necessary before releasing into waterbodies because the accumulation of PVA in water leads to number of environmental issues. Many degradation methods are applied on PVA-containing wastewater before releasing it. The COD and BOD values of PVA are threatening the aquatic life and are also affecting the agricultural land and production. PVA has significantly observable effects on human health by outbringing some serious issues. Therefore, use of PVA should be reduced or it should be completely degraded before releasing. Some suitable alternatives to PVA are available which are non-toxic and produce same result as PVA, so the use of these alternatives should be promoted.

References

1. Agarwal S, Sadegh H, Monajjemi M, Hamdy AS, Ali GAM, Memar AOH, Shahryari-Ghoshekandi R, Tyagi I, Gupta VK (2016) Efficient removal of toxic bromothymol blue and methylene blue from wastewater by polyvinyl alcohol. J Mol Liq 218:191–197
2. Ahmad TM, Sushil M, Krishna M (2012) Influence of dye industrial effluent on physico chemical characteristics properties of soil at Bhairavgarh, Ujjain, MP India. Res J Environ Sci 1(1):50–53
3. Aslam M, Kalyar MA, Raza ZA (2018) Polyvinyl alcohol: a review of research status and use of polyvinyl alcohol based nanocomposites. Polym Eng Sci 58(12):2119–2132
4. Baker MI, Walsh SP, Schwartz Z, Boyan BD (2012) A review of polyvinyl alcohol and its uses in cartilage and orthopedic applications. J Biomed Mater Res Part B Appl Biomater 100 B(5):1451–1457
5. Behera SK, Kim JH, Guo X, Park HS (2008) Adsorption equilibrium and kinetics of polyvinyl alcohol from aqueous solution on powdered activated carbon. J Hazard Mater 153(3):1207–1214
6. Ben Halima N (2016) Poly(vinyl alcohol): review of its promising applications and insights into biodegradation. RSC Adv 6(46):39823–39832
7. Bossmann SH, Oliveros E, Göb S, Kantor M, Göppert A, Braun AM, Lei L, Yue PL (2001) Oxidative degradation of polyvinyl alcohol by the photochemically enhanced Fenton reaction. Evidence for the formation of super-macromolecules. Progr React Kinet Mech 26(2–3):113–137
8. Carmen Z, Daniela S (2010) Characteristics, polluting effects and separation / elimination procedures from industrial effluents–a critical overview. Text Organ Dye 55–86
9. Chemicals & Laboratory Equipment SL co (2013) Material Safety Data Sheet n-Propyl alcohol MSDS 1–6. http://www.sciencelab.com/msds.php?msdsId=9924736
10. Chen Y, Sun Z, Yang Y, Ke Q (2001) Heterogeneous photocatalytic oxidation of polyvinyl alcohol in water. J Photochem Photobiol A 142(1):85–89
11. Chou WL, Wang CT, Huang KY (2010) Investigation of process parameters for the removal of polyvinyl alcohol from aqueous solution by iron electrocoagulation. Desalination 251(1–3):12–19
12. Cordes MG (2007) Polyvinyl alcohol. XPharm Comprehen Pharmacol Ref 24:1–2
13. Chiellini E, Corti A, D'Antone S, Solaro R (2003) Biodegradation of poly (vinyl alcohol) based materials. In: Progress in polymer science (Oxford), vol 28(6)
14. Dadi D, Stellmacher T, Senbeta F (2017) Environmental and health impacts of effluents from textile industries in Ethiopia : the case of Gelan and Dukem, Oromia regional state. Environ Monit Assess 189(1):11
15. Demerlis CC, Schoneker DR (2003) Review of the oral toxicity of polyvinyl alcohol (PVA)-PDF free download. Food Chem Toxicol 41(3):319–326
16. Etim UJ, Inam E, Umoren SA, Eduok UM (2013) Dye removal from aqueous solution using coconut coir dust extract-modified polyvinyl alcohol: a novel adsorbent. Int J Environ Bioenergy 5(2):62–79
17. Foo KY, Hameed BH (2010) Decontamination of textile wastewater via TiO_2/activated carbon composite materials. Adv Coll Interface Sci 159(2):130–143

18. Gaaz TS, Sulong AB, Akhtar MN, Kadhum AAH, Mohamad AB, Al-Amiery AA, McPhee DJ (2015) Properties and applications of polyvinyl alcohol, halloysite nanotubes and their nanocomposites. Molecules 20(12):22833–22847

19. Ghaly A, Ananthashankar R, Alhattab M, Ramakrishnan V (2014) Production, characterization and treatment of textile effluents: a critical review. J Chem Eng Process Technol 5(1):1–19

20. Ghemati D, Aliouche D (2014) Dye adsorption behavior of polyvinyl alcohol/glutaraldehyde/β-cyclodextrin polymer membranes. J Appl Spectrosc 81(2):257–263

21. Hebeish A, Higazy A, El-Shafei A (2006) New sizing agents and flocculants derived from chitosan. Starch Staerke 58(8):401–410

22. Imtiazuddin SM, Tiki S (2018) Impact of textile wastewater pollution on the environment. Pakistan Textile J 68(8):38–39

23. Odjegba VJ, Bamgbose NM (2012) Toxicity assessment of treated effluents from a textile industry in Lagos, Nigeria Afr. J Environ Sci Technol 6(11):438–445

24. Kong Y, Zhuang Y, Han Z, Yu J, Shi B, Han K, Hao H (2019) Dye removal by eco-friendly physically cross-linked double network polymer hydrogel beads and their functionalized composites. J Environ Sci (China) 78(2018):81–91

25. Lin CC, Lee LT, Hsu LJ (2014) Degradation of polyvinyl alcohol in aqueous solutions using UV-365 nm/S2O82- process. Int J Environ Sci Technol 11(3):831–838

26. Mahmoodi NM, Mokhtari-Shourijeh Z (2016) Modified poly(vinyl alcohol)-triethylenetetramine nanofiber by glutaraldehyde: preparation and dye removal ability from wastewater. Desalin Water Treat 57(42):20076–20083

27. Mallya YJ (2007) The effect of dissolved oxygen on fish growth in aquaculture 30

28. Alem M, Tarlani A, Aghabozorg HR (2017)Synthesis of nanostructured alumina with ultrahigh pore volume for pH-dependent release of curcumin. RSC Adv

29. Mara D (2020) Wastewater re-use in agriculture. In: Domestic wastewater treatment in developing countries

30. Marwari R, Khan TI (2012) Effect of textile waste water on tomato plant, Lycopersicon esculentum. J Environ Biol 33(5):849–854

31. Mo J, Hwang JE, Jegal J, Kim J (2007) Pretreatment of a dyeing wastewater using chemical coagulants. Dyes Pigm 72(2):240–245

32. Mohod AV, Gogate PR (2011) Ultrasonic degradation of polymers: Effect of operating parameters and intensification using additives for carboxymethyl cellulose (CMC) and polyvinyl alcohol (PVA). Ultrason Sonochem 18(3):727–734

33. Mok CF, Ching YC, Muhamad F, Abu Osman NA, Hai ND, Che Hassan CR (2020) Adsorption of Dyes Using Poly(vinyl alcohol) (PVA) and PVA-based polymer composite adsorbents: a review. J Polym Environ 28(3):775–793

34. Morrison P (1990) The merck index: an encyclopedia of chemicals, drugs, and biologicals eleventh edition. Scient Am 263(1):122–123

35. Muppalaneni S (2013) Polyvinyl alcohol in medicine and pharmacy: a perspective. J Developing Drugs 02(03):1–5

36. Nemours P, De Falls N, Yorlc N (1965) Poly(vinyl Alcohol) 3:4181–4189

37. Okhamafe AO, York P (1988) Studies of interaction phenomena in aqueous-based film coatings containing soluble additives using thermal analysis techniques. J Pharm Sci 77(5):438–443

38. Reddy N, Chen L, Zhang Y, Yang Y (2014) Reducing environmental pollution of the textile industry using keratin as alternative sizing agent to poly(vinyl alcohol). J Clean Prod 65:561–567

39. Roohi M, Riaz M, Saleem M, Muhammad S, Yasmeen T, Atif M, Tahir S (2016) Varied effects of untreated textile wastewater onto soil carbon mineralization and associated biochemical properties of a dryland agricultural soil. J Environ Manag

40. Hegazy SA, Abdei-AAL SE, Abdel-Rehim HA, Nevien A (2000). Removal of some basic dyes by poly (Vinyl Alcohol/Acrylic Acid) hydrogel

41. Schonberger H, Baumann A, Keller W (1997) Study of microbial degradation of polyvinyl alcohol (PVA) in wastewater treatment plants. Am Dyestuff Rep 86(8):9–18

42. Shaikh MA (2009) Environmental issues related with textile sector. Pakistan Textile J 58(10):36–38
43. Solaro R, Corti A, Chiellini E (2000) Biodegradation of poly(vinyl alcohol) with different molecular weights and degree of hydrolysis. Polym Adv Technol 11(8–12):873–878
44. Stegmaier T, Wunderlich W, Hager T, Siddique AB, Sarsour J, Planck H (2008) Chitosan-a sizing agent in fabric production-development and ecological evaluation. Clean: Soil, Air, Water 36(3):279–286
45. Sun W, Chen L, Wang J (2017) Degradation of PVA (polyvinyl alcohol) in wastewater by advanced oxidation processes. J Adv Oxidat Technol 20(2)
46. Sun W, Tian J, Chen L, He S, Wang J (2012) Improvement of biodegradability of PVA-containing wastewater by ionizing radiation pretreatment. Environ Sci Pollut Res 19(8):3178–3184
47. Sun W, Chen L, Tian J, Wang J, He S (2013) Radiation-induced decomposition and polymerization of Polyvinyl Alcohol in aqueous solutions. Environ Eng Manag J 12(7):1323–1328
48. Taghizadeh MT, Yeganeh N, Rezaei M (2015) The investigation of thermal decomposition pathway and products of poly(vinyl alcohol) by TG-FTIR. J Appl Polym Sci 132(25):1–12
49. Umoren SA, Etim UJ, Israel AU (2013) Adsorption of methylene blue from industrial effluent using poly (vinyl alcohol). J Mater Environ Sci 4(1):75–86
50. Wang C, Yang F, Liu, LF, Fu Z, Xue Y (2009) Hydrophilic and antibacterial properties of polyvinyl alcohol/4-vinylpyridine graft polymer modified polypropylene non-woven fabric membranes. J Memb Sci 345(1–2):223–232
51. Won Y, Baek S, Tavakoli J (2001) kinetics, catalysis, and reaction engineering wet oxidation of aqueous polyvinyl alcohol solution 60–66
52. Wu HF, Yue LZ, Jiang SL, Lu YQ, Wu YX, Wan ZY (2019) Biodegradation of polyvinyl alcohol by different dominant degrading bacterial strains in a baffled anaerobic bioreactor. Water Sci Technol 79(10):2005–2012
53. Yan J, Huang Y, Miao YE, Tjiu WW, Liu T (2015) Polydopamine-coated electrospun poly(vinyl alcohol)/poly(acrylic acid) membranes as efficient dye adsorbent with good recyclability. J Hazard Mater 283:730–739
54. Yu H, Gu G, Song L (1996) Degradation of polyvinyl alcohol in sequencing batch reactors. Environ Technol (United Kingdom) 17(11):1261–1267
55. Ye B, Li Y, Chen Z, Wu Q-Y, Wang W-L, Wang T, Hu H-Y (2017) Degradation of polyvinyl alcohol (PVA) by UV/chlorine oxidation: radical roles, influencing factors, and degradation pathway. Water Res 124:381–387
56. Zhu Z, Cao S (2004) Modifications to improve the adhesion of crosslinked starch sizes to fiber substrates. Text Res J 74(3):253–258
57. Zuber M, Zia KM, Bhatti IA, Jamil T, Rehman FU, Rizwan A (2012). Modification of cellulosic fabric using polyvinyl alcohol, part-II: colorfastness properties. Carbohyd Poly 87(4):2439–2446

Nanoparticles Functionalized Electrospun Polymer Nanofibers: Synthesis and Adsorptive Removal of Textile Dyes

Shabna Patel, Sandip Padhiari, and G. Hota

Abstract Textile wastewaters are highly toxic due to the occurrence of large amounts of organic contaminants as an example dyestuffs; their chemical and photostability have led to a serious problem for the ecosystem. The removal of textile dyes has been studied using a number of different methods. Adsorption methods, in particular, are frequently regarded as one of the most promising solutions for treating dye-contaminated water because of their practical and financial benefits. Overall, the adsorbent material's nature is strongly linked to the adsorptive elimination of colors from water's efficiency. Several materials have been studied as adsorbents, including raw and processed materials, but none have shown to be effective. The utilization of functional nanomaterials as effective adsorbents for wastewater treatment has been the focus of study over the past few decades. As an ideal adsorbent, nanoparticles functionalized electrospun polymer nanofibrous possesses extraordinary porosity with greater surface area. Therefore, for many applications, they have been recommended as excellent candidates, especially in the treatment of waste wastewater. The highlighting fetcher of this chapter is to understand the fundamental aspects of electrospun nanofibrous and their basic properties, as well as the potential of nanofibrous as enormous adsorbents for the textile dye remediation. We believe that the provided information about the fascinating electrospun functional nanofibrous materials will enhance the basis of scientific knowledge for their practical use as effective adsorbents for removing textile dyes.

Keywords Textile dye · Electrospun polymer nanofibrous · Adsorbent · Nanoparticles

S. Patel
Department of Mathematics and Science, UGIE, Rourkela 769004, Odisha, India

S. Padhiari · G. Hota (✉)
Department of Chemistry, NIT, Rourkela 769008, Odisha, India
e-mail: garud@nitrkl.ac.in

1   Introduction

In the recent past wastewater effluents containing contaminants of organic dyes have received immense concerns due to the rapid industrial developments like paper, plastics, pharmaceuticals, cosmetics, food processing, and textile finishing. Among the above sources, textile industries are a chief consumer of fresh water and are known to discharge colors and organic dyestuffs in considerable concentrations in their effluents. These dye pollutants discharged into the aqueous stream are mutagenic and carcinogenic and are the reason for several ecological and health problems in human beings [57]. Most of the dyes have high molecular mass, complex structures, and minimal biodegradability; meanwhile, even at a very low concentration, these dyes in water are visible. Moreover, they are resistant to light, heat, and chemical reagents and thus have the potential risk of bioaccumulation. Hence, before discharge into the environment, the removal of these dye contaminants is highly essential. Thus, to tackle this issue research must be focused on the creation of cost-effective and ecologically friendly products friendly techniques for the reduction of these contaminants in the wastewater. Many biochemical and physicochemical processes including flocculation-coagulation, biological process, electrochemical, photocatalytic degradation, chlorination, and adsorption have been investigated widely for the treatment of dye contaminants. Due to its effectiveness, cost efficiency, and ease of regeneration, adsorption is considered an efficient technique [17]. Since the cost and the properties largely determine the adsorption process of an adsorbent, so most attention has been focused on the designing of low-cost and highly efficient adsorbents. Taking the advantage of nanotechnology, as a substitute for traditional adsorbents, various nanomaterials have been reported to show promising efficiency for the adsorption of contaminants from wastewater.

Among the various nanoadsorbents, in the last two decades, EPNFS (electrospun polymer nanofibers) have gotten a lot of press in the field of adsorption [35]. In comparison to commercial resins or polymers, electrospun nanofibers offer good interfiber a significant specific surface area and porosity [2, 14, 44]. As a result, these nanofibers are becoming viable industrial and commercial materials in a variety of applications. The most significant technical and scientific knowledge on nanofibers, the electrospinning method, and product characterization is crucial, and the electrospinning technique has lately because of its reliable fiber production in the submicron range, it has gotten a lot of attention. Various natural and man-made polymers have been turned into nanofibers utilizing electrospinning processes after being dissolved in appropriate solvents. The polymer nanofiber by the application of suitable organic reagents can be functionalized by diverse groups [12, 56], while nanofibers with a cross-linked polymeric structure have a greater number of functional groups, they may offer mechanical benefits such as durable and resilient fibers for reuse and regeneration, as well as simplicity of operation. [31]. Furthermore, one-dimensional electrospun polymer nanofibers nanostructured material have exceptional characteristics like high aspect ratio, numerous sites, and greater affinity for adsorption.

These properties permit EPNFs as adsorbents for effective adsorption of toxic water contaminants.

Apart from electrospun polymer nanofibers, metal oxides like cerium oxides, iron oxides, manganese oxides, aluminum oxides, magnesium oxides, and manganese oxides, etc., due to their cost-effective, environmentally benign, and relatively high natural abundance, have been widely investigated as a powerful adsorbent for the removal of contaminants from polluted water [32]. Furthermore, these adsorbents surfaces can be altered by incorporation of functional groups like amine, carboxyl, or hydroxyl through surfactants, and polymers to improve their adsorption capacity to a significant extent [33].

2 Fabrication of EPNFS

Within the last two decades, electrospinning has become one among the most popular and widely utilized polymer nanofiber production techniques. This is due to its versatility and simplicity, as well as its capacity to manufacture nanofibers in a wide range of sizes and forms, ranging from less than one nanometer to several micrometres in diameter. [15, 33]. This approach has a simple setup that comprised of four fundamental components: a power source with a high voltage, a syringe pump, a collector, and a spinneret (Scheme 1). EPNFs are created by allowing the polymer to dissolve in an approroate solvent to create a viscosity-controlled solution, which is then spun

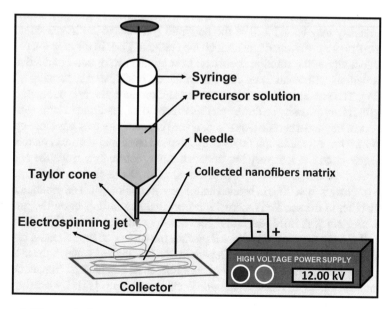

Scheme 1 Electrospinning setup

at high voltage using either or direct or alternating current. Due to surface tension, the viscous polymer solution is first pumped and controlled by the syringe pump before being released as spherical droplets from the spinneret. When a high voltage is supplied to the spherical droplets on the spinneret, electrostatic repulsive forces between them are created, which operate to overcome surface tension and transform into a Taylor cone [1]. Due to the curvature's instability, the jet first extends in a straight path after forming the Taylor cone, but subsequently suffers intense whisking motions while moving toward the collector. While the jet transforms into ultra thin fibers as it approaches the collector, the solvent evaporates quickly, resulting in nanofibers being deposited on the collecting surface. Electrospinning polymer 2021, 13, 20 3 of 38 processing conditions (e.g., flow rate (F.R), distance from the collector (TCD) to the spinneret, applied voltage (V), collector geometry, spinneret geometry, and diameter) polymer solution (e.g., viscosity, concentration, molecular weight, surface tension, and conductivity) and environmental circumstances (e.g., molecular weight, conductivity, surface tension, concentration, and viscosity) (e.g., temperature, humidity, and air speed). One factor that enhances the spinnability of polymer solutions is applied voltage, which must be adequate to overcome the polymeric solution droplet's surface tension. Droplets occur at the needle's tip due to insufficient or excessive voltage, resulting in beaded nanofibers. The solution flow rate has a significant impact on fiber formation. Because the fiber jet from the needle to the collection is only partially dry, at higher flow rates, beaded fibers occur [54]. To generate a beadless and uniform nanofiber, the needed rate of flow can be set as low as the minimal value [53]. Using polymeric systems for example polyethylene oxide (PEO), polycaprolactone (PCL), polyvinyl alcohol (PVA), and polyurethane, Theron et al. investigated the connection between applied and voltage flow rate. The charge density may be reduced as the flow rate and voltage are increased, causing nanofibers to merge before depositing on the collector [39]. In the creation of fibers, the conductivity of the solution is critical. Low conductive or non-conductive polymeric solutions will result in no charges on the droplet's surface, forming a Taylor cone [37]. The conductivity of a polymeric solution must rise to a particular point, causing the charge density on the droplet's surface to rise, resulting in the creation of a Taylor cone. Beyond a certain degree, increased conductivity may have an impact on fiber form. The polymeric solution's concentration has a major impact on the shape of bead-free nanofibers, at very low concentrations, before they reach the collector, the electric field that is used and surface tension break the chains of polymers into minute fragments, resulting in beaded nanofibers formation [9]. The separation from the collection to the needle is a crucial aspect that has an impact on the produced nanofibers quality. A little separation between the collector and the needle can aid in the full solvent evaporation before it reaches the surface of the collector, resulting in bead-free nanofiber production. Depending on the applied voltage, these lengths can be varied [36]. The morphology of nanofiber mats can have an impact be environmental factors as an example humidity and temperature [11]. By altering these parameters, the form and sizes of the created nanofibers may be optimized based on the planned usage of the nanofibers [5, 8, 13]. The structure of formed nanofibers can also vary depending on the used technique, for example needlefree electro-spinning [45], multi-jet electrospinning [43], bubble electrospinning [50], electro blowing [41], coaxial electrospinning [30], and emulsion electrospinning [47]. To make a single-layered nanofiber mat, the basic electrospinning technique is usually utilized. Core–shell structured nanofibers are made using either a typical coaxial method of electrospinning or a simple emulsion method of electrospinning [51]. In the coaxial approach, two needles are joined together coaxially for two distinct poly-meric solutions. The inner needle pushed the core material, while the coaxial spin-neret's outer needle pumped the shell material [22]. The fundamental characteristic for this process is that the shell is most likely made of a polymer solution that can be electrospinned, whereas the inner core is made of a chemical solution or another non-spinnable substance. The triaxial electrospinning setup, which comprised of three needles coupled to the intermediate, core of a spinneret device as well as sheath solutions, may produce multi-layered nanofibers. The intermediate solutions and core should be immiscible in this approach, however the intermediate solutions and outer sheath can be miscible [24]. Because of their large loading capacity and ease of loading poorly soluble drugs into the polymeric shell, core–shell and multi-layered nanofibers are mostly employed in biomedical applications [10].

2.1 Modification of EPNFS

In spite of the benefits of polymer nanofibers, such as porosity and high surface–volume ratios, certain virgin polymers still have adsorption restrictions. Some have insufficient adsorption capability to remove contaminants (e.g., nylon and polyacry-lonitrile (PAN)), others are not stable in solutions of water (e.g., polyacrylic acid (PAA), PVA, and polyvinyl pyrrolidone (PVP),) and still others, such as chitosan, have low mechanical qualities. Researchers have worked hard to enhance the charac-teristics of nanofibers by surface modification in order to overcome these issues. The surface modification seeks to improve nanofiber stability in solutions, enhance their wettability and hydrophilicity qualities, increase the number of surface adsorption sites, and improve their mechanical strengths [19]. A one-step treatment technique conducted during the electrospinning process (blends and nanocomposites) and a post-treatment process carried out after the electrospinning process may both modify the surface of nanofibers (e.g., wet-chemistry, plasma, coating, and grafting).

2.2 Blending with Other Polymers

The characteristics of virgin nanofibers from polymer may be improved by combi-nation one polymer with other polymer in particular proportions [4]. The ratio, addi-tional polymer molecular weight, and surface energy as well as the solvent, influence these qualities. Several hydrophobic polymers, including PVDF [25], polystyrene

[42], PET [20], PCL, and PES, have been improved using the mixing process. Before the electrospinning process, hydrophobic nanofibers that were utilized to remove ionic contaminants from aqueous solutions are commonly changed by addition of amphiphilic copolymer or hydrophilic homopolymer. Ghani et al., for example, blended different ratios of chitosan to modify the surface of polyamide-6 nanofibers. They discovered that raising the chitosan ratio enhanced the hydrophilicity of polyamide-6 nanofibers, which boosted their adsorption ability for anionic dye removal. In another work, Xu et al. used a one-step electrospinning procedure to create a mix of MMA-co-AA (PES/P) nanofibers [48]. They discovered that when the proportion of P(MMA-co-AA) grew, so did their hydrophilicity and adsorption ability for removing cationic (MB) dye.

2.3 Incorporation of Nanomaterials

Dispersing nanoparticles in a polymer solution and electrospinning them into polymer nanofibers is one of the best ways to change the surface features and polymer nanofibers mechanical strengths while also increasing the adsorption capacity. Included nanoparticles have a decent characteristic for dispersion in the solution containing the polymer and the capacity to separate nanofibers surface for the modification technique to be effective. The most often utilized materials for changing the polymer nanofibers surface and applying for various pollutants removal include carbon nanomaterials (e.g., MWCNTs-COOH and GO), nano-clay, MOFs, bacteria, and metal oxide NPs.

2.4 Incorporation of Functional Groups

Wet modification, which involves a reaction between the nanofibers' surface and an agent for functionalization in the solution, is an efficient method for functionalizing EPNFs' surfaces. The functional agent's nature determines the surface characteristics of the modified electrospun polymer nanofibers. This approach is based on the EPNFs' surface functional groups, which are vulnerable to interacting with various chemicals. The hydrolysis and aminolysis processes can produce polar functional groups on EPNFs. The most common wet-chemistry process for modifying the surface characteristics of nanofibers is alkali solution hydrolysis of polyester, PLGA, PCL, and PBGL. PAN nanofibers' hydrophilicity was improved by altering nitrile groups to more polar groups like carboxylic acids, amidoxime, and amino groups, which is an additional example of wet chemical modification.

3 Application of Nanoparticles Functionalized EPNFS for Removal of Textile Dyes

3.1 Silica/EPNFS Nanocomposites

Silica is a suitable nanofiller that was utilized to improve the nanofibers qualities such as mechanical capabilities, adsorption property, and surface area. It has surface functionalized with hydroxyl groups, and a high surface area [21] used electrospinning of thiol functionalized PVA to create absorbent nanofibrous membranes composite of PVA/SiO$_2$ and evaluated the removal of acid red and indigo carmine utilizing the produced fiber as an adsorbent. The adsorption maximums (Langmuir isotherm) for acid red and indigo carmine were 211.74 and 266.77 mg g^{-1}, respectively, due to high electrostatic and hydrogen bond membrane surface interaction and dye molecules and enlarged surface area (140.1 m^2 g^{-1}). In addition, the Redlich-Peterson model was more adapted to the adsorption process than the Langmuir and Freundlich isotherms. The pseudo-first-order and pseudo-second-order models were used to calculate the kinetics of the adsorption process. Mahmoodi et al. [27] used an electrospinning sol–gel technique to create mesoporous polyvinyl alcohol/chitosan/silica composite nanofibers (PCSCN). The tetraethyl orthosilicate boosts the electrospinning property of the gel by providing more active sites for the Red 80 dye. Dye adsorption is endothermic and follows pseudo-second-order kinetics, according to the Langmuir isotherm model. The impact of operating factors on the adsorption property was examined. The best adsorbent dose, beginning pH, and initial dye concentration were 50.015 g, 2, and 15 mg/L, respectively, whereas 322 mg g^{-1} was the best adsorption capacity. Teng et al. [38] used a combination of electrospinning and sol–gel procedures to create cyclodextrin mesostructure functionalized PVA/SiO$_2$ composite nanofibers. The FESEM and TEM pictures of the manufactured fiber are shown in Fig. 1. The indigo carmine dye was used as a model pollutant to look into the adsorption capability of the PVA/SiO$_2$ composite adsorbent. It was possible to achieve adsorption equilibrium after 40 min, and the highest adsorption capacity was 495 mg g^{-1}. Furthermore, the membranes have outstanding regeneration capabilities for practical use.

As shown in Fig. 2, hierarchical SiO$_2$/-AlOOH (boehmite) core/sheath nanofibers were effectively synthesized using a combination of hydrothermal and electrospinning methods, and were used to adsorb Congo red from aqueous medium [29]. In comparison to boehmite powder, this membrane that stands on its own is retrievable, simple to handle, and very flexible. The adsorption capacity of boehmite powder (10.4 mg g^{-1}) was nearly two times lower as compared to nanofibers (21.3 mg g^{-1}) due to the uniform distribution of boehmite nanoplatelets and the manufactured fiber's larger surface area.

The process of Congo red adsorption onto the surface of nanofibers was strikingly comparable to Li's study. By combining the electrospinning and sol–gel processes, Xu et al. [46] created a mesoporous PAA/SiO$_2$ modified with the vinyl group for malachite green removal from water. This newly designed membrane was discovered to

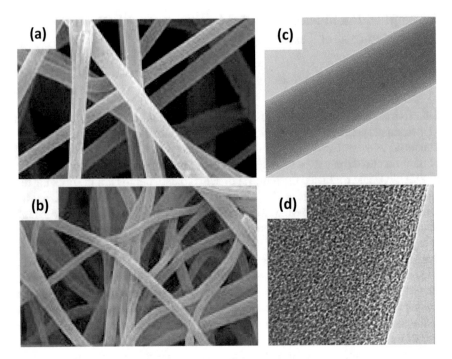

Fig. 1 SEM images of the silica/PVA fibers as they were electrospun (**a**) and following CTAB removal, and the mesoporous silica fibers (**b**). Scale bars: 1 μm. TEM pictures at different magnifications of the cyclodextrin-functionalized mesoporous fiber membranes (**c, d**)

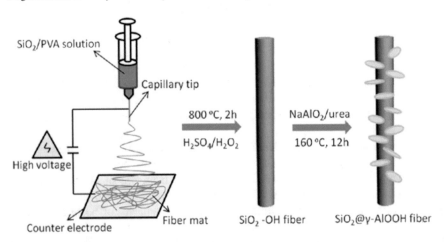

Fig. 2 Synthesis of SiO_2@γ-AlOOH (Boehmite) core/sheath fibers

have a large specific surface area ($523.84 \text{ m}^2 \text{ g}^{-1}$) and hence a high adsorption capability for malachite green removal. The PAA/SiO$_2$ membranes made of nanofibers were shown to have a 98.8% removal effectiveness after 240 min, with an equilibrium adsorption ability of 220.49 mg g^{-1}. The adsorption of malachite on nanofiber membranes was discovered to follow a pseudo-second-order model. The following three factors are primarily accountable for the developed nanofibers' high capacity for adsorption: To begin with, there is a significant electrostatic connection between the malachite green molecule ($-N^+$) positive surface and the nanofibrous membrane (-COO$^-$ of PAA and -Si$^-$O$^-$ of silica,) negative charge. Second, nanofibers' huge surface area ($523.84 \text{ m}^2 \text{ g}^{-1}$ and extraordinary energy on the surface may contribute significantly to physical adsorption dye; and, lastly, the conjugation action between the vinyl (or carbonyl group and the dye molecule's delocalized bonds. Figure 3 displays a schematic diagram of the mechanism. Furthermore, good regeneration characteristics of the mesoporous PAA/SiO$_2$ nanofibrous membranes were reported in the first three cycles.

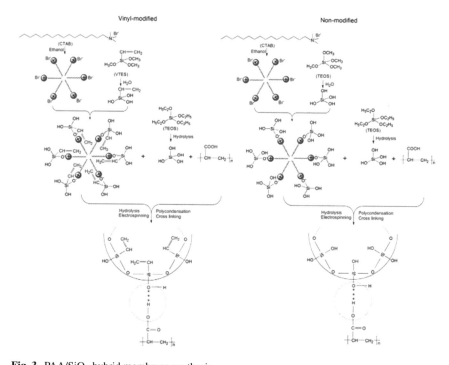

Fig. 3 PAA/SiO$_2$ hybrid membrane synthesis

3.2   EPNFs/Metal Oxides Nanocomposites

Metal oxides materials are known to be the best adsorbents that were utilized to
remove a variety of contaminants. However, due to chemical degradation and dissolution at low pH, the metal oxide NPs utilization for dye adsorption, particularly
in acidic conditions, is still limited. Metal oxides can be included in polymeric
nanofibers to tackle both the corrosion problem and the less mechanical characteristics of the polymer nanofibers. Fard et al. [3] created a -Fe2O3 nanofiber with
a hydroxyl group on its surface using electrospinning and in-situ polymerization
methods and explored its ability for cationic dye adsorption. The hydroxyl group
present on the α-Fe$_2$O$_3$ nanofiber surface was functionalized by polymerizing vinyl
acetate monomer on the nanofiber's surface, which was then alkaline hydrolyzed to
change the acetate group to hydroxyl group. The best monomer concentration and
initiator concentration were found to be 3.5 mL and 0.2 g, respectively. Figure 4
illustrates the effective polymerization process using FTIR and FE-SEM. The pore
size distribution study found that when the average diameter of nanofibers increased,
the pore size changed toward high values. The adsorption behavior of the produced
nanofiber was investigated by removing BR46, BB41, and BR18. The maximum
adsorbent efficiency was obtained at pH 8.5, as shown in the adsorption studies. By

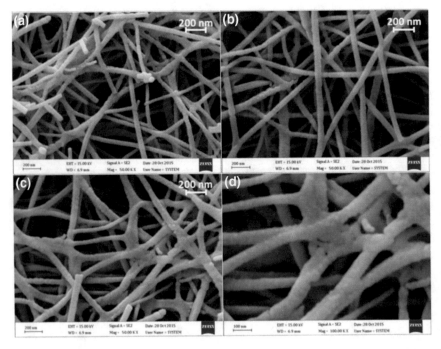

Fig. 4 a α-Fe$_2$O$_3$ nanofiber, **b** vinyl-functionalized α-Fe$_2$O$_3$ nanofiber, and **c, d** hydroxylated
α-Fe$_2$O$_3$ nanofibers, as shown in FE-SEM images

fitting the adsorption data to nonlinear kinetic and isotherm models, the rate and mechanism of the adsorption process were investigated. Chi-square test (χ 2), the summation of the squares of the errors (ERRSQ), and namely Average relative error (ARE) error functions were employed for Optimization of the procedure. The findings suggest hydroxyl group on the α-Fe_2O_3 nanofiber surface, which has a relatively high adsorption capacity, might be a good adsorbent for cationic dyes removal from polluted aqueous solutions.

Apart from Fe_2O_3, it has also been reported that ZnO embedded in nanostructures has outstanding capacity of adsorption for the elimination of different colors from contaminated water. Polyacrylonitrile (PAN), hinokitiol (HT), and zinc oxide composite nanofibers were developed by Phan et al. [34] and employed for bactericidal exertion and color removal. The ZnO-HT-PAN H nanofiber outperforms PAN nanofibers implanted with HT or zinc oxide in terms of dye removal capabilities. Because ZnO particles are loosely held on the PAN nanofibers surface, they disperse irreversibly in the aqueous environment, which is undesirable for the removal of dye. However, because of the glue effect, ZnO nanoparticles and HT were bonded securely together in ZnO-HT-PAN H, increasing dye-adsorption efficiency. The dye-adsorption effectiveness of reactive red 195 and reactive blue 19 over a 26-h period also indicated enhanced adsorption without any leaching. The highest capacity of adsorption for RB 19 and RR 195 for ZnO-HT-PAN H is 267.37 mg/g and 245.76 mg/g, respectively. Xu et al. [49] produced a Ti, Ag nanoparticle-embedded electrospun nanofiber composite with a well-dispersed structure that showed excellent MB adsorption ability.

Concentration of dye at the start, solution pH, dose, time for contact, temperature, and chemical modification all have an influence on adsorption capacity and process. The adsorption isotherms and kinetics of PAN-TA (PAN/Ag_3VO_4/TiO_2) composite nanofibers were determined to better understand their adsorption characteristics. The pseudo-second-order model obeyed the Langmuir isotherm model, in accordance with the data acquired from the adsorption investigation. At 25 °C, the maximal adsorption capacity is measured to be 155.4 mg g^{-1}. The adsorbent was shown to have good reuse characteristics in terms of practical use. Zhou et al. [59] effectively produced a functionalized nanofiber membrane with excellent methylene blue adsorption efficiency using electrospinning and a surface modification technique. First, titanium dioxide (TiO_2) nanoparticles (NPs) were deposited on the substrate poly (lactic acid) (PLA) nanofiber sheet using the sol–gel process. Polysiloxane was then placed on the surface of the modified nanofiber to minimize surface energy and enhance surface roughness. Furthermore, in an aqueous solution, the functionalized nanofibrous membrane demonstrated quick and reusable adsorption of hazardous dyes such as methylene blue. This opens up a new avenue for easily fabricating nanofibrous membrane materials that can effectively remove hazardous dyes.

Cellulose acetate (CA)/chitosan/single-walled carbon nanotubes (SWCNT)/ferrite (Fe_3O_4)/titanium dioxide (TiO_2) nanofibers were manufactured by ZabihiSahebi et al. [52]. By adsorption and photocatalysis, the nanofiber was employed to remove aqueous solutions of congo red and methylene blue. The adsorption performance of SWCNT/Fe_3O_4/TiO_2, SWCNT/Fe_3O_4, and TiO_2

were examined in a batch mode adsorption procedure. The effect of several factors as an example initial adsorbent concentration, contact length, and pH on Congo red and Methylene blue adsorption was examined. Nanofiber reusability was also tested for five adsorption–desorption cycles. Polyhydroxybutyrate/carbon nanotubes (CaAlg-c-PHB/CNT) electrospun nanofiber covered with coating calcium alginate hydrogel was created by Zhijiang et al. [58]. The CaAlg-c-PHB/CNT nanofiber filtration property was tested at operation pressures ranging from 0.1 to 0.7 MPa. To assess the antifouling capabilities, the separation efficiency of BSA solution and emulsified oil–water is also examined. The nanofiber with CNT of 1 wt% has the best tensile properties, according to the findings, with initial Modulus values of 6.14, 5.12, and 142.64, 127.5 MPa in dry and wet states, respectively. After covering with the calcium alginate hydrogel, the observed decreased in the contact angle from 83.6° to 17.8°. Operating pressure and flow of pure water (PWF) increased from 0.1 to 0.7 MPa and 68.61 to 150.72 L/m^2 h, respectively. Five dyes were tested and the molecular weights of which range from 400 to 900 g/mol and chosen for filtering. Under 0.5 MPa pressure, a 90 percent rejection rate was maintained for dyes with molecular weights greater than 600 g/mol. The composite nanofiber membrane has strong antifouling properties for protein and both oil. Magnetic one-dimensional (1D) nanomaterials have lately drawn interest for the separation of biomolecules and catalysts due to their enormous surface area, low density, superior chemical resistance, and outstanding charge transport capabilities.

Savva et al. [58] developed poly(ethylene oxide) (PEO) and poly(L-lactide) (PLLA) fibers containing magnetite Fe3O4 coated with oleic acid using an electrospinning procedure ($OA.Fe_3O_4$). The created fibers' adsorption technique is put to the test in order to eliminate malachite green oxalate dye (MG) from an aqueous medium. UV–Vis spectrophotometry was used to evaluate the adsorption process as a function of the dye content and pH of the solution. The equilibrium results revealed a monolayer adsorption mechanism. When an adsorption research of nanofiber without $OA.Fe_3O_4$ was conducted, the influence of $OA.Fe_3O_4$ on the adsorption performance was discovered. The obtained data show that the magnetite nanoparticles impair the adsorption effectiveness to some extent. Furthermore, the thermodynamic investigation conducted at three different temperatures shows that the process of adsorption is endothermic. Although the presence of Fe_3O_4 within the fibers hinders the adsorption process, the nanofiber may be readily retrieved from the solution using an external magnet. The MG is desorption from the surface of the nanofibers in an alcohol solution, allowing the adsorbent to be reused without losing its adsorption effectiveness. For dye adsorption from water, Ghourbanpour et al. [7] produced a poly (vinyl alcohol) (PVA)/chitin nanofiber cross-linked with Fe(III). The influence of chitin nanofibers (NFs) and Fe (III) concentration on the removal of dye effectiveness was examined. Initial dye concentration, contact duration, and solution pH were all evaluated as factors that impact the adsorption process. With the addition of chitin NFs, the removal efficacy of the dye of PVA/Fe(III) was significantly improved. Furthermore, the dye removal effectiveness in alkaline solution did not alter when chitin NFs were present. The thermodynamics of the process of adsorption, isotherms, and kinetics were also carefully examined. The Freundlich isotherm and a pseudo-second-order kinetics model were used to match the adsorption data. The PVA-chitin NFs-Fe(III) has 810.4 mg/g adsorption capacity. In addition, the reuse trials revealed five cycles of adsorption/desorption. Finally, the mechanism of the adsorption of dye was presented. The results obtained suggested that the created bio nanocomposites may be regarded as potential recyclable adsorbents for treatment of wastewater.

3.3   EPNFs/MOFs Nanocomposites

MOFs (metal–organic frameworks) are a potential type of material made up of organic linkers and metal–oxide units. MOFs have a lot of porosity, a lot of surface area, a lot of thermal stability, a lot of chemical stability, and a lot of mechanical characteristics [40]. MOFs are used in a large number of possible uses, such as hydrogen storage, catalysis, and water decontamination [18, 26]. Recently, there has been a lot of interest in incorporating MOFs are being turned into polymeric materials for water treatment [6]. Bio-MOF-1/PAN nanofibers were effectively synthesized by Li et al. [23] using co-electrospinning of polyacrylonitrile (PAN) and bio-MOF-1. The synthesized nanofiber was used to selectively adsorb methylene blue from an aqueous solution. They discovered that incorporating bio-MOF-1 into PAN nanofibers improved the mechanical and hydrophilicity characteristics of the material. The bio-MOF-1/PAN nanofibers showed improved selectivity in terms of adsorption and separation MB from aqueous solutions containing mixed colors. A synergistic impact between the nucleophilicity of -CN in PAN and the anionic charge of bio-MOF-1 was ascribed to the substantial selectivity in adsorption and separation. Mahmoodi et al. created a Chitosan (CS)/Polyvinyl alcohol electrospun (PVA) nanofiber incorporating the ZIF-8 Zeolite framework via electrospinning [28]. Depending on whether ZIF-8 is coated in the first, second, or third cycles, the composite fiber is designated as ZIF-8@CS/PVA-ENF(1), ZIF-8@CS/PVA-ENF(2), or ZIF-8@CS/PVA-ENF(3). The nanofiber that resulted was utilized to remove MG dye from tainted water. To find the most suitable conditions for adsorption, the results of the pH, adsorbent dosage, and starting MG concentration were examined. When the pH of the solution was 6, the efficiency of the removal of MG was improved by increasing the dosage of ZIF-8@CS/PVAENF. These two elements are the primary driving forces of MG adsorption. Because the difference in removal effectiveness between 0.004 and 0.003 doses was minimal, 0.003 g was chosen as the most effective dose for ZIF-8@CS/PVA-ENF. For the adsorption process, the pseudo-second-order model was observed. The ZIF-8@CS/ PVA-ENF also has a high level of reusability and stability (90%).

Zhan et al. [55] created ZIF-8/PAN nanofibrous mats with great success. For the removal of MB and MG, the dye-adsorption efficacy of the ZIF-8/PAN mats of the nanofibrous was investigated (Fig. 5). The following conditions were investigated for the optimal removal of MB and MG: The concentrations were 50 mg/L and 300 mg/L, respectively, with temperatures of 30 °C and 30 °C and pH levels of 11, 5, and 5

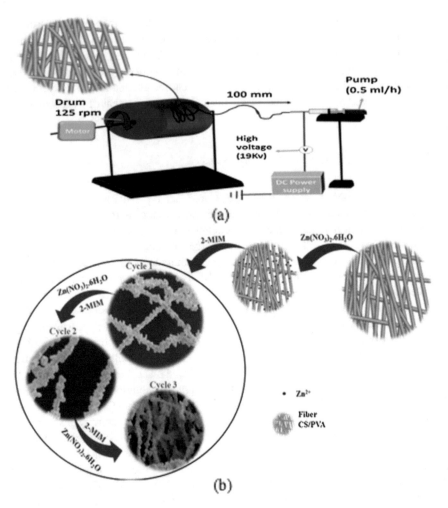

Fig. 5 a Fabrication of ZIF-8@CS/PVA-ENF and **b** Electrospinning apparatus for preparing nanofiber

correspondingly. The kinetics and adsorption isotherms for ZIF-8/PAN nanofibrous mats were developed. explored further to establish the adsorption capacity maxima and mechanism of the adsorption process for the elimination of MB and MG. The second-order kinetic equation and the Langmuir isotherm model were shown to be the best fit for MB and MG adsorption. Furthermore, employing an ethanol solution, the ZIF-8/PAN nanofibrous mat could be readily recovered. As a consequence of the foregoing findings, it can be inferred that the synthesized ZIF-8/PAN nanofibrous might be engaged for dye removal. Jin et al. [16] used a simple electrospinning approach to create a ZIF-67/PAN membrane that had outstanding MG adsorption capability. The maximal adsorption capacity was found to be 1305 mg g^{-1}, which is greater than previously reported materials. The ZIF-67/PAN can be renewed by washing it four times with ethanol. This adsorbent may be utilized to remove pollutants from water because of its strong adsorption capacity, easy manufacturing, high recyclability, and high effective adsorption of MG in water. Despite the function of MOFs in increasing certain EPNFs' ability to remove colors from aqueous solutions, including MOFs into the EPNFs matrix poses some obstacles. Due to the complicated MOFs composition, limited dispersion in solvents containing polymer, and their varied chemical characteristics with polymers, between polymers and MOFs has a low interface bonding is one of these difficulties, making it impossible to electrospun certain polymers directly.

4   Conclusions and Future Perspectives

EPNFs' distinctive qualities, including their high porosity and high surface–volume ratio, qualify them for a variety of applications, including water treatment. Furthermore, the addition of nanoparticles to its surface can improve its efficacy in the removal of contaminants from water, particularly anionic and cationic dyes. In this chapter, we discussed how electrospun polymer nanofibers and their modified versions may be used as nanoadsorbents from aqueous solutions to remove dyes. Because of the increased adsorption sites and improved mechanical characteristics, the modified EPNFs via functionalization of nanoparticles demonstrated greater efficacy in different dyes removal and aqueous environments stability than the unmodified EPNFs. The review is confined to in vitro studies that have been reported in the last two decades on utilizing EPNFs as novel adsorbents for dye removal in aqueous systems. Furthermore, the paper highlighted the most critical elements influencing EPNFs' adsorption capacity in dyes removal, the adsorption process, and several strategies of modification for improving their aqueous environments stability, as well as their mechanical qualities and efficiency. The key issue for these materials' uses, on the other hand, is their usage in the field via the treatment of actual wastewater including various sorts of contaminants. As a result, new perspectives for utilizing these materials can be established in a variety of ways. The first is an in vitro dye removal selectivity test. The second application is in the field, where they are used to clean real industrial effluent. The third goal is to increase the lab-scale production of electrospun polymer nanofibers, particularly by using inorganic materials as nanofillers. The fourth step involves using nonlinear three-parameter isotherm models to characterize the process of adsorption, and doing column studies rather than batch studies. Another area of future study that should be considered is the development of modification techniques to improve EPNF surface characteristics (e.g., porosity, surface area, and functional groupings that are active), mechanical characteristics, and aqueous solution stability.

Acknowledgements The authors are grateful to NIT Rourkela (Odisha), India, for providing the research facilities to carry out this work.

References

1. Agarwal S, Greiner A, Wendorff JH (2013) Functional materials by electrospinning of polymers. Prog Polym Sci 38:963–991. https://doi.org/10.1016/j.progpolymsci.2013.02.001
2. Ahmed FE, Lalia BS, Hashaikeh R (2015) A review on electrospinning for membrane fabrication: challenges and applications. Desalination 356:15–30. https://doi.org/10.1016/j.desal.2014.09.033
3. Chizari Fard G, Mirjalili M, Najafi F (2017) Hydroxylated α-Fe2O3 nanofiber: optimization of synthesis conditions, anionic dyes adsorption kinetic, isotherm and error analysis. J Taiwan Inst Chem Eng 70:188–199. https://doi.org/10.1016/j.jtice.2016.10.045
4. Diani J, Gall K (2006) Finite strain 3D thermoviscoelastic constitutive model. Society 1–10. https://doi.org/10.1002/pen
5. Doshi J, Reneker DH (1993) Electrospinning process and applications of electrospun fibers. Conf Rec IAS Annu Meet IEEE Ind Appl Soc 3:1698–1703. https://doi.org/10.1109/ias.1993.299067
6. Elrasheedy A, Nady N, Bassyouni M (2019) Matrix membranes : review on applications in water purification. Membranes (Basel) 9:1–31
7. Ghourbanpour J, Sabzi M, Shafagh N (2019) Effective dye adsorption behavior of poly(vinyl alcohol)/chitin nanofiber/Fe(III) complex. Int J Biol Macromol 137:296–306. https://doi.org/10.1016/j.ijbiomac.2019.06.213
8. Greiner A, Wendorff JH (2007) Electrospinning: a fascinating method for the preparation of ultrathin fibers. Angew Chem Int Ed 46:5670–5703. https://doi.org/10.1002/anie.200604646
9. Haider S, Al-Zeghayer Y, Ahmed Ali FA et al (2013) Highly aligned narrow diameter chitosan electrospun nanofibers. J Polym Res 20. https://doi.org/10.1007/s10965-013-0105-9
10. Han D, Steckl AJ (2013) Triaxial electrospun nanofiber membranes for controlled dual release of functional molecules. ACS Appl Mater Inter 5:8241–8245. https://doi.org/10.1021/am402376c
11. Hardick O, Stevens B, Bracewell DG (2011) Nanofibre fabrication in a temperature and humidity controlled environment for improved fibre consistency. J Mater Sci 46:3890–3898. https://doi.org/10.1007/s10853-011-5310-5
12. Homaeigohar S, Elbahri M (2014) Nanocomposite electrospun nanofiber membranes for environmental remediation. Materials (Basel) 7:1017–1045. https://doi.org/10.3390/ma7021017
13. Huang ZM, Zhang YZ, Kotaki M, Ramakrishna S (2003) A review on polymer nanofibers by electrospinning and their applications in nanocomposites. Compos Sci Technol 63:2223–2253. https://doi.org/10.1016/S0266-3538(03)00178-7
14. Islam MS, Ang BC, Andriyana A, Afifi AM (2019) A review on fabrication of nanofibers via electrospinning and their applications. SN Appl Sci 1. https://doi.org/10.1007/s42452-019-1288-4
15. Jian S, Zhu J, Jiang S et al (2018) Nanofibers with diameter below one nanometer from electrospinning†. RSC Adv 8:4794–4802. https://doi.org/10.1039/c7ra13444d
16. Jin L, Ye J, Wang Y et al (2019) Electrospinning Synthesis of ZIF-67/PAN fibrous membrane with high-capacity adsorption for malachite green. Fibers Polym 20:2070–2077. https://doi.org/10.1007/s12221-019-1196-7
17. Kausar A, Iqbal M, Javed A et al (2018) Dyes adsorption using clay and modified clay: a review. J Mol Liq 256:395–407. https://doi.org/10.1016/j.molliq.2018.02.034
18. Khan NA, Hasan Z, Jhung SH (2013) Adsorptive removal of hazardous materials using metal-organic frameworks (MOFs): a review. J Hazard Mater 244–245:444–456. https://doi.org/10.1016/j.jhazmat.2012.11.011
19. Kurusu RS, Demarquette NR (2019) Surface modification to control the water wettability of electrospun mats. Int Mater Rev 64:249–287. https://doi.org/10.1080/09506608.2018.1484577
20. Li G, Zhao Y, Lv M et al (2013) Super hydrophilic poly(ethylene terephthalate) (PET)/poly(vinyl alcohol) (PVA) composite fibrous mats with improved mechanical properties prepared via electrospinning process. Colloids Surf A Physicochem Eng Asp 436:417–424. https://doi.org/10.1016/j.colsurfa.2013.07.014

21. Li M, Wang H, Wu S et al (2012) Adsorption of hazardous dyes indigo carmine and acid red on nanofiber membranes. RSC Adv 2:900–907. https://doi.org/10.1039/c1ra00546d

22. Li M, Zheng Y, Xin B, Xu Y (2018) Roles of coaxial spinneret in taylor cone and morphology of core-shell fibers. Ind Eng Chem Res 57:17310–17317. https://doi.org/10.1021/acs.iecr.8b04341

23. Li T, Liu L, Zhang Z, Han Z (2020) Preparation of nanofibrous metal-organic framework filter for rapid adsorption and selective separation of cationic dye from aqueous solution. Sep Purif Technol 237:116360. https://doi.org/10.1016/j.seppur.2019.116360

24. Liu W, Ni C, Chase DB, Rabolt JF (2013) Preparation of multilayer biodegradable nanofibers by triaxial electrospinning. ACS Macro Lett 2:466–468. https://doi.org/10.1021/mz4000688

25. Ma F, fang, Zhang D, Zhang N, et al (2018) Polydopamine-assisted deposition of polypyrrole on electrospun poly(vinylidene fluoride) nanofibers for bidirectional removal of cation and anion dyes. Chem Eng J 354:432–444. https://doi.org/10.1016/j.cej.2018.08.048

26. Ma L, Abney C, Lin W (2009) Enantioselective catalysis with homochiral metal-organic frameworks. Chem Soc Rev 38:1248–1256. https://doi.org/10.1039/b807083k

27. Mahmoodi NM, Mokhtari-Shourijeh Z, Abdi J (2019) Preparation of mesoporous polyvinyl alcohol/chitosan/silica composite nanofiber and dye removal from wastewater. Environ Prog Sustain Energy 38:S100–S109. https://doi.org/10.1002/ep.12933

28. Mahmoodi NM, Oveisi M, Taghizadeh A, Taghizadeh M (2020) Synthesis of pearl necklace-like ZIF-8@chitosan/PVA nanofiber with synergistic effect for recycling aqueous dye removal. Carbohydr Polym 227:115364. https://doi.org/10.1016/j.carbpol.2019.115364

29. Miao YE, Wang R, Chen D et al (2012) Electrospun self-standing membrane of hierarchical SiO 2atγ AlOOH (Boehmite) core/sheath fibers for water remediation. ACS Appl Mater Inter 4:5353–5359. https://doi.org/10.1021/am3012998

30. Moghe AK, Gupta BS (2008) Co-axial electrospinning for nanofiber structures: preparation and applications. Polym Rev 48:353–377. https://doi.org/10.1080/15583720802022257

31. Najafi M, Frey MW (2020) Electrospun nanofibers for chemical separation. Nanomaterials 10. https://doi.org/10.3390/nano10050982

32. Pereao OK, Bode-Aluko C, Ndayambaje G et al (2017) Electrospinning: polymer nanofibre adsorbent applications for metal ion removal. J Polym Environ 25:1175–1189. https://doi.org/10.1007/s10924-016-0896-y

33. Pham QP, Sharma U, Mikos AG (2006) Electrospinning of polymeric nanofibers for tissue engineering applications: a review. Tissue Eng 12:1197–1211. https://doi.org/10.1089/ten.2006.12.1197

34. Phan DN, Rebia RA, Saito Y et al (2020) Zinc oxide nanoparticles attached to polyacrylonitrile nanofibers with hinokitiol as gluing agent for synergistic antibacterial activities and effective dye removal. J Ind Eng Chem 85:258–268. https://doi.org/10.1016/j.jiec.2020.02.008

35. Ray SS, Chen SS, Li CW et al (2016) A comprehensive review: Electrospinning technique for fabrication and surface modification of membranes for water treatment application. RSC Adv 6:85495–85514. https://doi.org/10.1039/c6ra14952a

36. SalehHudin HS, Mohamad EN, Mahadi WNL, Muhammad Afifi A (2018) Multiple-jet electrospinning methods for nanofiber processing: a review. Mater Manuf Process 33:479–498. https://doi.org/10.1080/10426914.2017.1388523

37. Sun B, Long YZ, Zhang HD et al (2014) Advances in three-dimensional nanofibrous macrostructures via electrospinning. Prog Polym Sci 39:862–890. https://doi.org/10.1016/j.progpolymsci.2013.06.002

38. Teng M, Li F, Zhang B, Taha AA (2011) Electrospun cyclodextrin-functionalized mesoporous polyvinyl alcohol/SiO2 nanofiber membranes as a highly efficient adsorbent for indigo carmine dye. Colloids Surf A Physicochem Eng Asp 385:229–234. https://doi.org/10.1016/j.colsurfa.2011.06.020

39. Theron SA, Zussman E, Yarin AL (2004) Experimental investigation of the governing parameters in the electrospinning of polymer solutions. Polymer (Guildf) 45:2017–2030. https://doi.org/10.1016/j.polymer.2004.01.024

40. Tranchemontagne DJ, Tranchemontagne JL, O'keeffe M, Yaghi OM (2009) Secondary building units, nets and bonding in the chemistry of metal–organic frameworks. Chem Soc Rev 38:1257–1283. https://doi.org/10.1039/b817735j
41. Um IC, Fang D, Hsiao BS, et al (2004) Electro-spinning and electro-blowing of hyaluronic acid 1428–1436
42. Valiquette D, Pellerin C, Hc QC (2011) Miscible and core—sheath PS / PVME fibers by electrospinning 2838–2843
43. Varesano A, Rombaldoni F, Mazzuchetti G et al (2010) Multi-jet nozzle electrospinning on textile substrates: observations on process and nanofibre mat deposition. Polym Int 59:1606–1615. https://doi.org/10.1002/pi.2893
44. Wang SX, Yap CC, He J et al (2016) Electrospinning: a facile technique for fabricating functional nanofibers for environmental applications. Nanotechnol Rev 5:51–73. https://doi.org/10.1515/ntrev-2015-0065
45. Wang X, Niu H, Lin T, Wang X (2009) Needleless electrospinning of nanofibers with a conical wire coil. Polym Eng Sci 49:1582–1586. https://doi.org/10.1002/pen.21377
46. Xu R, Jia M, Zhang Y, Li F (2012) Sorption of malachite green on vinyl-modified mesoporous poly(acrylic acid)/SiO_2 composite nanofiber membranes. Microporous Mesoporous Mater 149:111–118. https://doi.org/10.1016/j.micromeso.2011.08.024
47. Xu X, Zhuang X, Chen X et al (2006) Preparation of core-sheath composite nanofibers by emulsion electrospinning. Macromol Rapid Commun 27:1637–1642. https://doi.org/10.1002/marc.200600384
48. Xu Y, Bao J, Zhang X et al (2019) Functionalized polyethersulfone nanofibrous membranes with ultra-high adsorption capacity for organic dyes by one-step electrospinning. J Colloid Interface Sci 533:526–538. https://doi.org/10.1016/j.jcis.2018.08.072
49. Xu Z, Wei C, Jin J, et al (2018) Development of a novel mixed titanium, silver oxide polyacrylonitrile nanofiber as a superior adsorbent and its application for MB removal in wastewater treatment. J Braz Chem Soc 29:560–570. https://doi.org/10.21577/0103-5053.20170168
50. Yang R, He J, Xu L, Yu J (2009) Bubble-electrospinning for fabricating nanofibers. Polymer (Guildf) 50:5846–5850. https://doi.org/10.1016/j.polymer.2009.10.021
51. Yarin AL (2011) Coaxial electrospinning and emulsion electrospinning of core-shell fibers. Polym Adv Technol 22:310–317. https://doi.org/10.1002/pat.1781
52. ZabihiSahebi A, Koushkbaghi S, Pishnamazi M et al (2019) Synthesis of cellulose acetate/chitosan/SWCNT/Fe3O4/TiO2 composite nanofibers for the removal of Cr(VI), As(V), Methylene blue and Congo red from aqueous solutions. Int J Biol Macromol 140:1296–1304. https://doi.org/10.1016/j.ijbiomac.2019.08.214
53. Zargham S, Bazgir S, Tavakoli A et al (2012) The effect of flow rate on morphology and deposition area of electrospun nylon 6 nanofiber. J Eng Fiber Fabr 7:42–49. https://doi.org/10.1177/155892501200700414
54. Zeleny J (1935) The role of surface instability in electrical 219:1314–1345
55. Zhan Y, Guan X, Ren E et al (2019) Fabrication of zeolitic imidazolate framework-8 functional polyacrylonitrile nanofibrous mats for dye removal. J Polym Res 26. https://doi.org/10.1007/s10965-019-1806-5
56. Zhang W, He Z, Han Y et al (2020) Structural design and environmental applications of electrospun nanofibers. Compos Part A Appl Sci Manuf 137:106009. https://doi.org/10.1016/j.compositesa.2020.106009
57. Zhao R, Wang Y, Li X et al (2015) Synthesis of β-cyclodextrin-based electrospun nanofiber membranes for highly efficient adsorption and separation of methylene blue. ACS Appl Mater Inter 7:26649–26657. https://doi.org/10.1021/acsami.5b08403
58. Zhijiang C, Cong Z, Ping X et al (2018) Calcium alginate-coated electrospun polyhydroxybutyrate/carbon nanotubes composite nanofibers as nanofiltration membrane for dye removal. J Mater Sci 53:14801–14820. https://doi.org/10.1007/s10853-018-2607-7
59. Zhou Z, Liu L, Yuan W (2019) A superhydrophobic poly(lactic acid) electrospun nanofibrous membrane surface-functionalized with TiO_2 nanoparticles and methyltrichlorosilane for oil/water separation and dye adsorption. New J Chem 43:15823–15831. https://doi.org/10.1039/c9nj03576a

Printed in the United States
by Baker & Taylor Publisher Services